BIRD BLOW FLIES (*Protocalliphora*) IN NORTH AMERICA (DIPTERA: CALLIPHORIDAE), with notes on the Palearctic species

BIRD BLOW FLIES (*Protocalliphora*) IN NORTH AMERICA (DIPTERA: CALLIPHORIDAE), with notes on the Palearctic species

Curtis W. Sabrosky,
Gordon F. Bennett, and
Terry L. Whitworth

Smithsonian Institution Press
Washington, D.C., and London

Printed in the United States of America

Library of Congress
Cataloging-in-Publication Data

Sabrosky, Curtis, W.

Bird blow flies (protocalliphora) in North America (diptera: calliphoridae), with notes on the palearctic species / Curtis W. Sabrosky, Gordon F. Bennett, and Terry L. Whitworth.

p. cm.

Bibliography: p.
Includes index.
ISBN 0-87474-865-8
1. Protocalliphora--North America. I. Bennett, Gordon F.
II. Whitworth, Terry L. III. Title.
QL537.C2423 1989

595.77'4--dc20 89-30488

91 90 89 4 3 2 1

∞ The paper used in this publication meet the minimum requirements of the American National Standard for Permanence of Paper for Printed Library Materials Z39.48 1984.

Publisher's note: For reasons of speed and economy, this book is published from camera-ready copy prepared electronically by the authors, who assume full responsibility for the contents and form.

Dedication

The authors take pleasure in dedicating this work to the many biologists, chiefly entomologists and ornithologists, who have furnished material for this study and who have been most patient in awaiting the result. In particular we honor three superb field naturalists whose interest in *Protocalliphora* supplied us with abundant data and material, and who encouraged and assisted in the study. We regret that Professor Spencer did not live to see the finished product.

William L. Jellison, Jr. (Hamilton, Mont.)
George J. Spencer, 1888-1966 (Vancouver, B.C.)
Edward S. Thomas (Columbus, Ohio)

Contents

Preface

This long-delayed study of bird blow flies (*Protocalliphora*) has grown steadily since its inception by the senior author about 1950 on the adult flies. Graduate study and field studies by Bennett at the University of Toronto and at the Wildlife Research Station in Algonquin Park, Ontario, provided a good foundation on the immature stages that complemented and greatly aided the study of the adults. Later graduate studies by Whitworth at Utah State University and Gold at the University of California at Albany extended knowledge of these flies for the Intermountain Region and California and added species to the known fauna.

This book has been essentially ready for some years, although modified and improved from time to time, and the senior author is responsible for most of the delay. In partial extenuation, the long development of the project has undoubtedly brought us nearer to--although it has not necessarily ensured--complete knowledge of the fauna as far as the number of known species and their distribution are concerned. But any possible incompleteness does not justify further postponement, and we present our best efforts at this time as a record of the present status of knowledge and a basis for further observations.

In addition to the basic taxonomic framework the book includes a review of the history and geographic distribution of *Protocalliphora* and a summation of other aspects such as life history, host relations, and ecology. Maps have been prepared on Goode's Base Map Series No. 102 to show the known distribution of most of the species.

In this work, which deals with the species of *Protocalliphora*, specific names are often used by themselves to avoid too frequent repetition of the long generic name *Protocalliphora* or its abbreviation *P*. All other generic or subgeneric names are spelled in full.

The host birds are referred to by their common or English names throughout the text. The Latin names are given in the list of natural hosts (Appendix), which covers all North American species mentioned as known hosts of *Protocalliphora.* The references contain annotations by Sabrosky on the identities of the species referred to.

The book is a contribution from various organizations as follows, thus identifying the authors and acknowledging the support given them:

Systematic Entomology Laboratory, Agricultural Research Service, U.S. Department of Agriculture, Washington, D.C. (Sabrosky). Mailing address: Systematic Entomology Laboratory, USDA, % U.S. National Museum of Natural History (NHB 168), Washington, D.C. 20560.

Department of Parasitology, Ontario Research Foundation, Toronto, Ont., Canada, and (present address of author) International Reference Centre for Avian Haematozoa, Memorial University of Newfoundland, St. John's, Nfld., Canada A1B 3X9 (Bennett). A portion of the work was made possible by a research grant to the Ontario Research Foundation from the Department of Economics and Development of the Province of Ontario, and research fellowships to Bennett from the Research Council of Ontario and the National Research Council of Canada.

Department of Zoology, Utah State University, Logan, Utah (Whitworth). Present address: Whitworth Pest Control Inc., 3707 - 96th St. E., Tacoma, Wash. 98446.

Curtis W. Sabrosky
Gordon F. Bennett
Terry L. Whitworth

Introduction

Knowledge that certain calliphorid larvae ingested the blood of nestling birds dates back at least to the middle of the 19th century. Dufour (1845) in France described the hematophagous habit of the larvae in his paper on a new species that he named *Lucilia dispar*, now considered a synonym of *Protocalliphora azurea* (Fallén), and Rossi (1848) reared "*Musca azurea*" from a large number of larvae found on young birds. In America, Walsh (1866a) mentioned large larvae found attached to the head and body of a young swallow, and he identified these as belonging to the "*OEstrus* family," later corrected by Osten Sacken to "*Musca* family in the vicinity of *Musca* or *Sarcophaga*" (Walsh 1866b). Undoubtedly these were larvae of *Protocalliphora.*

Subsequent years saw the accumulation of many records of bloodsucking maggots on nestling birds, in both entomological and ornithological literature. Although 19th century workers made some observations, much more attention was paid to the group in the 20th century, particularly within the last few decades. The numerous past records have been discussed and summarized, in part at least, by Rodhain and Bequaert (1916), Bezzi (1922), Séguy (1929), Hall (1948), Owen (1954), Hicks (1959, 1962, 1971), and Zumpt (1965). There have been two revisions of the Nearctic species prior to the present one (Shannon and Dobroscky 1924, Hall 1948), and for the Palearctic species there have been three revisions (Zumpt 1956, Gregor and Povolný 1959, Peus 1960), a series of papers by Grunin (1966-75), chiefly on Russian species, and a revision of *Trypocalliphora* by Rognes (1985).

Books on parasitic insects, parasitology, and birds make little or no mention of *Protocalliphora.* For example, there is no mention whatsoever of the genus or any of its synonyms in Marshall's recent *The Ecology of Ectoparasitic Insects* (1981), although there is considerable attention to such minor genera as *Carnus* (Carnidae) and the only recently described *Mystacinobia*, "which may be parasitic" (Mystacinobiidae, later placed in Calliphoridae). Textbooks on ornithology and books on the biology and the life history of birds make little if any mention of *Protocalliphora* despite the facts that the maggots infest a sizeable proportion of birds nests, especially of passerine birds, and that they have sometimes been charged with the death of nestlings, or with weakening them so as to contribute to their death from other causes. In the great series by A. C. Bent (21 volumes, 1919-68) on life histories of North American birds, *Protocalliphora* is mentioned in only seven--one of those only as speculation--and usually only as a brief mention or citation of one or a few references. The omission or scanty mention by ornithologists seems especially curious in view of the numerous papers by C. W. Johnson in ornithological journals, the extensive rearings reported by Shannon and Dobroscky (1924), and the papers of Henshaw, Mason, McAtee, C. W. Miller, Plath, Stoner, Storer, and others, which antedated most of Bent's volumes and the recent books and texts on birds and their lives. In the periodical literature of ornithology there is rarely more than a mere mention of the bird blow flies and their maggots; *Protocalliphora*, if mentioned, is commonly not even indexed. Even papers specifically on the insect faunas of birds nests all too often have lists of lice, fleas, mites, and ticks with no mention of *Protocalliphora.* Wing's (1956) book on the *Natural History of Birds: A Guide to Ornithology* is one of the few bird books to say much about *Protocalliphora*, but even so this is not indexed.

A common name is perhaps unnecessary for these flies. The adults are seldom net collected, and the maggots are usually found by entomologists, for whom the generic name *Protocalliphora* is readily understood, or by ornithologists. Various publications have used such names as "blood-sucking larva flies," "parasitic bird-flies," "nestling screwworms," and "bird-nest screwworms," but these seem awkward. The last two are also misleading because the larvae do not have the characteristic appearance of screwworms. For that matter, "blow fly" is also inaccurate because *Protocalliphora* larvae do not "blow" carcasses, but it is the common name for flies of the family Calliphoridae. The German literature uses the appropriately suggestive name "Vogelblutfliegen" (bird blood flies), although this might imply, incorrectly, that the flies themselves suck blood, and the term would also fit some Hippoboscidae. There are also other flies with ornithophilic blood-sucking larvae, such as *Philornis* (Muscidae) and *Neottiophilum* (Neottiophilidae). But if a common name is desired, "bird blow flies" would be short and euphonious; blow flies (calliphorids) that attack birds. This was adopted by Zumpt (1965).

Difficulties in the identification of Nearctic *Protocalliphora* stimulated Sabrosky about 1950 to reexamine the classification of the genus. Soon after, Bennett, based on observations by D. M.

Davies in 1949, undertook a study of the ecology, life history, and immature stages of the genus, principally at Lake Sasajewan (45° 35' N., 78° 30' W.) in southern Algonquin Park, Ontario, at the Wildlife Research Station of the Ontario Department of Lands and Forests. The parallel studies on adults and immature stages, although initially not equally comprehensive, proved to be so mutually advantageous and complementary that they soon developed concomitantly. The ecological and life history studies have also contributed significantly to understanding of the genus. These are summarized from an unpublished doctoral dissertation (Bennett 1957) that was generously made available to other workers and has been cited by them, and from the M.S. thesis and doctoral dissertation by Whitworth in Utah (Whitworth 1971, 1976, 1977).

As a result of our studies, covering the extensive Algonquin Park and Utah material, plus collections from all major museums in this continent and other sources in the United States and Canada, the present revision recognizes 26 species (15 new and one newly recognized as Holarctic), compared with 10 recognized by Hall (1948) in the last revision of the genus in North America. Furthermore, unlike most previous publications, immature stages are also covered. Table I presents a summary that shows the stages that are known and described in this paper. The studies are based on a total of 9,292 adult specimens (4,649 males, 4,643 females), and large numbers of larvae and puparia. Of the 16 names already published and available, the senior author has studied the types of 13, one is lost, and the other two are European names now being used in North America.

Much remains to be learned about *Protocalliphora*, in spite of the nearly complete record for mature larvae, puparia, and adults (Table I). The biology and ecology of most species are unknown, at most inferred from host records. The complete life history has not been observed for any species. The distribution of most of the species is spottily known, although the scattered records indicate wide ranges. Presumably, *Protocalliphora* occurs throughout the nesting ranges of the bird hosts, with the exception of the far south as judged from the typically northern or higher altitude range of the genus (cf. Map 1). However, a few specimens from Mexico and southwestern United States suggest that attention to the Sonoran fauna would yield interesting results. Few areas have been studied thoroughly, and isolated available specimens hint at other undescribed species.

It is apparent from some puzzling complexes that much work remains to be done before the far western fauna is fully understood. Certain species, such as *asiovora* and *chrysorrhoea*,

Table I. The described stages of Nearctic *Protocalliphora*

Species	Egg	Larval Instars 1st	2nd	3rd	Puparium	Adults Male	Female
Subgenus Trypocalliphora							
braueri			X	X	X	X	X
Subgenus Protocalliphora							
aenea	X		X	X	X	X	X
asiovora				X	X	X	X
avium	X		X	X	X	X	X
beameri						X	X
bicolor				X	X	X	X
brunneisquama					!X	X	X
chrysorrhoea				X	X	X	X
cuprina				*X	X	X	X
deceptor				X	X	X	X
fallisi				X	X	X	X
halli				*X	X	X	X
hesperia					X	X	X
hesperioides				X	X	X	X
hirundo			X	X	X	X	X
interrupta				*X	X	X	X
lata				X	X	X	X
metallica	X	X	X	X	X	X	X
parorum				X	X	X	X
sapphira						X	X
seminuda				*X	X	X	X
shannoni		X	X	X	X	X	X
sialia	X	X	X	X	X	X	X
spatulata				X	X	X	X
spenceri				*X	X	X	X
tundrae				*X	X	X	X

* = Important characters described from puparium.
! = Atypical; only two undersized puparia known.

are clearly marked. In species such as *hesperia*, however, one finds it difficult to determine whether one is dealing with one or a few variable species in which the local populations differ slightly, or whether there are a number of closely related species. Such factors as mixed infestations, lack of outstanding characters in both sexes, lack of conspicuous differentiation in the immature stages, and immaturity of many reared series make the interpretation extremely difficult.

Perhaps some hybridization is taking place, or the western region is one of active speciation and the local populations show tendencies that have not yet stabilized as consistent characteristics. On the other hand, reasonable stability may be present, but obscured by unfortunately narrow distinctions between species that may occur together in mixed infestations. Purposeful field work like that of Bennett in the Algonquin Park area, Whitworth in Utah, and Gold and Dahlsten in California is essential to clarify the distinctions and relationship. In the meantime, a tentative arrangement is offered as the best that can be deduced at this time from the available material. Certainly in the West it is even more necessary than in the East that firm identifications be based on series containing both sexes, plus mature larvae or puparia whenever possible. Separation of these western forms from the numerous known species will probably be difficult and will require increasingly careful attention to details.

Keys and detailed descriptions are given for males, females, and immatures. For each species there is a brief statement of the known distribution, followed by the detailed records. For already described species, the identification records have been somewhat abbreviated, if numerous; the detailed records will be filed with the Systematic Entomology Laboratory, U.S. Department of Agriculture, Washington, D.C., or in the Archives of the Smithsonian Institution, so that future workers may locate and recheck the material if later developments require it.

Voucher specimens for published records have been checked whenever possible, and unconfirmed published records are few in number. For a few species, such as the occurrence of *avium* in crows' nests and in the ears of hawks, or records of *braueri* (*hirudo*) reared from larvae found feeding subcutaneously, the records can be accepted as almost certainly correct.

Dates of reared specimens are of little significance for *Protocalliphora*. Life history studies show that the breeding of *Protocalliphora* species is confined to and continuous throughout the nesting seasons of their bird hosts. Moreover, dates on most reared specimens do not state whether they are the dates of nest collection or of emergence of adults. Where dates are given with host records, they are those of emergence of adult flies unless otherwise specified. Dates of hand-caught specimens may be significant, however; early and late dates give evidence of overwintering by adults.

The collector's name, if known, is stated for type series and for uncommon species, but not for the numerous records of common species already described, such as *aenea*, *avium*, and *sialia*. The collector's name is sometimes supplied from published information, notably from the papers of C. W. Johnson (1925-32).

For each species there is a list of known hosts and a brief statement of the known or inferred ecology. The latter is based on the field studies plus analysis of other available records and of published information where the species identifications have been checked. Ecological data on the host birds have been derived from the publications by Bent (1919-53), Peterson (1947), and Taverner (1934), as well as the field studies by Bennett and Whitworth.

Host records refer to flies reared from larvae and puparia found in the nests of the hosts, but any records of direct attack upon the birds themselves, e.g., myiasis in the ears, or larvae in abscesses under the skin, are so specified and given in the detail available.

Location of paratypes and other material is indicated by the following abbreviations for the frequently cited collections:

CAS	California Academy of Sciences, Golden Gate Park, San Francisco, Calif. 94118
CNC	Canadian National Collection, Agriculture Canada, Ottawa, Ontario. K1A OC6
MCZ	Museum of Comparative Zoology, Harvard University, Cambridge, Mass. 01238
UBC	University of British Columbia, Spencer Entomological Museum, Vancouver, B.C. V6T 2A9
USNM	U.S. National Museum of Natural History, Washington, D.C. 20560
USU	Utah State University, Department of Biology, Logan, Utah. 84322

Infrequently cited collections have either been cited in full or somewhat shortened by standard and readily understood abbreviations, e.g., U. Del. = University of Delaware; U. Calif., Berkeley = University of California at Berkeley. University material is usually in the collection of the Department of Entomology.

Acknowledgments

The writers gratefully acknowledge their indebtedness to numerous individuals and institutions for the loan of material and for other help during the course of this study. Particular thanks are due to A. Murray Fallis, former Director, Department of Parasitology, Ontario Research Foundation, Toronto, Ont., and Wilford J. Hanson, Department of Biology, Utah State University, Logan, Utah, under whom Bennett and Whitworth, respectively, carried out field and laboratory investigations for their doctorate programs, for their sympathetic interest and sound counsel; to the Ontario Department of Lands and Forests for making available the facilities of their Wildlife Research Station, Algonquin Park, Ontario, and to the staff and other colleagues at the Station for assistance and information during the field work; to D. M. Davies, McMaster University, for initiating the field studies in Algonquin Park; to Kenneth J. Capelle, Brigham City, Utah, (retired, Fish and Wildlife Service, U.S. Department of the Interior), for suggesting Whitworth's study and for his assistance throughout the project; to the late G. J. Spencer, University of British Columbia, for the generous and long-time loan of his large collection of *Protocalliphora* gathered over a period of some years; to W. L. Jellison, Rocky Mountain Laboratory, U.S. Public Health Service, for making available the large collection of *Protocalliphora* accumulated by himself, C. B. Philip, and others in the course of their field work; to the late J. C. Bequaert, Museum of Comparative Zoology, Harvard University, and Arthur G. Humes, Boston University, for the loan of important collections accumulated and published on by Charles W. Johnson; to Edward S. Thomas, Ohio State Museum, the late M. T. James, Washington State University, E. L. Kessel and the California Academy of Sciences, and Clifford S. Gold and Donald L. Dahlsten of the University of California at Albany, for the loan of particularly important collections; to George E. Watson, Smithsonian Institution, and Richard Banks, Fish and Wildlife Service, U.S. Department of the Interior, for many courtesies in connection with ornithological literature and the nomenclature of the bird hosts, and to Knut Rognes of Norway for generously sharing with us his results of dissections of types and his descriptions from his manuscript on Calliphoridae for the *Fauna Entomologica Scandinavica.*

We are indebted to Deborah Roney for the drawings and the makeup of the plates, to Richard Banks, Michael Schauff, Guy E. Shewell, and Thomas Pape for careful and critical reading of the manuscript and excellent comments, to two anonymous referees, to Vera Lee for typing of the manuscript, to Elizabeth Klafter for the preparation of camera ready copy, and to Linda Lawrence for adjusting plates to final copy.

Our appreciation for material and information is also due to the following individuals, some of whom are deceased: John F. Anderson, Paul H. Arnaud, Jr., John R. Baker, Russell P. Balda, Sister Barbara Ann (All Saints' Convent, Catonsville, Md.), R. H. Beamer, W. W. Becklund, John N. Belkin, Robert Bohm, Gary R. Bortolotti, J. W. Boyes, A. E. Brower, George W. Byers, Robert A. Cannings, S. G. Cannings, K. J. Capelle, John A. Chapman, Frank R. Cole, B. E. Cooper, Monique Coulloudon, Scott Crocoll, C. H. Curran, Vivie E. Davis, Vasiliki Demas, Henry Dietrich, H. R. Dodge, Richard P. Dow, W. L. Downes, Jr., Emmet R. Easton, Howard E. Evans, W. G. Evans, D. C. Ferguson, R. H. Foote, Woodbridge A. Foster, Norman R. French, John George, J. D. Gregson, G. C. D. Griffiths, A. H. Grewe, Jr., K. J. Grunin, Harvey L. Gunderson, G. E. Haas, David G. Hall, Jr., Jeffrey A. Halstead, Mrs. Frances Hamerstrom, Joseph J. Hickey, Cluff Hopla, Donald S. Horning, Jr., Paul D. Hurd, Jr., C. D. Johnson, W. W. Judd, Rokuro Kano, Ben Keh, E. E. Kenaga, Ke Chung Kim, Frank Kuhlman, W. E. LaBerge, Paul K. Lago, Robert D. Lee, Hugh B. Leech, Robin Leech, B. Lindeberg, Tom Lund, J. E. H. Martin, Wayne N. Mathis, J. F. McAlpine, A. T. McClay, E. T. McKnight, John F. Mehner, Heinz Meng, Donald H. Messersmith, Virgil I. Miles, Robert T. Mitchell, Val Nolan, Jr., Pekka Nuorteva, Don R. Oliver, L. L. Pechuman, P. I. Persson, Fritz Peus, T. David Pitts, Adrian C. Pont, Kim G. Poole, Dalibor Povolný, H. W. Prescott, Frank W. Preston, Robert L. Rausch, H. J. Reinhard, C. L. Remington, David C. Rentz, Chandler Robbins, Vincent D. Roth, E. P. Rouse, R. E. Ryckman, R. I. Sailer, Earl C. Schriver, G. G. E. Scudder, William M. Shields, David R. Smith, E. Graywood Smyth, Robert M. Stabler, Sarah H. Stabler, L. R. Steeves, George C. Steyskal, Frederic M. Stiner, Jr., Wallace A. Tarpley, Jack D. Tiner, Peter Tirrell, Charles A. Triplehorn, Neely Turner, William J. Turner, S. L. Tuxen, Nicolaas A. M. Verbeek, Kenneth A. Walker, Judith Stenger Weeden, R. L. Wenzel, A. Wetmore, Nixon Wilson, D. M. Wood, D. L. Wray, Russell E. Wright, Lawrence Zeleny, and F.

Zumpt. Inasmuch as detailed records are not given except for type series and special cases, the individual contributions are ordinarily not elaborated. We humbly regret if any persons in the many years of this project have been overlooked.

History of the Classification of Bird Blow Flies

The genus *Protocalliphora* was proposed in North America by Hough (1899a) for *Musca azurea* Fallén (designated as type species) and *M. chrysorrhoea* Meigen. Both names were originally proposed for European species but then also used in North America. Hough unquestionably described a genus of bird blow flies, but his designation of *azurea* has led to arguments and problems. What is *azurea*? Does it belong to a different genus? Is it really *Protophormia terraenovae* (Robineau-Desvoidy)? Inasmuch as *azurea* is the type species of *Protocalliphora*, we must digress from the narrative to discuss its status even though we know now that the specific name does not apply to a Nearctic species. Relevant publications will be considered chronologically.

Fallén (1817): *Musca azurea* named and described, from "♂♀," sent to him by Gyllenhal from Westergöthland (modern Västergötland) Sweden. The number of specimens was not indicated.

Fallén (1821): *M. azurea* redescribed essentially as in 1817 but with a few additional details consistent with *Protophormia terraenovae* ("Squama nigricans," etc.).

Meigen (1826: 63): stated clearly that Fallén's *azurea* included two species, and that only the male is *azurea.* The squamae (i.e., the calypteres) are described as white, and the description fits *Protocalliphora* as we know it. If he was dealing with original material, Meigen is the first reviser, in limiting *azurea* to males with white squamae. He visited Fallén at Lund and at least twice, in prefaces to his volumes, recorded his indebtedness to Fallén: "Die meisten von dem berühmten Professor Fallen in Lund in Schweden öffentlich gemachten Arten, haben wir durch des Verfassers Freigebigkeit gleichfalls zur Ansicht und Vergleichung gehabt" (Meigen, 1818: xix); and Meigen, 1824: vi, in which he described his visit to Lund in 1823 in the company of the dipterist Wiedemann. They were received, Meigen stated, "with open arms" by Professors Fallén and Zetterstedt, who made their collections readily available for study by the visitors. Meigen undoubtedly saw original material, and his recognition of a mixed series is reinforced by Zetterstedt.

Zetterstedt (1838: 657) revised the Fallén material of *azurea* and recognized two species: *azurea* in the sense of a species of *Protocalliphora* and *Musca groenlandica* n.sp., now a synonym of *Protophormia terraenovae*. His description of *azurea* fits *Protocalliphora* for the most part, although he still says "squama nigricante." He specifically mentions Fallén's material, "which that author had before his eyes when he was about to describe his species." Later he divided the collections and sent the Fallén Collection to Stockholm. Van Emden (1954) accepted Zetterstedt (1838) as first reviser.

Villeneuve (1918), under the combination *Phormia azurea*, noted that Zetterstedt's description of *Musca azurea* was based on "le type même de Fallén," and Stein (1924) referred to his study of "Die Type in Lund." Neither labeled any specimen of *azurea* as type or lectotype, nor gave any way of recognizing the specimen referred to.

Townsend (1931), who visited the collections at Stockholm and Lund, recorded the holotype in Stockholm, but no specimen there bears Townsend's label. Curiously, he called the species *Protocalliphora*, even though the specimens in the Fallén Collection at Stockholm are *Protophormia terraenovae* (Ringdahl 1937, 1945; Hennig 1939; Sabrosky 1956).

Ringdahl (1937) stated that the two males and a female of *Musca azurea* in the Fallén Collection in Stockholm are identical with *Phormia terraenovae* Robineau-Desvoidy and *groenlandica* Zetterstedt.

Hennig (1939) borrowed a male of *azurea* from Stockholm, which he referred to as "der Typus." He dissected and figured the male genitalia, and recognized that the specimen belonged to the saprophagous genus *Protophormia*. He did not label the specimen as type or lectotype, perhaps assuming that the specimen sent him was *the* type. One of the two males in the Fallén Collection is obviously the specimen dissected by Hennig; it bears a printed label "hiervon micr. Präp./Kopulat.-Apparat," and a small pink label with a number that indicates it was loaned to Hennig.

Ringdahl (1945) also synonymized "the true *Musca azurea*" with *Phormia groenlandica* Zett., after "the type specimen in the Riksmuseum" in Stockholm. Again, he neither labeled nor identified a type or lectotype.

Sabrosky (1956) reviewed the problem of "The nomenclature of *Protocalliphora*," concluded that Meigen was really the first reviser, and stated that in his opinion "action of a first reviser takes precedence over later actions, even if one of the latter involves selection of a lectotype or neotype." He then designated a lectotype consistent with Meigen's revision, choosing an old unlabeled male in the collection at Lund.

With the advantage of hindsight and the subsequently developed International Code of Zoological Nomenclature (1961, 1985), we now conclude as follows:

1. Lectotype designation has been given precedence over restriction by revisers (ICZN, Art. 74a.ii, 3rd edition; cf. also Rec. 74A). Sabrosky's 1956 point of view in favor of reviser action did not prevail. Unfortunately, as a result, a heedless or inadvertent designation of lectotype can upset usage long established by revisers.

2. The material sorted by Zetterstedt contained two species, one a *Protocalliphora*, the other a *Protophormia*. From the labels, most of the former now in the collection at Lund were collected after *Musca azurea* was published and therefore cannot possibly be the syntypes, or part of the original series. One specimen, unlabeled as very old specimens often are, might have been an original specimen and at least cannot be demonstrated not to have been. This was the specimen selected as lectotype by Sabrosky (1956), who then believed that the first reviser (Meigen 1826) limited the species to a specimen of *Protocalliphora*. The other specimens that also appear to be original material, two males and one female, are in Stockholm; these are *Protophormia terraenovae*.

3. Once the first reviser approach is rejected, then the question of lectotype must be reexamined critically. Did any of those who referred to "the type" or its equivalent in any language actually designate a lectotype? None of those so labeled a specimen, but labeling is not specifically required by the Code. If not labeled the specimen must surely be described in some way so that it can be recognized as the type; otherwise the so-called designation is ambiguous and imprecise and cannot be regarded as valid. Hennig, who dissected the male genitalia, appears to qualify because the specimen he studied is labeled as dissected, and it also bears a numbered pink slip of paper that indicates it was loaned to Hennig. True, Hennig did not see all of the original material and in a sense did not really *revise*, but under Article 74b of the Code his inference that the specimen he studied was "*the* type" is deemed to have been lectotype designation should other syntypes be discovered. This assumes, of course, that the lectotype was one of the original syntypes, and it appears to have been, or at least cannot be demonstrated not to have been.

4. We are thus led to the unfortunate conclusion that Hennig's (1939) figured specimen is the lectotype, and that accordingly *Musca azurea* in the strict sense is a *Protophormia*, with priority over the relatively common and widespread species long known as *Protophormia terraenovae* (Robineau-Desvoidy).

5. All authors are agreed on preserving the generic name *Protocalliphora* on the basis of misidentified type species (ICZN, Article 70), because Hough's description clearly applies to the bird blow flies and the name has always been used in that way. Peus (1960) agreed that the type species was misidentified, but considered that Hough based his genus on American species then identified incorrectly under the European names. There is nothing in Hough's paper, however, to suggest that he saw only American examples. Furthermore, there is evidence in a paper he published the same year (Hough 1899b) that European material was available to him ("I have compared my American with European specimens from Prof. G. Strobl and Dr. O. Schmiedeknecht"). We can therefore accept his designation of *azurea* Fallén as referring to a European species.

6. The question of the correct specific name for *azurea* sensu Hough and European authors is not so easily solved. The next oldest name that has been associated with the bird blow flies is *Phormia caerulea* Robineau-Desvoidy, 1830, but no type material is known to exist and the description is not identifiable. In view of the extensive usage of the name *azurea* Fallén for the bird blow flies, *Protocalliphora*, we believe the best course would be to suspend the rules and fix the name *azurea* for a species of those flies, thus also leaving the widely used name *Protophormia terraenovae* undisturbed.

Sabrosky (1984) applied to the International Commission on Zoological Nomenclature for an amendment to the International Code that would relax the present requirement that cases of misidentified type species must be referred to the Commission. In actual practice many taxonomists have proceeded unilaterally to recognize the species actually before the original author, when the circumstances are considered clearcut that it was misidentified. Although no official decision has yet been published, we understand that reaction was favorable. We therefore continue to use the name *Protocalliphora* for the bird blow flies in the belief that ultimately this name will be approved, whatever the mode of achieving it. Meanwhile a formal application to the Commission has been prepared (Sabrosky 1988), because it is still worthwhile to fix the use of *azurea* for a bird blow fly and thus to avoid the confusing transfer that would substitute *azurea* for the widely used name *terraenovae* in *Protophormia*.

Hendel (1901) erected a new genus, *Avihospita*, for *azurea* and relatives, but Aldrich (1901) promptly pointed out that it was a synonym of *Protocalliphora* Hough, with the same type species. The genus *Protocalliphora* was recognized as distinct in both the Nearctic (Aldrich 1905) and Palearctic catalogues (Bezzi and Stein 1907). On the other hand, one of the most respected of European specialists on muscoid Diptera, Villeneuve (1911), rejected the genus as useless and pronounced it a synonym of *Phormia*. In later publications (1928, 1931) he referred to it as at best no more than a subgenus. The genus has usually been maintained as distinct by American dipterists, however, as in the well-known manuals by Williston, Curran, and Townsend, the revisions of Nearctic Calliphoridae by Shannon (1923, 1926) and Hall (1948), and in the new Manual of Nearctic Diptera (Shewell 1987), and it is recognized as distinct by modern European dipterists. The distinctive biology and larval morphology are important factors in its recognition.

Aside from occasional records, there was little attention to the genus in North America in the first two decades after its proposal, and it was regarded as rather uncommon, probably because specimens are seldom taken by ordinary net collecting. Harbeck (1907), probably based on an identification by Coquillett, apparently first referred to *Protocalliphora* a described Nearctic species, *Calliphora splendida* Macquart. Later, Townsend (1919) described *Phormia metallica* in this group, using *Phormia* instead of *Protocalliphora* because of a nomenclatural confusion.

The first revision of the North American species was by Shannon and Dobroscky (1924). The Palearctic names *azurea* and *chrysorrhoea* were dropped and three Nearctic species were recognized, *avium* and *hirudo* (now *braueri*) as new species, and *splendida* (Macquart) (syn., *metallica* Townsend), together with one variety of *avium*, three varieties and a subspecies of *splendida*, and one variety and one subspecies of *hirudo*. The male terminalia were utilized to some extent, but except in *avium* with its unusually broad surstyli (outer forceps) the differences were not fully appreciated and were dismissed as being too intangible for the differentiation of species.

The second revision of the Nearctic species was that of Hall (1948) in his book on *The Blowflies of North America*. Hall recognized ten species, raising four of Shannon and Dobroscky's "varieties" to specific rank and describing three new species. He also proposed a new genus, *Apaulina*, for the Nearctic species, based in part on certain differences from the Palearctic species pointed out by Shannon (1923) and by Shannon and Dobroscky (1924). *Apaulina* is considered by us to be a synonym of *Protocalliphora* (cf. later discussion), and the synonymy has already been published by several authors on the authority of Sabrosky in litt.

The present revision, the third for North America in 60 years, includes a total of 26 species, including 15 described as new and two (*chrysorrhoea* and *braueri*) recognized as Holarctic. True *chrysorrhoea* is here recorded in North America for the first time, but *braueri* was previously known as a Nearctic species, *hirudo*. Some additional species appear to be new, but these are left undescribed because of inadequate material. A general discussion of this taxonomic revision and of the field and laboratory work was presented both as an invitational paper and a demonstration at the Tenth International Congress of Entomology at Montreal in 1956 (abstract, Sabrosky and Bennett 1958), and as a contributed paper at the Thirteenth Congress at Moscow in 1968 (short abstract, Sabrosky and Bennett 1971).

Evidence from field and laboratory studies of both adults and immature stages supports the view that the genus is composed of a number of distinct species albeit closely related and confusingly similar, and not of a few variable species with color forms or ecological races. Additional evidence has been furnished by a study carried out by Prof. J. W. Boyes of McGill University, who found good specific differences in the somatic chromosomes of six eastern species, *aenea*, *avium*, *hirundo*, *metallica*, *shannoni* (as n. sp. near *sialia*), and *sialia*, made available to him at Algonquin Park, Ontario (Boyes 1961, Boyes and Brink 1965, Boyes and Shewell 1975).

Geographic Distribution

The genus *Protocalliphora* is Holarctic and predominantly northern (Maps 1, 2). In North America it occurs in Alaska, throughout Canada, Greenland, in most of the contiguous states of the United States, and in the Nearctic part of Mexico. The flies seem to occur chiefly at higher altitudes in the southern extensions of the range. There are only a few scattered records from the southeastern states, the southern Mississippi Valley, and throughout the Great Plains states and provinces. In the East, specimens have been collected in the Appalachian Mts. south to northern Georgia. The southernmost eastern record, from south central Georgia (Norris 1958), is based on a single puparium found in the nest of a brown-headed nuthatch. In the West, specimens are known from California as far south as San Diego and from Baja California Norte, from various localities in Arizona and New Mexico, especially in mountain areas, and from six localities on the Mexican Plateau (the southernmost: in Morelos, Puebla, and Tlaxcala states south and southeast of Mexico City). The few Mexican localities, with a total of 11 net-collected specimens, highlight the extensive Nearctic territory in Mexico that is terra incognita for *Protocalliphora.*

It remains to be seen whether the absence of records in some states reflects merely lack of collecting. We have made numerous efforts to find material from some of those states, and to encourage collecting of birds' nests in a search for *Protocalliphora,* but thus far with little or no success. Lack of collecting would be especially noticeable in a group like *Protocalliphora,* which requires rearing of immature stages rather than ordinary net collecting of adult flies.

The map (Map 1) of course does not necessarily show the abundance of *Protocalliphora.* A dot on the map may represent only a single net-collected individual, or only one species, or it may represent hundreds of specimens of several species, or repeated recoveries over a period of years. The major areas of collecting (marked with large dots), in terms both of number of specimens and of species, are Kamloops, B.C. (G. J. Spencer and others), Ravalli County, Mont. (W. L. Jellison and C. B. Philip), northern Utah (T. L. Whitworth), Algonquin Park, Ont. (G. F. Bennett), Ithaca, N.Y. (R. C. Shannon, I. D. Dobroscky and others), eastern Massachusetts (many localities close together, the records published by C. W. Johnson 1925-32), and California localities (university staffs and students).

In the Palearctic Region (Map 2) the genus also appears widespread, with numerous records from Europe (not individually spotted for present purposes), but only widely scattered records from temperate Asia. North Africa and vast stretches of Palearctic Asia are virtually bare of records. There are published records from Algeria, from the Altai Mountains in central Asia, from the Ussuri Region near eastern Manchuria, and from Japan. There are hitherto unpublished records in the collection of the National Museum of Natural History from Korea, China (Shanghai, and Harbin in Manchuria), and from the Wa-Hu Pass, 16,400 ft., on the border of Sichuan (Szechuan) and Tibet. *Protocalliphora azurea* was recorded from Iceland in 1889 and 1890, but the records are now doubted and the species has not since been recovered (Nielsen et al. 1954). This may reflect lack of collecting from birds' nests. For a half century *Protocalliphora* was known from Greenland from a single specimen collected in 1892, and the next specimens (reared) did not turn up until 1952 (cf. *P. tundrae*).

Any general statements must of course be qualified by "as far as known." The conspicuous lack of records from some areas and regions (Map 1) often reflects lack of collecting, or lack of rearing from birds' nests. Net collections yield very few specimens, and the full extent of a fauna can only be determined by the time-consuming method of locating birds' nests and rearing any mature larvae or puparia found after the young have left the nest. Unverified published records of species cannot usually be relied upon in view of the complexity now known to exist in the genus, although identifications by competent dipterists can be accepted as showing the presence of the genus *Protocalliphora*.

One odd record should be noted, only to dispose of it: *Protocalliphora azurea* (Fallén) from Kona, Hawaii (Grimshaw 1901: 27, as *Calliphora azurea*). Both Adrian C. Pont of the British Museum (Nat. Hist.) and Sabrosky have examined the specimen and find that it is actually the black blow fly, *Phormia regina* (Meigen). The genus *Protocalliphora* has never been found in the Hawaiian Islands. Hardy (1981) was in error in assigning this specimen to *Chrysomya megacephala* (Fabricius).

The Palearctic Fauna

Before dealing with the taxonomy of the Nearctic species, a brief review of the Palearctic fauna will be useful for comparison. Although the generic distinctness of the bird blow flies was first recognized by an American, some knowledge of the Palearctic species long antedated that of the Nearctic fauna. In the last few decades, knowledge of *Protocalliphora* in the Palearctic Region has expanded much like that in the Nearctic Region. For many years, European dipterists recognized only one or two species, *azurea* Fallén (*sordida* Zetterstedt) and *chrysorrhoea* Meigen, the latter often called a variety or subspecies of *azurea*. In the older literature, usage of *azurea* has been so varied, and so many species have been described since *azurea* was used in a broad sense, that it would seem hopeless to straighten out the published literature without a thoroughgoing revision based on good reared series of both sexes supported by larvae and/or puparia, followed by reexamination of all existing types and of voucher specimens for published records.

Table II. Palearctic species of Protocalliphora

Subgenus Protocalliphora	Specimens	Rearing	Provenance
1817. azurea (Fallén)(Musca)	♂♀(number?)		Sweden
violacea Meigen 1838 (Musca)[See note]	♂(number?)		Germany
dispar Dufour 1845 (Lucilia)	♂♀(many, at least 80)	reared	France
surcoufi Séguy 1928 (N. syn., Rognes, in litt.)	♂♀(number?)		France
1826. chrysorrhoea (Meigen) (Musca)	♀(number?)		Austria
sordida Zetterstedt 1838 (Musca)	2♀		Sweden(Lapland)
1928. falcozi Séguy (as azurea var.) (species by Séguy 1929)	♂♀(number?)	part reared	Austria
1956. asiatica Zumpt	1♂		USSR (Kirghizia)
1959. peusi Gregor and Povolný	2♂ ,1♀		Czechoslovakia
1960. amblyogma Peus	1♀		"Siberia"
1960. isochroa Peus	1♂ ,1♀	reared	se. Russia
1965. lii Fan	1♂	reared	China
kaszabi Grunin 1971 (syn. by Grunin 1975)	10♂		Mongolia
1966. maruyamensis Kano & Shinonaga	8♂ ,1♀		Japan
1966. distincta Grunin (as asiatica ssp.)	6♂ ,2♀		USSR(Kazakh SSR)
1966. frontosa Grunin	1♂		USSR(Tadzhik SSR)
1966. proxima Grunin	4♂		USSR(Maritime Terr.)
1970. sabroskyi Grunin	1♂		USSR(Central Siberia)
1972. nuortevai Grunin	7♂ ,7♀	reared	Finland
Subgenus Trypocalliphora			
1901. braueri (Hendel) (Avihospita)	♂♀ (3♂ ,2♀ rec. by Peus 1960)	reared	Austria
lindneri Peus 1960 (Trypocalliphora)	5♂, 4♀	reared	Germany
compacta Grunin 1966 (Trypocalliphora lindneri ssp.)	5♂, 12♀	reared	USSR (Leningrad Region)

Table II, end
Specific names of uncertain or disputed identity

** 1780. *reviso* Harris (*Musca*). Questioned as synonym of *Protocalliphora azurea* by Pont 1976.

** 1829. *carnarida* Stephens (*Musca*). Nomen nudum, synonym of *P. sordida* by Pont 1976.

1830. *caerulea* Robineau-Desvoidy (*Phormia*). Senior synonym of *Protocalliphora sordida* by Séguy 1929, but synonym of *P. azurea* by Zumpt 1956, 1965, and Schumann 1986.

* 1838. *violacea* Meigen (*Musca*). Listed by Rodhain & Bequaert 1916, from Meigen's type, as male of *Protocalliphora sordida* Zett., and by Schumann 1986 as a nomen dubium in Calliphorinae. Villeneuve wrote to Aldrich on Feb. 2, 1923 (card in U.S. National Museum of Natural History, Washington, D.C.) that the male type is undoubtedly *Phormia* (*Protocalliphora*) *azurea* (Fallén), confirmed by Knut Rognes, n. syn. in litt., but without dissection of the type. Whatever its true identity, the name *violacea* cannot be used in any case because of primary homonymy; it is preoccupied several times over in *Musca*, by Scopoli 1763, Panzer 1804, and Fabricius 1805.

1863. *nigripalpis* Robineau-Desvoidy (*Phormia*). Synonym of *Phormia caerulea* by Bezzi & Stein 1907, but synonym of *Protophormia terraenovae* (Robineau-Desvoidy) by Hall 1948, Zumpt 1956, 1965 (both as *nigripalpus*), and Schumann 1986.

* 1863. *corusca* Robineau-Desvoidy (*Phormia*). Synonym of *Phormia caerulea* by Bezzi & Stein 1907, but of *Protophormia terraenovae* by Schumann 1986.

* 1867. *nidicola* van Heyden *in* Nowicki (*Calliphora*). Nomen nudum, synonym of *Protocalliphora azurea* in Bezzi & Stein 1907, but of *Trypocalliphora braueri* in Schumann 1986.

* = Not mentioned by Zumpt 1956.
** = Not mentioned by Schumann 1986, nor Zumpt 1956.

Although a few of the recent species have been based on series of specimens, in general most species were described and named on the basis of very inadequate material (cf. Table II). Of the 22 available names (15 presently recognized species and seven synonyms; Schumann 1986), the type material of only seven was reared, including one in part. For the older species, the number of specimens was not specified, but in most cases was probably only one or two, judging from the surviving material in the well-preserved Fallén, Meigen, and Zetterstedt collections. The exception is *Lucilia dispar* Dufour, of which at least 80 were reared. Most disturbing, eight of the 14 *Protocalliphora* described in and since 1956 were based on a total of 14 specimens (11 males, 3 females, with only four of the 14 reared), and seven were described from one sex only.

Three names were first proposed as subspecies or varieties, in addition to the frequent treatment of *chrysorrhoea* as a subspecies of *azurea*. In three of the four cases, our experience in the Nearctic fauna convinces us that the characters given mark distinct species. Recognizing these as good species, and accepting some published synonymy--although with realization that a broadly based revision is needed, the presently known Palearctic species of *Protocalliphora* (including *Trypocalliphora*, see later discussion) can be listed as in Table II (species listed chronologically to show the historical sequence of the development of knowledge).

The revisions by Gregor and Povolný (1959) and by Peus (1960) contain keys to the adults of the Palearctic species and figures of male and female terminalia. Other figures of male terminalia were published by Séguy (1929), Hennig (1939), Zumpt (1956), Kano and Shinonaga (1966), Grunin (1966), and Rognes (1985). Figures of the immature stages were published by Rohdendorf (1957) and Ishijima (1967), although the immature stages are neither described nor figured in the detail of those in the present paper. The cited publications together with available Palearctic material of adults of nine species and puparia of five have made possible comparisons of the Palearctic and Nearctic species. Two species are recognized here as Holarctic, one, *chrysorrhoea*, virtually restricted to nests of a Holarctic bird, the sand martin (of European literature) or bank swallow, *Riparia riparia* (L.), in a distinctive ecological niche. Another species, *braueri* (syn., *hirudo*), is now recognized as Holarctic (Rognes 1985), although Peus (1960) had a different opinion and described the European form as a distinct species, *Trypocalliphora lindneri* (see discussion under *braueri*).

The literature on Palearctic species is usually not relevant to the present study and it has been referred to sparingly. It has been summarized by various authors, including the revisions cited in the preceding paragraph and the authors cited in the second paragraph of the introduction. There are numerous host records, but the Palearctic species of *Protocalliphora* have not been studied with the detailed ecological approach reported in the present study. It seems to us quite probable that more intensive biological and ecological studies, and attention to the characters of the immature stages, at least of puparia or 3rd-instar larvae, might reveal additional species in the extended Palearctic Region. Noteworthy also is the dearth of information about some of the species already named but known from single or few net-caught specimens, and association of sexes might even be questioned in some cases. The immature stages are unknown or undescribed for most of the species. These are difficulties that we are in no position to solve, but as far as possible we have compared the Nearctic and Palearctic species before publishing our new species.

Life History

The entire life history has not been followed for a single species of *Protocalliphora* in North America, and perhaps not anywhere in the world, but enough is known of parts of the cycle and the factors affecting it that inferences can be drawn and reasonable statements made. Observations and experiments were recorded by Bennett (1957), Whitworth (1976), and Gold and Dahlsten (1984).

Species of *Protocalliphora* apparently overwinter as adult flies. Overwintering in the egg or larval stages would be impracticable in most cases, even if they were cold tolerant, inasmuch as most birds do not utilize old nests, and such nests exposed to the weather are usually destroyed by the following season. Overwintering in the pupal stage would theoretically be possible, but evidence from Bennett's studies suggests that the pupae of these flies are also not cold hardy. Other evidence also points to the adult stage: (1) Adult flies emerge in mid or late summer from puparia in the nests of the last brood of birds although no more hosts will be available that season; (2) the length of life of adults can be long; (3) only empty, dead, or parasitized puparia are found in old nests taken in fall or winter (e.g., Dobroscky 1925); and (4) there are numerous late and early season records of adult flies, and even a few winter records.

Johnson (1930) listed a number of fall, winter, and spring captures of live adult *Protocalliphora* as indicating that "there is no doubt that *Protocalliphora* hibernates as an adult and awaits the arrival of the birds in the spring to oviposit in their nests." The winter dates of live specimens are especially significant. One specimen, a male *P. bicolor*, was collected at Rumney, N.H., on December 22 by P. J. Darlington, Jr., in chopping up a dead pine. Another specimen, a female *P. shannoni*, was collected at Concord, Mass., Jan. 10, 1925. Stiner (1969) recorded a frayed-winged male *P. metallica* found under bark on April 22, 1968 at Newark, Del., and from abundant evidence on nesting songbirds in the area he concluded that it could not have developed in a 1968 nest. From the timing plus its "old" appearance, the individual appeared to have overwintered. In Norway, Sømme (1961) recorded finding live *P. azurea* in crevices in mountain huts at 900 and 950 meters in March and early April. Adults of *Protocalliphora* can be long lived, as shown by Bennett's studies on their longevity, in which four species averaged 75-83 days, with maxima of 170-281 days. Whitworth (1976) found some adults still alive after eight months, in a cage exposed to winter weather but with sawdust offering available protection for the flies.

The available records indicate overwintering by both sexes, rather than by fertilized females alone as in other Calliphoridae known to overwinter as adults--e.g., the related *Phormia regina* (Meigen) and *Protophormia terraenovae* (Robineau-Desvoidy). Sperm apparently remain viable for long periods. Bennett found that sperm removed from the seminal receptacles of *Protocalliphora avium* 119 days after the last insemination appeared to be as motile as those removed only 10 days after insemination. Overwintering females might still be fertilized, however, and overwintering males only occasional survivors or infertile or ineffective. It would seem a more efficient adaptation in the life history of *Protocalliphora* for overwintering females to be fertile and ready to oviposit at once in the nests of the first broods of the early-nesting birds. An especially relevant bit of evidence here was contributed by the late G. J. Spencer (personal communication), who informed us that live puparia of *Protocalliphora* were found in magpie nests in British Columbia in February, the magpies nesting when snow was still on the ground. Obviously, gravid adult females were present in the winter and became active enough on warm days to deposit eggs in the new nests. It is of course still possible, with both sexes overwintering, that mating could occur when adults become active on warm days.

There is some evidence that males of *Protocalliphora* may gather at high, bare aggregation sites as do some other muscoid Diptera. Dodge and Seago (1954) recorded two "spp.", now known to be *sialia* and *braueri* (*hirudo*), on mountaintops in northern Georgia. For *P. spatulata*, records from Montana, Wyoming, and New Mexico specify the top of a peak or of a range. A number of other specimens are labeled from specific mountain peaks, and some of these may also have meant the mountaintop.

Female flies apparently oviposit in or on nests containing very young nestling birds. Although Zumpt (1965) stated that it is not known whether *Protocalliphora* is oviparous or larviparous, numerous observations now show that these flies are oviparous. In those studies, of course, the nesting birds were usually not disturbed, and in order to find mature larvae and puparia of *Protocalliphora* the nests were not collected until after the young had fledged. Bennett (1957) and Whitworth (1976) had numerous records and experiments with eggs deposited by females caged in the laboratory. In Europe there have also been studies on the oviposition of *P. azurea* on the great tit, *Parus major* L. (Eshuis-van der Voet 1972, 1975, and with others, especially Eshuis-van der Voet and Kluyver 1971, who established that the eggs were laid "in the nest-material, close to the nestlings.").

A unique observation is that of Meng (1954), recorded in an obscure and hard-to-find paper that summarized observations on *P. avium* during an intensive 4-year study of the Cooper's hawk in New York State, that "as soon as the hawk eggs start hatching, the adult flies, which look very much like bluebottle flies, deposit their eggs along the edges of the nests. Soon after being laid the fly eggs hatch, and the larvae find their way into the ear openings of the young hawks." Professor Meng (personal communication) has assured us that on several occasions he saw the flies laying eggs. A suggestive supporting observation is that of Law (1929), who observed a large fly looking like *Protocalliphora*, which, after some circling, finally entered a nest of the pygmy nuthatch, *Sitta pygmaea* (San Bernardino Mts., Calif.). Law further stated that "On many occasions I have seen these blue flies circling the entrance to nests containing the young of nuthatches, chickadees, and bluebirds."

Rausch (1972), in discussing cutaneous myiasis by *P. hirudo* (= *braueri*) in the head of a fledgling Wilson's warbler, speculated that the aggregation of larvae in a single locus in the head indicated the likelihood that "the eggs are deposited directly upon the head." Rognes (1985), in reviewing the literature and records on *Trypocalliphora braueri*, chiefly in Europe, concluded that the eggs were "probably laid in clusters directly upon nestlings of various ages, apparently also on newly hatched ones." Whitworth (1976), who often observed adults of *chrysorrhoea* about bank swallow burrows, recorded that "on one occasion, an adult was observed in a nest, walking from one nestling to another while dipping its abdomen. Eggs were laid on several nestlings in small clumps and were attached to the feathers. No eggs were found deposited in the nest material." On the other hand, the observation of Meng (1954) suggests that even in the case of burrowing larvae, eggs might be laid on the nest and the young larvae could find their own way to an appropriate feeding site.

Tirrell (1978) reported that females of *P. avium* "apparently laid eggs directly" on nestling red-tailed hawks. He observed flies "entering the ears of two nestlings from which larvae had been removed. Subsequently, each ear cavity was infested within one day by approximately 40 uniformly sized larvae, each 2.5 mm in length." In earlier correspondence (Feb. 17, 1976), Tirrell had elaborated on his field notes during hours of observations: "Typically, there were 5-20 adult flies around the nest. One or two flies continuously attempted to settle on a nestling's head and to walk to an aural cavity. Their success per attempt was very low, but at least several times an hour, a fly attained the edge of an aural cavity. These flies usually moved their heads in and out of the ear opening, then turned and held the posterior near the edge of the cavity, or walked around the edge, stopping frequently, and touched the end of the abdomen on the edge." Although eggs were not then observed, he "interpreted this as egg-laying behavior," especially when re-infestation occurred so quickly in cases where he removed all the larvae and reexamined within a day. Also, on several occasions he found scattered clumps of eggs near the axillary areas of the nestlings, showing that indeed eggs had been laid on the body.

The egg stage is short, probably 24-48 hours, although this is an inference because of uncertainty about the exact time of oviposition. In the laboratory, Bennett found that eggs laid by *P. sialia* hatched within 24 hours, but Whitworth (1976) recorded 38-43 hours for *P. chrysorrhoea* and 72 hours for *P. asiovora.* In nature, eggs laid on the nests of smaller birds with short nestling period would necessarily have had to hatch quickly for larval development to be completed before the young birds fledged and left the nest.

Details of the life history and biology will be given in a later publication. There are some differences in the data from the respective studies, probably affected by differences in temperature, the species of hosts, and the species of flies, and also affected by uncertainty of the time of oviposition and the duration of the egg stage and the larval instars, with some inferences necessary. The time in the 1st and 2nd larval instars is relatively short, with a longer period for the 3rd instar, the three totaling 7-15 days. The prepupal period is short, varying from 1-4 days, and the puparial period has a wide range, 9-36 days or perhaps even more.

There may be very few larvae in a nest, especially on small birds, or there may be surprisingly large numbers. Séguy (1955), in a general review of *Protocalliphora*, no doubt chiefly from the European literature, stated that "one finds from one to thirty larvae per nest," but that is an understatement in our experience. In contrast, Bennett found 970 in one nest (Sabrosky and Bennett 1958), with other records of 400-800 per nest, and Whitworth recorded larval populations of 300-1200 per nest. In general, the numbers of larvae and/or puparia reported in the literature are much lower, although occasionally in the hundreds (e.g., Dobroscky 1925, 343 in a nest of an American crow; Mason 1944, 442 in a nest of an American tree swallow; Kenaga 1961, 206 in a nest of an eastern bluebird; and Gold and Dahlsten 1984, 273 in a nest of a chestnut-backed chickadee). Whitworth (1976) has given a detailed discussion of a number of factors that affect larval populations in nests. Large numbers of larvae may result primarily from large nests with suitable hiding places rather than from large birds. Whatever the bird size, small or fragile nests are inadequate shelter for very many larvae.

Various authors have observed that the larvae are "tough". One dramatic bit of evidence was provided by Plath (1919a, b). He had placed some larvae in a fixing fluid for six hours, washed them in 50% alcohol, and placed them in 90% alcohol, but two days later they were still alive! Moreover some remained alive for two or three days in "very strong insect powder," which makes one wonder at some of the control recommendations that have been made.

The larvae of *Protocalliphora* are obligatory blood-sucking parasites. Bennett closely observed feeding by the larvae, and observations were also made by Whitworth (1976). Bennett further details the feeding as follows: The anterior part of the body, with mouth hooks thrust out, turns like a person turning a paring knife, with a cutting motion. It penetrates the skin a little way, then thrusts in the prothoracic fringe, which points backward and apparently holds the maggot in the wound. Then the maggot cuts farther in with the same cutting motion, followed by another grip with the fringe. At first the maggot braces itself against the nest wall, sticks, or other material. When it is well into the skin it relaxes, and starts to suck blood. It takes a long time for this, about 45 minutes to get the crop one-third full. The amount of blood can be seen through the body wall. Fully-fed larvae are distended with bright red blood, which gradually darkens and decreases in size as digestion and excretion proceed. Some larvae burrow into the skin and may embed themselves (*P. braueri*, apparently typically), but even these larvae are basically blood-sucking. Rognes (1985), discussing *braueri* in Europe, stated that "the digestive tract of the larvae mostly contains blood," even though the larvae "make permanent subcutaneous burrows" and "extensive tissue destruction may occur." Arnold (1919) had recognized this long ago, pointing out that "the maggot eats a burrow or chimney into the flesh of its victim, remaining stationary and feeding upon the fluids." Although larvae of *braueri* are often found attached to their host, some larvae containing a blood meal have also been found by Bennett in nest material.

Third-instar larvae take two or three blood meals for maturity to normal-sized individuals. Apparently they can complete development on only one blood meal, but the resulting adults are undersized, or "runts."

Raptorial birds in particular seem characteristically to have *Protocalliphora* larvae in the ear cavities, and numerous authors have recorded this (e.g., Burtch 1920, Sargent 1938, Hill and Work 1947, Hamerstrom and Hamerstrom 1954, Meng 1954, Hickey 1969, Tirrell 1978). The Hamerstroms and Meng, who have studied myiasis in hawks in considerable detail, differ on whether or not the larvae remain attached until mature. The Hamerstroms believed that the larvae "spend the entire time between hatching from the egg until pupation within the hawk's ear," whereas Meng stated flatly that "the maggots are intermittent feeders and do not remain in the ears continuously." Certainly the ears are not large enough to contain fully fed larvae in the numbers that have been reported. The Hamerstroms have graphically described what can happen: "We doubt whether it makes much difference how many maggots there are in the ear. We have never seen just a few; there have always been either a goodly number or none at all. As the young maggots grow larger, they become crowded in the limited space of the ear cavity.

Breathing space is probably even more at a premium than feeding space. One by one the surplus maggots dropped out of the ears of our Cooper's Hawk's and fell to the floor of the cage where they died, until at last each ear contained what looked like the maximum number of maggots that could fit into the distended opening. These did not leave the ear prematurely and pupated successfully." In nature, surplus maggots would probably have dropped into the nest material and survived to feed elsewhere on the body.

A few of the other published records are detailed enough to bear repeating. Meng (1954) wrote that "When the [Cooper's] hawks are very small the only indications one has that they are infected are the droplets of dried blood around the ear openings. The larvae grow rapidly on their diet of blood, and by the time the eyases are three weeks old their ear opening may be completely plugged by the protruding ends of the screw-worm fly larvae" [meaning in this case the bird nest screw-worm, *P. avium*]. Hill and Work (1947) found two young sparrow hawks [i.e., American kestrels] with a black crust around the openings of the ears: "When this had been removed, it became apparent that the ear canal was markedly swollen and filled with fat, squirming, grayish larvae. The right nostril and both ears of each bird contained larvae; none was found elsewhere. The nostrils were deformed, as the diameter of the larvae found there far exceeded the width of the normal nasal passages."

Do *Protocalliphora* Larvae Kill Young Birds?

This is a challenging question, often asked, and of special interest to the many people interested in birds and their welfare. A flat yes or no answer is impossible. Opinion is divided, probably depending on the particular circumstances or the experiences of each author. In certain cases, death of the nestlings can clearly be ascribed to the maggots of *Protocalliphora*, as when larvae in the head penetrate to the brain. In other cases, however, nestlings seemed to be relatively unaffected in any way noticed by the observers. There is always the possibility that the larvae contributed to stress or weakness that resulted in death later from other causes, such as unfavorable weather, malnutrition in periods of food shortage, and other parasites or diseases such as arborviruses, *Plasmodium*, *Leucocytozoon*, or *Haemoproteus*. For example, Wing (1956:405) in his book on the *Natural History of Birds* concluded that "In inclement weather when little food may be brought to the young, the attacks of the parasites upon weakened nestlings may cause death." One can understand that this could happen especially in the case of heavy infestations, hundreds of larvae, on small birds. Some authors have also recorded nestlings smaller or weaker than normal, and a longer than normal nestling period, in the presence of large numbers of maggots. Whitworth (1976) has pointed out the likelihood that birds losing blood daily to *Protocalliphora* "would be somewhat anemic and perhaps less able to withstand additional stresses."

It is our impression that the literature contains much anecdotal material on which assumptions have been based, rather than factual evidence. Statements seem to reflect a belief that because the larvae are hematophagous, they must ipso facto be considered lethal or at least potentially life threatening. In many cases, the feeding of *Protocalliphora* larvae may be no more serious to the host than the feeding and irritation of fleas on a dog. On the other hand, comments in the literature as to little or no evident injury might have reflected slight or relatively low numbers of maggots, or large numbers of maggots but in times of favorable weather and abundant food when the nestlings were vigorous.

Protocalliphora may also be, on occasion, unjustly accused of causing the death of nestling birds. In at least one case that we have been able to investigate (Munro 1949), the presumed *Protocalliphora* larvae taken from dead fledglings proved, from reared adults, to have been the scavenger calliphorids *Phaenicia sericata* (Meigen) and *Phormia regina* (Meigen), which had invaded the carcasses subsequent to their death from other causes (or *regina* might have been directly involved in myiasis). This is further borne out by the observation of Coutant (1915), who investigated the death of a nestling American crow infested by *Protocalliphora* larvae and found that the crow actually died of a "malarial fever" and that the erythrocytes contained "enormous quantities of *Halteridium*" [=*Haemoproteus*, protozoan blood parasites]. In another case, "*Protocalliphora*" larvae submitted for identification as having killed barn swallows, undermining the back skin and "clear down next to the ribs in a bloody necrotic mess in one bird," proved to be larvae of *Phormia regina.*

Some authors have recorded dead birds along with infestation by *Protocalliphora*, although not assigning definite responsibility for the mortality, appearing to leave guilt by association. For example, Shannon and Dobroscky (1924) listed the type series of *hirundo* as 64 specimens from cliff swallow nests, adding "Many nests examined had dead remains of young." Of material received for identification, two larvae and 12 puparia were found at Auburn, Ala., in the nest of a wren "where young birds failed to mature."

Eliminating pure speculation and mere repetition of the views of others, we will summarize the expressed opinions and evidence, with appropriate quotations.

Little or no injury to the nestlings was reported by a number of authors, with various comments, e.g., Jellison and Philip (1933) and Jellison (1949), no fledgling crows or magpies ever found dead in the nest; Stoner (1936), bank swallows "suffering no serious handicap"; Krug (1941), in bluebirds and tree swallows "the mortality is almost negligible"; Zeleny (1970), bluebirds are "usually able to survive," and blow fly infestation, unless heavy, "does no noticeable harm to the birds"; and Seidensticker and Reynolds (1971), "we would attribute no mortality [of hawks] to these larvae." Many authors expressed the caveat, however, that even though they observed no injury or mortality, heavy infestations of larvae might have those effects, or (Gold and Dahlsten 1984) might so affect the young that they fledged "in a weakened condition." McAtee (1929) stated unequivocally, however, that "heavy infestations [of *Protocalliphora* in bird houses] are not especially destructive to nestling birds." Whitworth (1976), while acknowledging that high numbers of *Protocalliphora* larvae might be lethal, concluded from his studies of larval populations that "near-lethal numbers are uncommon under natural conditions." Bennett (1957) concluded that "in general it appeared that the *Protocalliphora* did not seriously harm their avian hosts," and Gold and Dahlsten (1984) found nestling mortality of chickadees "negligible," although they hypothesized that heavy infestations would result in weakened fledglings.

Except for references to *braueri*, discussed later, many European authors have not regarded *Protocalliphora* as unusually serious. Owen (1954) in a review in *British Birds* concluded that "it would appear that ... mortality of the host is exceptional and possibly always linked with other factors." A year later, Owen and Ash (1955) published additional British records of *Protocalliphora* infestation and again suggested "that any resultant mortality among nestlings is most exceptional." On the other hand, Boyd (1935, 1936) concluded that prevalence of *Protocalliphora* larvae "may materially affect the size of the broods and account for the broods of one or two young that are often reported" and further that the larvae "probably cause many deaths among the young [barn] swallows."

Death or serious injury has been reported by a number of authors, although it is not always clear what is observation and what is presumption based on what others have written or on the large numbers of larvae found in nests. Sometimes the presence of larvae is linked with other factors as contributory causes of death. For example, Allen and Nice (1952) on purple martins reported heavy infestations of *Protocalliphora* larvae coincident with cold wet weather and concomitant shortage of food, resulting in the death of many nestlings. Zeleny (1970) found that while in general *Protocalliphora* infestation does no noticeable harm to young bluebirds, the blood loss from heavy infestations "seriously weakened" young birds and "may be a contributory cause to the death" for both nestlings and fledglings, and "Very heavy infestations ... may kill the young birds outright."

Infestation of the ears of birds by *Protocalliphora* larvae is more obvious than the somewhat hidden feeding on the under side of the body, and when severe it attracts special attention. Sargent (1938) reported so many maggots in the ears of a red-tailed hawk that the ear openings were stretched to twice normal size and "completely plugged by the caudal ends of the maggots." Tirrell (1978) and the Hamerstroms (1954) both noted that infested hawks shook their heads intermittently or scratched them, apparently irritated by the larvae, although apparently there was no permanent damage. Hagar (1969, on peregrine falcons), Crocoll and Parker (1981, on broad-winged hawks), and Bortolotti (1985, on bald eagles) found, respectively, "no apparent ill effect," "no major deleterious effects," and "no indication of any ill effects" from *Protocalliphora* infestations. Meng (1954) stated that none of the observed Cooper's hawk eyases "seemed to be weakened appreciably." On the other hand, Bent (1937) recorded the results by a correspondent of skinning and dissection of two young red-shouldered hawks, in which the maggots had not only disfigured the outer ear cavities but had destroyed the ear drums. We conclude that, except for the occasional instance when such serious damage has been done, in general the large young of raptors can support sizeable numbers of maggots, the young birds have long nestling periods, the maggots mature and leave the birds, and the birds recover from any ill effects and fledge successfully. As the Hamerstroms remarked, when larvae left the ear, "By then our hawks were as large as branchers and made a good recovery."

Whitworth observed larvae commonly in the noses, ears, and inside feather sheaths of nestlings of its common hosts, especially raptors and corvids (crows, ravens, magpies). Whitworth (1976) noted that "the presence of these larvae is often indicated only by the scablike accumulations of larval excreta around cavity openings."

Subcutaneous infestation by *Protocalliphora braueri* (*P. hirudo* of Nearctic literature) is another feature of *Protocalliphora* that deserves special mention. A few authors have recorded serious damage by larvae of this species. R. C. Miller (1929) described a young yellow warbler with numerous pockets containing larvae: "Two of the lesions entered the pleural cavity, and the lungs had been partly devoured," which he considered evidence that when the larvae were numerous enough "they may literally devour the young birds alive." Rausch (1972) detailed the damage done to a fledgling pileolated [now Wilson's] warbler by *Protocalliphora* larvae in subcutaneous cavities: "Soft tissues on both sides of the beak, including the right nostril, had been destroyed, and the resultant two dorsal grooves were continuous with subcutaneous cavities extending below the eyes bilaterally, ... The bilateral cavities, each occupied by a single larva, extended posteroventrad below the eye to the level of the external opening of the ear." The bird, perched motionless, was easily captured, refused food, and soon died, after which dissection revealed the damage. Bedard and McNeil (1979) recorded that a nestling savannah sparrow had lost an eye from larval action. The many records listed under *P. braueri* (q.v.) show the subcutaneous sites of the larvae, variously described as in boils, swellings, cysts, cavities, purulent sores, or lesions, and obviously some locations would have resulted in severe damage or even death. Shannon and Dobroscky (1924) listed a fragmented adult fly from Ontario, misidentified then as *P. splendida aenea* but now recognized as *braueri*, labeled as "from brain of a living fledgling of sparrow kind." Some years ago, the senior author identified an adult *braueri* reared by Murray L. Johnson of Tacoma, Wash., from a juvenile barn swallow, and Dr. Johnson (personal communication) kindly supplied the following information: "It was obviously very ill and unable to fly, though the feathers were well developed. There was a large cavity extending cranially from the right naris, containing two large larvae. The upper beak was weakened and movable; the right eye was closed." The bird died about a day later. Many labels and published information state that the larvae were on the head, and larvae in such a location could easily have done serious damage.

In Europe, authors have also noted that infestation by *braueri*, especially if heavy, may kill young birds. Rognes (1985), in his recent revision of *Trypocalliphora braueri*, reviewed the European publications and records and concluded that "infestation weakens the host and is frequently lethal."

In conclusion, we believe that *Protocalliphora* larvae do not ordinarily kill nestlings or fledglings, but under certain conditions they may do so, or they may so weaken the young that other factors will be lethal or will combine with larval feeding to cause death.

Host Relations and Ecology

All species of *Protocalliphora* are obligatory bloodsucking parasites of nestling birds, and they need more than one blood meal for maturation. Field studies and abundant available records suggest that, with the possible exception of species with wet and messy nests, all species of birds with altricial young (confined to a nest for some period of time) in the range of *Protocalliphora* will eventually be recorded as hosts of one or more species of *Protocalliphora*. If such birds within the range of *Protocalliphora* are not now known to be attacked, the dearth of records may result from (1) lack of collecting of the nests of those species, (2) low percentages of parasitism and reduced chances of finding infested nests, or (3) the nests may be absent or rare in certain ecological niches. Many species of birds recorded from North America occur in geographical areas (e.g., Florida, Texas) outside the known range of *Protocalliphora*. On the other hand, precocial young birds would not be expected to be parasitized by *Protocalliphora* larvae, which need the relative permanence and stable food source of nestlings confined to a nest. Of course, even birds with precocial young could be hosts of *Protocalliphora* because of direct oviposition by *P. braueri* (e.g., there are records from domestic chickens and a budgerigar). Probably precocial young are not, or not often, as closely examined as nestlings.

The absence or virtual absence of host specificity, and the demonstrated occurrence of some species of *Protocalliphora* in the nests of a number of species of birds, make a list of hosts of less significance than formerly. For individual species of *Protocalliphora*, however, the host records could gradually contribute to a typical or characteristic ecological picture for each species of fly, or of host preference if not host specificity.

Hall (1948) listed 51 species of birds reported as natural hosts of *Protocalliphora* in North America. To this list may be added 31 from later published literature and 57 (including three subspecies) formally published here for the first time (including dissertation records by Bennett and Whitworth), making a total host list at the present time of 139 for North America. In addition to these natural hosts, there are published records of *Protocalliphora* on two domesticated birds, the domestic chicken and the budgerigar, and from a nest of caged ringed turtle-doves used in experimental studies in Colorado. A list of the common and scientific names of the natural hosts will be found in the Appendix.

The ordinal distribution of the known hosts, shown in Table III, reveals the great preponderance of parasitism by *Protocalliphora* on the perching birds or Passeriformes. In terms of the percentage of species presently known to be parasitized by *Protocalliphora*, the passerine birds and the raptorial birds (Falconiformes and Strigiformes) are almost equally good hosts, with 41% for the Passeriformes (114 species out of 280) compared with 37% for the raptors (20 out of 54 species breeding in our region). Precociousness of young virtually precludes parasitism by *Protocalliphora*, hence the water and shore birds and the Galliformes are free of such parasitism, except for two unverified records of *Protocalliphora* (probably *braueri*) larvae on domestic chicks (Hearle 1938, Munro 1949). However, the young of water and shore birds are not usually handled and observed as closely as those of nesting birds with altricial young, and if *P. braueri* deposits eggs directly on young birds as has been suggested (Rausch 1972, as *P. hirudo*), parasitism of the young of water and shore birds might occur in nature.

Table IV shows the distribution of records of *Protocalliphora* infestation among the families of perching birds, and it reveals the widespread occurrence of the parasitism and the sampling that has been done, either actively undertaken or achieved by the accumulation of scattered and chance rearings. If we disregard the skewed percentages that result from calculations involving only one to four species, the Table shows a range of between 17% and 47% for most families, very high for Hirundinidae (87.5%). The figures undoubtedly reflect the fact that most families of birds have some common species with readily observable or available nests that are easily sampled. Uncommon or secretive or inaccessible species are rarely sampled, and rare samples can easily be negative, especially if the percentage of parasitism is low for that species of bird or in that area or region. Swallows are common and readily observed, and hence most of our species have been sampled for *Protocalliphora.* The type and site of nest are factors: tight or solid cup nests or nests in cavities retain larvae and puparia of *Protocalliphora* more readily than "loose-weave" or few-stick nests, which may not provide suitable resting sites for larvae between blood meals, or from which larvae may drop to the ground and be missed by collectors or observers.

Mexico has the potential of a number of additional host species, but since no rearing records are known from that country Table IV is limited to hosts nesting north of Mexico, either as residents or as nesting migrants. In reality the number of available hosts should be lower and the percentages higher, because a number of chiefly Neotropical or subtropical birds occur along the Gulf Coast or in southern Florida and southern Texas, areas that are outside of what appears to be the normal range of *Protocalliphora* (cf. Map 1).

Bennett's ecological study (Bennett 1957, Sabrosky and Bennett 1958) was the first broadly based investigation of habitat preference by species of *Protocalliphora.* His studies indicate that most species of the genus show habitat or ecological preference rather than host preference or host specificity. Different species of *Protocalliphora* tend to be found in different types of environment, and even in different levels or strata within those environmental types. Whitworth (1971, 1976) also studied the nest ecology of *Protocalliphora* but in such a diverse area that the data did not show habitat preference as much as nest site preference.

Authors differ on whether some situations should be referred to as host preference, habitat preference, or nest site (niche?) preference. Several species of *Protocalliphora* that have been extensively studied in the field (e.g., *avium*, *asiovora*, *metallica*, *sialia*) have a number of hosts but the hosts usually nest in the same type of habitat, e.g., in the canopy, near the ground, or in cavities. Probably it is technically a misnomer to speak of habitat preference; the birds prefer the habitat, the flies prefer the birds. Or do the flies also show preference for nest sites? Perhaps we may be pardoned if we continue to speak of habitat preference to avoid any circumlocution. For *P. halli* four hosts are known but barn swallow predominates. Whitworth (1976) reported that *halli* (as species IV) commonly infested barn swallow nests

Table III. Ordinal distribution of known natural hosts of *Protocalliphora* in North America north of Mexico

Orders of birds	Total Spp.*	Known as Hosts	% Known
Water and shore birds	211	0	0
Galliformes (grouse, quail, etc.)	21	0**	0
Raptorial birds: Falconiformes (hawks, etc.), Strigiformes (owls)	54	20	37
Columbiformes (pigeons, doves)**	11	1	9
Cuculiformes (cuckoos, etc.)	6	1	17
Miscellaneous orders: Psittaciformes (parrots), Caprimulgiformes (nighthawks, etc.), Apodiformes (swifts, hummingbirds), Trogoniformes (trogons), Coraciiformes (kingfishers, etc.)	30	0**	0
Piciformes (woodpeckers, etc.)***	22	3	14
Passeriformes (perching birds: blackbirds, buntings, sparrows, warblers, etc.)***	280	114	41
Totals	635***	139***	22

* Based on the AOU "Check-list of North American birds", 6th edition, plus Greenland species. The numbers include the species listed as "breeding" or "resident," including species non-native but established and therefore possible hosts of *Protocalliphora*. Extinct species are not included.

** Records from budgerigar, baby chicks, and ringed turtle-dove are not counted here, because they are unnatural hosts (captive audience!).

*** The figures include three recently classified as non-typical subspecies: red-shafted flicker in Piciformes, Oregon junco and Audubon's warbler in Passeriformes (Cf. Appendix).

Table IV. Distribution of known hosts of *Protocalliphora* among the families of perching birds (Passeriformes)

Families	Total Spp.*	Known as Hosts	%Known
Aegithalidae (bushtits)	1	1	100
Alaudidae (larks)	1	1	100
Bombycillidae (waxwings)	2	1	50
Certhiidae (creepers)	1	1	100
Cinclidae (dippers)	1	1	100
Corvidae (crows, jays, magpies, ravens)	15	6	40
Emberizidae (buntings, blackbirds, juncos, sparrows, towhees, warblers, etc.	130	47	36
Fringillidae (finches, siskins, etc.)	15	7	47
Hirundinidae (martins, swallows)	8	7	87.5
Laniidae (shrikes)	2	1	50
Mimidae (catbirds, mockingbirds, thrashers)	9	3	33
Motacillidae (pipits)	6	1	17
Muscicapidae (bluebirds, robins, solitaires, thrushes, all in sub-family Turdinae, formerly Turdidae)	19	8	42
Paridae (chickadees)	10	4	40
Passeridae (house sparrow)	2	1	50
Sittidae (nuthatches)	4	4	100
Sturnidae (starlings)	1	1	100
Troglodytidae (wrens)	9	4	44
Tyrannidae (flycatchers, kingbirds, pewees, phoebes)	33	13	39
Vireonidae (vireos)	11	3	27
Totals	280	115	41

* Tabulated as in Table III.

bridges but did not infest cliff swallow nests in the same habitat (or nest site). *P. parorum* is a similar case, in that although it is known from nine different hosts, it is overwhelmingly a parasite of chickadees (*Parus* spp.), if we may lump three species of one genus as a kind of host preference. At one time, host specificity also seemed to be true for *chrysorrhoea* on bank swallows, although that might also have been interpreted as habitat preference (in a narrow sense) or nest site preference. Now that four other species of *Protocalliphora* are known from bank swallows, host specificity does not appear so reasonable. One might argue for either host preference (by selective oviposition) or nest site preference, and either would seem equally plausible, there being (with rare exceptions) only one species of bird in that one nest site or niche. Areas such as marshes, forest canopy, etc., are habitats in the broad sense of a certain type of environment, and they might also be nest sites in a narrow sense, although we prefer to think of a nest site as a particular place, or type of place, or niche, within a habitat.

Some species of *Protocalliphora* are known to parasitize many species of birds, but others are restricted to one or a few species of birds. The highest totals to date are by *braueri* on 42 species of birds (plus domestic chicks and a budgerigar), *metallica* on 35, and *sialia* on 33. These are common species often encountered in rearings, and all are parasites chiefly of small and ground- or low-nesting birds where the numbers of species of such birds as sparrows, warblers, and the Tyrannidae (flycatchers, kingbirds, phoebes) make possible the large totals. In contrast, *avium* is known from only 12 species because it is a parasite of large birds, chiefly raptors and high-nesting birds such as crows, with young confined to a nest for a long period of time, long enough for the relatively slow-growing *avium* to mature.

On the reverse side, some birds are parasitized by only one or two species of *Protocalliphora*, although some of this probably reflects lack of collecting. Again in contrast, three common birds with easily accessible nests that are most frequently sampled show 10 species on barn swallows and 9 each on American robins and common grackles.

Mixed infestations occur occasionally, and these can cause trouble in the correct association of sexes and of larvae or puparia. Bennett found 12.5% of 669 nests and Whitworth (1976) found 7.1% of 869 nests contained mixed infestations.

The host relations of *braueri* are especially interesting, as the species is often recorded in "tumors" or swellings beneath the skin, not only in nestlings and sometimes in fledglings, but at least once in a juvenile bird. It has been found in a wide range of hosts, from house finch to golden eagle. Presumably it may occur on almost any species of young bird. For further details, see the section on the adults of *braueri*. A similar special habit (see discussion under the species) is the infestation of ears and nostrils by *avium* and its western relative, *asiovora*, and sometimes by *chrysorrhoea* and *sialia*.

Parasites of *Protocalliphora*

Parasites and predators can be summarized briefly here. The most common is the polyphagous pupal parasite *Nasonia* [or *Mormoniella*] *vitripennis* (Walker) (Pteromalidae in Hymenoptera Chalcidoidea), which sometimes kills high percentages of *Protocalliphora*, as it does of many other muscoid flies. *Morodora armata* Gahan, another pteromalid, has been reared only from *Protocalliphora*, but it is seldom recorded. A third pteromalid, *Dibrachys cavus* Walker, was recorded as reared from *Protocalliphora* sp. by Burks (1979), but we have been unable to locate a North American record; possibly the entry was based on a published record in Europe. Finally, the encyrtid *Tachinaephagus zealandicus* Ashmead (Chalcidoidea) was recently recorded from *Protocalliphora* n.sp. by Gold and Dahlsten (1981).

Two other chalcidoid Hymenoptera have been reared from nests containing *Protocalliphora* but without definite association with it. The pteromalid *Muscidifurax raptor* Girault and Sanders was reared from a nest containing *Protocalliphora asiovora*, and three specimens of a eulophid, *Pediobius* sp. possibly *alcaeus* (Walker), were reared at Portage, Alaska, in 1978 (G. E. Haas) along with *Protocalliphora hesperia* from a bird nest, probably that of the gray jay, *Perisoreus canadensis* (L.). From the known hosts of the chalcidoids, it seems likely that they emerged from some other inhabitants of the nest, not *Protocalliphora*.

Evolution and Phylogeny

In the present state of knowledge of the group, with rapidly growing information about species in both Nearctic and Palearctic Regions, speculation on phylogeny seems premature. Attention has been concentrated on the difficult and presently more important problem of better understanding of the species involved, with more precise definition for purposes of identification and recording of data.

As for the origin of the genus, it seems likely that *Protocalliphora* arose from a carrion-feeding calliphorid or calliphorid-like ancestor, and became adapted to feeding on living hosts with the availability of young birds in soiled nests. As James (1969) has pointed out in discussing the origin of parasitism, there is a logical transition from scavenging to preying on other organisms in the same habitat to feeding on young birds in the soiled nests. Within the calliphorid tribe Phormiini, there are suggestions of these stages, culminating in the obligatory hematophagy of larval *Protocalliphora*. Even within this genus itself there are stages from simple intermittent external bloodsucking, to penetration of external openings such as the ears and nares, to existence in subcutaneous cysts, though still bloodsucking.

The wide range of hosts and the usual lack of host specificity preclude any consideration of parallel evolution of hosts and parasites, such as has been done, for example, in parasitic mites and Mallophaga. Maggots of *Protocalliphora* are - with perhaps rare exceptions - not permanent parasites, only intermittent feeders, but they are obligatory parasites, and they will apparently feed on any birds with nests and nestlings. The availability of such a "captive audience" is the determining factor. Feeding on the blood of young birds may have developed relatively recently, probably from saprophagous larvae in birds' nests to occasional or partial hematophagy and finally to obligatory hematophagy, but with specific association with certain host species. The parasites have by now become completely adapted to getting their blood meals from nestling birds and require blood meals for development. However, even though evolution has progressed to the point that parasitism on birds is obligatory and blood meals absolutely necessary for the parasites, experimental evidence shows that *Protocalliphora* larvae will also complete development on mammalian blood, including that of man, if a mammalian host is made available to them. Perhaps this is an indication that this last evolutionary step of obligatory parasitism on birds is relatively recent and has not yet become so firmly fixed but that the larvae can still develop on other bloods. Probably the isolating mechanism here is whatever attracts the female *Protocalliphora* to a nest for oviposition, whether the hatching of the eggs, or the oxidation products from the egg remains, or the odor of the new-born young.

The close similarity of most species of *Protocalliphora*, the apparently high degree of habitat preference, and the occurrence of mixed infestations in nests suggest the possibility that hybridization might have at times played some part in speciation in the group. Perhaps the phylogenetic picture is partly reticulate evolution, rather than a dichotomous branching. But this is only speculation at this time, and moreover experiments by Bennett, while not exhaustive, indicated that hybridization was unlikely.

Collecting, Rearing, and Preserving *Protocalliphora*

It is surprising that many papers on parasites of birds, and even directions for collecting the parasites, make no mention of *Protocalliphora*. Attention is given to lice, fleas, louse flies (Hippoboscidae), ticks, and mites, but *Protocalliphora* larvae, in spite of the fact that they are fairly common and sometimes abundant, seem to escape notice.

Adults of *Protocalliphora* are seldom collected with a net, and then never in numbers. The most practical method of securing adults involves rearing them from larvae and/or puparia found in birds' nests. This method usually makes possible a good series of specimens, samples of the immature stages, and a positive identification of the host, especially if the nest is located while the adult birds are still available.

Anyone interested in taking bird nests for scientific purposes, or in the removal of nests and use of substitute nests for parasite control (see next section on Control), must obtain a permit from the appropriate United States agency, the Canadian Wildlife Service at Ottawa, or a regional office of the Fish and Wildlife Service of the U.S. Department of the Interior. State or provincial permits may also be required.

Under most conditions, a nest should not be taken before the young birds have left it, although for experimental studies, or under certain circumstances where parasite control was desired in regularly observed nest boxes, nests have been removed and substitute nests provided. From the standpoint of the collector of *Protocalliphora*, early removal of a nest is both unnecessary and undesirable. With most birds, the period in the nest is short enough that the larvae of *Protocalliphora* require the full period to reach maturity. Ideally, an active nest should be located, the host identified, and the nest marked or noted for observation and collection at a later time. The nest should be examined for immature *Protocalliphora* as soon as possible after the fledglings have left, because some larvae may drop to the ground to pupate. This is particularly true of birds that build small, loosely woven nests; in large or compact nests, most or all larvae will pupate in the nesting material. The nest should be torn into small pieces and searched thoroughly for the immature stages. In cup nests such as those of robin and barn swallow, the puparia may even be cemented into the mud of the nest. With loosely constructed nests such as those of the mourning dove, or with nests on the ground such as those of the white-throated sparrow, the ground immediately under and around the nest should be searched to a depth of three or four inches.

The recorded data should always include the locality, name of collector, and name of host. Other useful and desirable data are date of nest collection, date of adult emergence, habitat in which the nest was found, and height of nest above the ground.

Larvae of *Protocalliphora*, unlike those of most calliphorids, do not survive in damp surroundings. They are best maintained in dry sawdust until they pupate. Large larvae about to do so are recognizable by the lack of contents in the alimentary tract, and by a uniform gray, cream, or yellow color. Other large larvae (10-15 mm) will pupate also in three to four days. A sample of the larvae should be preserved for study by placing them in a KAAD solution (1 part kerosene, 10 parts 95% ethyl alcohol, 2 parts acetic acid, 14 parts dioxane) for 24 hours, and storing in 70% ethyl alcohol + 10% glycerine. If KAAD solution is not available, the larvae should be killed in hot water before storing in preservative.

Puparia being held for emergence of the adults are best maintained in a mixture of sawdust and boric acid (about 1% of boric acid powder), slightly wetted. Ideally, some or all of the puparia should be segregated in separate vials, so that each emerging adult can be associated with its own puparium. Individual adults should be kept alive for 24-48 hours after emergence, for maturation and hardening of the integument before they are pinned. Adult and entire puparium should be kept together. The puparium and the two halves of the cap can conveniently be stored together in a gelatin capsule, which can then be pinned beneath the adult. Empty puparia should be stored dry.

The accurate association of adult and immature stage is of great significance. It may be critical in the differential analysis of mixed infestations, and in the study of incompletely known species and species complexes, such as the western fauna at the present time. Here the taxonomist is at the mercy of the collector. If adults and puparia from a given nest rearing are heterogeneously associated, but only one species happens to be present, no confusion will result. On the other hand, improper association in mixed infestations, especially if only a few puparia are preserved, can lead to erroneous conclusions. If accurate association of puparia and eclosed adults is not possible or practicable, then puparia should be preserved separately so that the taxonomist will be aware of the lack of individual relationship. He can thus deal with a *series* of adults associated with a *series* of puparia. If both series are homogeneous and of reasonable length, he has a useful association upon which to base conclusions and further study, despite the lack of individual associations.

Control of *Protocalliphora*

Although our studies have not been concerned with control of blow fly infestations, we give here a review of suggestions that have appeared in the literature. Obviously, control in nature for such flies as *Protocalliphora* is impracticable in most cases, as well as unnecessary. Furthermore, migratory birds ordinarily would not and should not be disturbed in their nests and the need for control would not be apparent. Where nest boxes are maintained and cleaning operations show that the birds were heavily attacked, some might wish to try to control the blow flies in order to alleviate the worst of the attacks and to promote survival of the young birds in the next brood or the following season.

Regular removal and destruction of nests immediately after the young birds have fledged would destroy larvae and pupae of *Protocalliphora* and might tend to reduce the local population and perhaps hold it down to reasonable levels, although we question any significant or lasting effect on the total area population of the flies. Unfortunately the procedure would also destroy the chalcidoid parasites, which are very common and often kill 25-100 percent of the puparia in a nest. Nest destruction would seem to have the greatest impact on hymenopterous parasites as the season advances, when the percentage of parasitism usually increases. The commonest parasite, *Nasonia vitripennis*, is common everywhere on a wide variety of muscoid hosts, and probably removal of a comparatively few would have only a slight local effect and no serious or lasting inhibition on the area population of the chalcidoids.

Another method of combating parasitism by *Protocalliphora* larvae was reported by Johnson (1932), based on information from one of his correspondents. The nest of the first brood of birds was removed and a handmade nest was substituted. This removed any larvae as well as any fly eggs that had just been laid, and gave the young birds a clean start. Such a method, however effective, would be time-consuming and could only be used in a limited way on nests or nest boxes that were being closely observed or being used in experiments or life history studies. It would also not deal with eggs laid directly on the nestlings.

Zeleny (personal communication) recounted to the senior author his experiences with nest removal, which were the basis of his recommendations in Zeleny (1970). In a closely observed nest, young bluebirds 10-12 days old were weak, and something seemed to be amiss. Inspection showed the nest to contain over 100 maggots of *Protocalliphora*, and it was therefore removed and a clean nest of dry grass was substituted. The young perked up and fledged successfully. Another infested nest was likewise successfully treated.

Some authors have recommended dusting the birds' eggs with a mild insecticide, a mild flea powder (Highhouse 1963), pyrethrum powder (Brower 1966), or rotenone dust (Zeleny 1970, 1976), in order to control the fly larvae before the chalcidoid parasites are present. The timing would depend partly on the duration of effectiveness of the insecticide. For example, pyrethrum dust has a residual life of about one hour, perhaps two to four hours in a dark nest box. If fly eggs are not laid until after the birds' eggs hatch (cf. life history studies), there would be no fly larvae present to be killed if the nests were dusted too early; if fly larvae were present, dusting would have to be very thorough because only larvae actually contacted with the insecticide would be likely to die. Note the "tough" larvae mentioned in the section of life history studies.

Zeleny (1976) has given a well considered discussion of the possible use of insecticides, listing 1% rotenone powder, pyrethrum powder, and pyrethrins. Pyrethrin formulations especially for use on caged birds could be used in nest boxes. All writers caution that although the insecticides mentioned do not appear to harm the birds, they might have obscure or deleterious effects that are not presently known, and hence they should be used sparingly and not routinely. Zeleny stated that "the dusting may be done before or after the eggs have been laid, but before they have hatched," although no fly larvae will be present then and any effectiveness will depend on the duration of the insecticide's potency. He does offer a further thought: "If it is necessary to treat a nest after the young birds have hatched, the nestlings should be lifted gently from the nest before the application, and the insecticide should be worked carefully into the nesting material with the finger so that as little of it as possible will come into contact with the nestlings when they are put back." No doubt this would be effective, because the larvae leave the nestlings after feeding and withdraw into the nest material where they could come into contact with the insecticide. We believe that under most conditions and for most people, however, birds in nest boxes should be left undisturbed and unexamined for parasites. If examination of the nests after the young have fledged showed serious blow fly infestations, then the nests could be destroyed and the boxes cleaned out, and perhaps the following season or for the following brood dusting could be done in the manner suggested by Zeleny.

A reader desiring to control *Protocalliphora* larvae in nest boxes or under experimental conditions would do well to check with a veterinarian specializing in or at least concerned with bird care. Sevin (carbaryl) is commonly used by veterinarians to control parasites on adult domestic birds, with no ill effects, and perhaps this would be effective in a nest box although we do not know that this has been tested. This product has about a one-week residual effect, so its effectiveness would be longer lasting than pyrethrum, for example, if it proved to have no ill effects.

Two factors need to be kept in mind. First, infestation by blow fly larvae may not be detected without undesirable disturbance to the birds. Second, larvae of *P. braueri* may stay attached to the host and would not be removed with the nest, if that method were used. Larvae that may stay in the nares or ears, such as those of *Protocalliphora avium* and *asiovora*, would also remain on the birds, of course, but birds so infested (raptors, corvids, bank swallows, starlings) are not those involved in nest boxes.

A side effect of all the methods suggested would be the reduction or even elimination of mites, fleas, and lice, no doubt to the delight and better health of the nestling birds.

Taxonomic Section
Genus *Protocalliphora* Hough

Protocalliphora Hough, 1899a: 65-66. Two species. Type species, *Musca azurea* Fallén, by original designation.
Avihospita Hendel, 1901: 29. Five species. Type species, *Musca azurea* Fallén, by original designation.
Protocalliphora (syn., *Avihospita*); Aldrich, 1901: 68; Aldrich, 1905: 523.
Apaulina Hall, 1948: 179. Ten species. Type species, *Protocalliphora avium* Shannon and Dobroscky, by original designation.
Protocalliphora (syn., *Apaulina*, ex Sabrosky in litt.); Hamerstrom and Hamerstrom, 1954: 5; James, 1955: 24; Zumpt, 1956: 94.
Protocalliphora, subgenus *Orneocalliphora* Peus, 1960: 198. Four species. Type species, *Musca chrysorrhoea* Meigen, by original designation.
Trypocalliphora Peus, 1960: 199. Three species. Type species, *Avihospita braueri* Hendel, by original designation [Sabrosky 1967: 122 was curiously and inexplicably in error in citing *T. lindneri* Peus as the type species].
Protocalliphora (syns., *Apaulina*, *Orneocalliphora*, *Trypocalliphora*); Zumpt, 1965: 83.
Trypocalliphora; Rognes, 1985: 371-382. [Revision].
Protocalliphora; Shewell, *in* McAlpine et al, 1987: 1134-5, 1140, 1143.

The generic diagnoses of Hall (1948, as *Apaulina*) and Zumpt (1956) will serve except for the few characters in which *Apaulina* was said to differ from *Protocalliphora*. Briefly, *Protocalliphora* can be characterized as follows in the family Calliphoridae:

Predominantly metallic blue to bluish purple or bluish black, with a few species partly aeneous or cupreous, all somewhat gray-microtomentose; epistoma strongly warped forward but not elongate, ending only barely below level of vibrissae; scutum in most species more or less flattened centrally, and with long, strong, erect bristles, including several pairs of well developed presutural acrosticals; greater ampulla (subalar know) with naplike pile, not long haired; stem vein (remigium, basal section of R) ciliate posteriorly on upper surface (Fig. 4a); subcostal sclerite setulose (Fig. 4b); surface of calypteres with short to minute pile but without long hairs, the lower calypter bare and narrowly rounded; hind coxa bare posteriorly. The side view of the head and the wing venation, which are quite uniform for the genus, have been figured by Hall (1948, plates 4C and 9H).

The flies are readily recognized as typical calliphorids (blow flies), the color (except in a few species) being suggestive of *Calliphora* but the body shape narrower as in *Phormia regina* (Meigen). Detailed species descriptions of color are usually difficult and unsatisfactory because of the range of condition of the specimens. By far the largest part of the available material has been reared, and the specimens were killed in different degrees of maturity and condition. The thorax and abdomen are dark metallic bluish or purplish black in males, the females likewise in some species but green in others (Figs. 33-35). Males are always more shining than the females, the latter being more heavily microtomentose, at least on the scutum, with the microtomentum as viewed from behind demarcating three broad but weakly defined shining mesonotal stripes. These stripes are especially evident when the thorax is viewed from behind at a low angle. The head is chiefly dark, but the frontal vitta and genal grooves are usually lighter, reddish to dark reddish or brown. The lower ends of the facial ridges and the vibrissal angles are usually reddish yellow, and the palpi, narrow bases of 3rd antennal segments, and the 2nd antennal segments at least apically, are reddish to reddish yellow in most species.

The general color and structure are surprisingly uniform throughout the genus, a uniformity that has led to numerous misidentifications. Hall (1948) described in great detail the characters held in common by all calliphorids. His characterization of adult *Protocalliphora* (as *Apaulina* in his work) will serve for all species of *Protocalliphora* herein treated, except for a few characters, discussed elsewhere, in which we do not agree that *Apaulina* is distinct from *Protocalliphora*. The general characterization has not been repeated, and the keys and descriptions in this revision are focused on features that have shown useful differences. For ready reference by users of this work, Figures 1-4 illustrate the major structures and measures that are used.

One character, the strong mid ventral bristle on the middle tibia, is present in all known Nearctic species and has not been mentioned in the descriptions. *Protocalliphora maruyamensis* Kano and Shinonaga (1966), described from Japan, is unique among all known species in lacking this bristle. Also, females of *Protocalliphora* regularly have one pair of reclinate upper orbital bristles (Fig. 1), and the character is mentioned only for *braueri*, in which the bristles are absent.

The wing venation is uniform throughout the genus and is not referred to in the descriptions. Hall (1948) has described it, and figured the wing of *P. metallica* (his plate 9H). We note only that the 3rd vein (R_{4+5}) is actually setose from a third to halfway to the small crossvein (r-m), not merely at base.

An occasional author has synonymized *Protocalliphora* under the older name *Philornis* Meinert 1890, probably because both attack nestling birds. The latter is a genus of Muscidae quite different from the calliphorid genus *Protocalliphora*, and references to those authors have not been included here. *Philornis* is chiefly Neotropical, with outliers in Florida and southern Texas.

Synonymy of *Apaulina*

Apaulina was proposed on the basis of supposedly significant characters of chaetotaxy, reinforced by zoogeography, so that it is important to examine the basis for our synonymy, especially since *Protocalliphora* was then said not to occur in North America. Additional material has revealed that the characters used for *Apaulina* are actually somewhat variable, that there are intermediate species, and that a typical species of *Protocalliphora* even in Hall's restricted sense does in fact occur in North America, in the nests of a Holarctic bird, the bank swallow. The points of supposed differentiation are discussed seriatim.

1. Two posterior bristles on the fore tibia in *Protocalliphora*, usually one in *Apaulina*: Occasional specimens of *Protocalliphora* have only one, and conversely, certain species of *Apaulina* (especially *hesperia*, *sapphira* and *tundrae*) quite often have two, at least on one side. Occasionally two have been seen even in *avium* and *asiovora*.

In view of the considerable reliance placed on the character of the posterior bristles on the fore tibia, the variability in an obviously homogeneous series of *P. tundrae* reared from snow bunting nests on Ellesmere Island is significant and enlightening. The series consists of 70 specimens (32 males, 38 females), of which 36 (19 males, 17 females) have two posterior bristles on each fore tibia, and 22 others (2 males, 20 females) have two on one side or the other, but 12 (1 male, 11 females) have only one on each fore tibia.

2. Tympanic membrane (Fig. 3b) haired or setulose in *Protocalliphora*, bare in *Apaulina*: Typical species of the former, such as *azurea* and *chrysorrhoea*, do have a number of strong black setulae, and *Apaulina* spp. usually none, but some species included in the latter regularly have some hairs on the tympanic membrane. For example, out of 71 specimens of Nearctic *braueri*

checked for this character, 28 (39 %) showed from one to 13 hairs on one or both sides. One individual from Seattle, Wash., had 13 on the left side and 10 on the right, all coarse, black, and conspicuous. Even in such a distinctly *Apaulina*-like species as *metallica*, occasional individuals have up to four or five tympanic hairs. A reverse combination occurs in the Nearctic species *lata*, which has the two posterior bristles on the fore tibia, typical of Palearctic *Protocalliphora*, but only a few (2 or 3) weak hairs on the tympanic membrane.

3. Postalar wall (Fig. 3b) with a median tuft of hairs in *Protocalliphora*, bare in *Apaulina*: Hall (1948) regarded this as an important character for *Apaulina*, and Peus (1960) considered that the question of whether *Protocalliphora* can be used for Nearctic species depended on the taxonomic value assigned to this very character. Both believed, however, erroneously as we now know, that the postalar wall is bare in all the Nearctic species. Many if not most specimens of *Apaulina* actually have one to a few hairs on the wall, although these admittedly are usually short, pale, and inconspicuous. In some individuals part of these are dark, and occasionally long. Moreover, *P. chrysorrhoea*, a widely distributed species in Europe with a conspicuous tuft of hairs on the postalar wall, is now known to be Holarctic, a fact unknown to both Hall and Peus. This further weakens the argument that Nearctic and Palearctic species can be neatly divided by this character.

4. Scutellum with apical bristles (Fig. 3a) in *Protocalliphora*, without apicals in *Apaulina*: Apparently the same pair of bristles is present in almost all specimens, but varying in position from nearly apical to subapical, and sometimes even appearing to be discal scutellars.

5. Stem vein (basal section of R, remigium) with some cilia on ventral surface in *Protocalliphora*, none in *Apaulina*: Generally true, but *sapphira* and *chrysorrhoea* in North America often have the setae, and *braueri* in Europe does not, so there are exceptions on both sides. The cilia have also been seen on occasional specimens of other Nearctic species.

6. Immature stages: There are no generic differences in the immature stages between the species of *Protocalliphora* and *Apaulina* Hall (1948). Comparison of species results in even closer ties. For example, the puparia of *chrysorrhoea* are similar to those of *sialia*, but adults of the latter are typical *Apaulina* sensu Hall and those of the former *Protocalliphora* in the strict sense.

7. Zoogeography: The difference once thought to exist

--*Protocalliphora* Palearctic and *Apaulina* Nearctic--has disappeared as more material has become available. Two species have been found to be Holarctic: *braueri*, formerly called *hirudo* in North America and *lindneri* in Europe, and *chrysorrhoea*, which occurs in bank swallow nests in both Palearctic and Nearctic Regions. (In fairness, both *chrysorrhoea* in North America and the *hirudo*-like species in Europe were unknown to Hall when he proposed *Apaulina*). Hall's new species *sapphira* was described in *Apaulina*, but the tympanic pit has coarse black setae and the holotype and one paratype have two posterior bristles on the fore tibia. The long series of *tundrae* now available shows that these characters are typical of many--but not all--examples of the species. *P. lata* from North America has virtually bare tympanic pit but two posterior bristles on the fore tibia, hence a combination of the characters supposedly characteristic of *Apaulina* and *Protocalliphora*, respectively. The reverse combination is found in *P. asiatica* Zumpt in the Palearctic Region, which has numerous black setae on the tympanic pit but only one posterior bristle on the fore tibia. It is known from only a single specimen, and the tibial chaetotaxy could be variable, as often noted in the Nearctic *hesperia*, for example.

In conclusion, then, the variation or intergradation of the characters of chaetotaxy, and the absence of the supposed geographical separation, together with the fundamental similarity of the anatomy, male terminalia, immature stages, ecology, and host relationships of the Palearctic and Nearctic species all lead us inevitably to the conclusion that *Apaulina* is synonymous with *Protocalliphora*. It cannot even be recognized as a subgenus.

Synonymy of *Orneocalliphora*

Peus (1960) proposed to divide the genus *Protocalliphora* into two subgenera: *Orneocalliphora* for those species (e.g., *chrysorrhoea*) in which the two sexes are essentially alike in size, color, and microtomentum, and *Protocalliphora* in the strict sense for those species (*azurea* et al.) with strong sexual dimorphism in size and microtomentum, and usually in color. In the numerous Nearctic species there are different degrees and combinations of characters, and we do not believe this subgeneric division has any validity. Zumpt (1965) and others have also rejected the division as untenable.

Status of *Trypocalliphora*

Peus (1960) also proposed a new genus *Trypocalliphora* for three presumably distinct species, the Palearctic *lindneri* and *braueri* and the Nearctic *hirudo*. As with *Apaulina* and *Orneocalliphora*, our review of the numerous Nearctic species of bird blow flies leads us to question the taxon as a genus. A number of the characters used are differences only of degree, of the amount of reduction from typical *Protocalliphora*. For example, the range in amount of flattening of the scutum is so great in some series that one is led to suspect the influence of degree of maturity, especially since so many available specimens are reared, and because the relatively few wild-caught late-season specimens, obviously mature, are quite convex. This is especially true in the series of *metallica*, the species most often net collected late in the season. The postalar wall is said to be bare in *Trypocalliphora*, and it usually is, but specimens have been seen with one to three short black hairs, including several specimens from Schlorer's original rearing of *lindneri* at Diersheim, Germany. Furthermore, in ordinary *Protocalliphora* there is a range from a strong and distinct tuft of rather long hairs to few or none, and *Apaulina* was said to be characterized by having the postalar wall bare. The tympanic tuft varies through the same range (cf. previous and also later discussion of that character). The surstyli of the male terminalia range from extremely narrow, long and curved, to the broadly obovate type of several American species (cf. Figs. 7 - 32). The frontal width in the male likewise ranges from very narrow to broad (Figs. 33-41), in different combinations with the male terminalia. Species may or may not be dichromatic. Puparia, and hence the mature larvae, are all of the same general type (Figs. 42-64), but the integument ranges from smooth and almost without spines--the spines being reduced to minute tubercles as in *metallica*--to densely long spined, with different degrees of rugosity and cuticular folds. Some larval characters thought by some European authors to be characteristic of *Trypocalliphora* appear to us to be only degrees of reduction from *Protocalliphora*, extreme though they are.

Grunin (1970a) in his generic key characterized the calypteres of *Trypocalliphora* as yellow with yellow edge but those of *Protocalliphora* as white with light edge, at most the thoracic calypter brown. That was probably a character of convenience for the species with which he was dealing rather than a significant generic character. There is again a range in the North

American species that casts doubt on the generic value of the character. In *lata* and *brunneisquama*, the calypteres are very dark, black or blackish brown. In *cuprina* and *hesperia* they are brown. In *aenea* and *halli*, they are white or at most yellowish white in the females but deep yellow (*aenea*) to light brown (*halli*) in the males. In *braueri* the calypteres of the males are darker than those of the females, a deep yellow to light brown, compared with yellow for the females.

Some European authors have maintained that *Protocalliphora* and *Trypocalliphora* are distinct because species of the former are external bloodsucking parasites whereas those of the latter are subcutaneous; however, the subcutaneous habit is neither uniformly nor exclusively characteristic of the species referred to *Trypocalliphora.* It is important first to note that the species of *Trypocalliphora* are also bloodsucking, even though they commonly embed themselves when they feed. In Nearctic *braueri*, a typical *Trypocalliphora*, Bennett has found many larvae in the nest material, engorged with blood, like typical species of *Protocalliphora.* It is true that records of subcutaneous larvae in North America frequently prove to be *braueri*, but not always. *P. avium* and *P. asiovora* are often found in the ears of raptorial birds and may even penetrate to the brain, but they are also found embedded in the flesh in other areas (cf. notes under *avium* in particular; also Hamerstrom and Hamerstrom 1954). Specimens of the new species *seminuda*, a typical *Protocalliphora*, were reared along with *braueri* from larvae found in purulent sores on a fledgling horned lark in New Mexico (Walton 1914). Specimens of *chrysorrhoea* were reared in Alberta "from larvae that were removed from under the skin and breast muscles of nestlings."

Generic distinction between *Trypocalliphora* and *Protocalliphora* breaks down notably in *P. deceptor*, which is intermediate. The habitus of this species is immediately suggestive of *braueri.* It is strongly dimorphic in the color of males and females. The scutum posteriorly is convex. The frons is unusually narrow in both sexes, little over half the width of an eye in the female and so narrow in the male that the parafrontals almost touch. The basicosta is yellow to bright orange. Unlike *Trypocalliphora*, however, the reclinate upper orbital bristle ("Frontalborste" of Peus) is present in the females. This alone might not be considered a serious divergence from *Trypocalliphora*, but the male terminalia are of the type found in *avium* and *asiovora*, with broadly obovate surstyli (Fig. 9) and aedeagus typical of subgenus *Protocalliphora* (Fig. 6a).

Furthermore the puparium is covered with numerous fine spines, quite unlike the relatively bare puparium of *braueri*.

Protocalliphora aenea likewise shows, at least in mature specimens, a convex scutum, and it has a strong sexual dimorphism in color, but the reclinate upper orbital bristle is present in females, the front is wider in both sexes, the male terminalia are more like those of *chrysorrhoea*, and the puparium is well covered with spines.

In conclusion, the various combinations of characters lead us to doubt that more than one genus can be recognized. We do agree that the species described in or referred to *Trypocalliphora*-- originally three, *lindneri*, *braueri*, and *hirudo*, now one, *braueri*, with the others as synonyms--form a species group characterized notably by the distinctive aedeagus and in the female by the lack of a reclinate upper orbital bristle. In our view, variation and the range of characteristics in the now greater known number of species greatly decrease the apparent distinctiveness of *Trypocalliphora* although it can be argued that the distinct aedeagus is still justification for its recognition. In order to preserve the usefulness and meaningfulness of the name *Protocalliphora* in general biological literature but at the same time to segregate the species group, we choose to recognize *Trypocalliphora* as a subgenus of *Protocalliphora*. Sabrosky (1967) has already suggested this. The remaining species form the typical subgenus *Protocalliphora*.

Rognes (1985) published a detailed study of *Trypocalliphora* (as a genus) in which he concluded that there is only one species, wide-ranging and Holarctic. The oldest available name is Hendel's *Avihospita braueri* (1901), with synonyms *Protocalliphora hirudo* Shannon and Dobroscky 1924, *T. lindneri* Peus 1960, and *T. compacta* Grunin 1966. This synonymy returned to the identification by Sabrosky (Heinz 1954) of *hirudo* for European specimens, which was accepted in several publications such as the revision by Zumpt (1956). (At that time, the specific identity of *braueri* was unknown). Peus (1960) rejected the identification of *hirudo* and he described European specimens as *lindneri* n.sp., which was accepted by Zumpt (1965). Rognes (1985) has concluded that *lindneri* and *hirudo* are in reality conspecific, that *lindneri* was based on variable characters, and that both are synonyms of the older but poorly described *braueri*. We cannot contribute on the last named, but Rognes thoroughly studied the extant type series. We do concur in the conspecificity of European and North American material, and we have adopted the name *braueri* in place of the familiar and rather often cited *hirudo*.

Generic Relationships of *Protocalliphora*

There are few clearcut generic distinctions in the adults of *Protocalliphora*, as one quickly realizes when trying to construct a satisfactory generic key. However, the characteristic biology, the obligatory bloodsucking parasitism on nestling birds, and the distinctive larvae mark the group as a distinct entity, although some might argue that it is merely a specialized subgenus of *Phormia*.

The larvae are more distinct generically than the adults. Larvae of *Phormia* resemble those of *Cochliomyia* and other chrysomyine calliphorids in having anterior bands of spines on the abdominal segments (screwworms). In *Protocalliphora*, on the contrary, the spines are distributed over most of the surface, even in a few species in which the spines are extremely reduced (*metallica*, *braueri*). Furthermore, all species of *Protocalliphora* have in all larval instars the characteristic and unique feature--occasionally much reduced but detectable--of a fringe of long hairs on the anterior margin of the prothoracic segment (Hall 1948, plate 38D). These larval features and the characteristic biology thus justify generic recognition, even though the adult characters may not seem impressive.

Protophormia and *Boreellus*, which contain species also formerly included in *Phormia*, are easily separated from *Phormia* and *Protocalliphora*, but the latter two are not so easily separated from each other. Hall (1948) keyed *Phormia regina* as having "Mesonotum convex; eyes in male very narrowly separated or contiguous; mesothoracic spiracle with bright orange hair," whereas the other phormiines, including all *Protocalliphora* (as *Apaulina* in his book), were said to have "Mesonotum conspicuously flattened on disc; eyes in male distinctly separated (except in several species of *Apaulina*); mesothoracic spiracle with dark hair (dark orange brown in *Apaulina hirudo*)." Unfortunately, none of these characters will consistently differentiate *Protocalliphora* from *Phormia*. More species of *Protocalliphora* are now known in which the eyes of the male are only narrowly separated, and in which the mesothoracic spiracle has brightly colored hairs, not as bright as in *Phormia regina* but enough to be confusing. As for the first character, some difference of interpretation is evident. Hall, following Townsend (1935), keyed *Protocalliphora* as having the postscutum, or posterior portion of the mesonotum, "conspicuously flattened on disc," but under this they were including *P. hirudo*, which was included by Peus (1960) in his genus *Trypocalliphora*, which he characterized as having the mesonotum convex. In practice, the character is difficult to apply because specimens are so often pinned through that part of the body, either in the center or to one side of the midline. Obviously also there is some difference in the degree of flattening. Further, when one compares reared material--all too often more or less teneral--with specimens caught in nature, one sees that the flattening of the mesonotum is often more obvious in reared material and scarcely or not at all evident in hand-caught specimens. Then when one remembers that most specimens in collections are reared, because these flies are seldom collected otherwise, one realizes that most specimens in collections will show flattening. There may be slight differences between species in the degree of convexity of the mesonotum, if one could get perfect specimens and measure the curvature properly, but we have concluded that much of the observed flattening is an artifact, or is more apparent than real, and that in practice it cannot be used to separate *Trypocalliphora* from *Protocalliphora*. This conclusion is reinforced when one compares *braueri* with *deceptor*: both with convex scutum but radically different male terminalia, and female with reclinate upper orbital bristle in *deceptor*.

Discussion of the status of the other genera of the tribe Phormiini is irrelevant here; it is sufficient to point out in the following statement the distinctive nature of *Protocalliphora* compared with related phormiines:

Protocalliphora: Upper (alar) calypter with at most minute pile on the upper surface, any long hairs confined to the rim; anterior margin of prothoracic segment of larvae with a complete band of spines, forming a characteristic fringe, which is long in most species (Hall 1948: plate 38D), greatly reduced in *braueri*; larva dorsally and laterally thickly set with spines over the entire surface, except in a few species in which they are greatly reduced; larvae obligatory bloodsucking parasites on nestling birds; dorsum of thorax and abdomen more or less microtomentose, the scutum viewed from behind (at least under a microscope) showing three dark stripes, more conspicuous in females than in males; mesonotal bristles strongly developed, including complete rows of acrostical bristles and 3 or 4 strong post-intraalar bristles (Fig. 3a); postalar wall often bare or with relatively few inconspicuous hairs about midway, especially in Nearctic species (Fig. 3b).

Phormia, *Protophormia*, and *Boreellus*: Upper (alar) calypter with numerous long, erect hairs on upper surface of the outer half or more (pale and easily overlooked in *Phormia*, black in the others); larvae without prothoracic fringe; most larval segments with distinct anterior or anterior and posterior bands of spines, the intervening integument bare; larvae normally saprophagous, occasionally involved facultatively in myiasis; thorax and abdomen in both sexes highly shining, only a thin, obscure microtomentum on the anterior slope of the scutum, the scutum not vittate; acrostical bristles absent *(Protophormia*, *Boreellus)* or weakly developed *(Phormia)*; two post-intraalar bristles, rarely *(Protophormia)* a weak third; middle to lower half of postalar wall with numerous, long, coarse, conspicuous hairs.

Possible confusion of *Protocalliphora* with the blue bottle flies, *Calliphora* or *Cynomyopsis*, because of the blue color of the abdomen, is easily resolved by a glance at the upper surface of the stem vein (basal section of R, remigium), which is entirely bare in the calliphorine genera but long ciliate posteriorly in the phormiine genera (Fig. 4a).

The place of *Protocalliphora* in the blow fly family Calliphoridae and its relationship to the genera previously discussed may best be shown by a key, somewhat abbreviated for present purposes. The characters used for the adults, although not appearing of great significance, will serve to separate *Protocalliphora* from *Phormia*. Superficially, *Phormia* is easily recognized by having a dark and much more shining and unstriped thorax, hence its common name, black blow fly. Species of *Protocalliphora*, even in the relatively shining males, have a more microtomentose thorax, the microtomentum especially evident in two broad submedian stripes that delineate three broad shining stripes, particularly when viewed from behind. *Protocalliphora* is also much more bristly, and the bristles are larger and more erect.

In passing, we note that Lehrer (1970) proposed a new tribe Protocalliphorini, but we agree with Shewell (1987) that this tribe is not justified. The fundamental similarity of *Phormia*, *Protophormia*, and *Protocalliphora* indicates to us that the last named is properly placed in the Phormiini.

Abbreviated keys to genera and higher categories of Nearctic Calliphoridae

Adults

1. Stem vein (remigium) ciliate posteriorly on upper surface of wing (Fig. 4a) (Subfamily Chrysomyinae) 2

-- Stem vein (remigium) bare above . . . . Subfamilies Calliphorinae (blue- and greenbottle flies) and Polleniinae (cluster flies)

2. Hind coxa bare posteriorly; lower calypter at most with short, naplike pile above; head predominantly black (Tribe Phormiini) 3

-- Hind coxa setose posteriorly; in Nearctic forms, lower calypter with long erect hairs above on basal half; head predominantly yellow . . . . Tribe Chrysomyini

3. Presutural acrostical bristles present, strong (Fig. 3a); head not elongate below, the vibrissal angles not strongly projecting and lower margin of head in profile appearing rounded 4

-- Presutural acrosticals absent or vestigial, at most only faintly thicker than the long, fine, erect scutal hairs; head strongly elongate below, the vibrissal angles projecting and vibrissal axis obviously longer than antennal axis, lower margin of head straight . . . . *Protophormia* . . . Townsend and *Boreellus* Aldrich and Shannon

4. Three postsutural intraalar bristles (Fig. 3a); postalar wall (Fig. 3b) bare or with few minute hairs in most Nearctic species (long, but still fine and weak, in *chrysorrhoea*); dorsum of abdomen more or less heavily microtomentose; larvae are bloodsucking parasites of nestling birds. . . . . *Protocalliphora* Hough

-- Two postsutural intraalars, the first absent; postalar wall with strong tuft of long bristly hairs; dorsum of abdomen entirely shining; larvae normally saprophagous but may have healthy tissue in wound myiasis . . . . *Phormia* Robineau-Desvoidy

Mature larvae and puparia

1. Prothoracic fringe (Hall 1948, pl. 38D) present (very short in *braueri*); integument dorsally and laterally densely covered with spines in most species, minute (like fine sandpaper) in *metallica*, minute and reduced to anterior spine band in *braueri*; posterior surface (Fig. 42ff) with numerous spines except in *braueri* and *metallica* . . . . *Protocalliphora* Hough

-- No prothoracic fringe; most segments with anterior band of strong spines, a few may have anterior and posterior bands, rest of integument bare (screwworm appearance); posterior surface bare of spines . Other Calliphoridae

Taxonomy and Taxonomic Characters of the Adults

In the descriptions of the species, the following arrangement has been adopted: Citations and synonymy, brief diagnosis, description of male, description of female, description of immature stages, type series, specimens examined (figures for males and females), distribution (general statement, followed by records, which are detailed if only few, more generalized if many), hosts, ecology and biology, variation, remarks and etymology (for new species).

References under each species are chronological, except in one or two cases where minor references were grouped under different combinations, each group chronological within itself. For citations other than original references, the name of the author of the species is omitted; the name of the species is then followed by .--, and the author and citation of the reference.

In the descriptions of the adults, all measures were based on 25 specimens of each sex except where stated otherwise. Occasionally, damage to a specimen made some measure(s) unavailable, and for some new species the limited available material is admittedly an inadequate sample. All measurements were made by means of an ocular micrometer, at the same magnification (27x). The mean ratio between any two measures is based upon the sums of the measurements for each. Each mean ratio is followed by a statement in parentheses of the range of the individual ratios. When fewer than 25 specimens were measured, ranges are followed by a slant and the number of specimens measured. Thus for males of *beameri*, the ratio of frons: head width is 0.12 (.11-.14/10), based on 10 specimens. The same method is used when 25 specimens were measured but one ratio was based on a smaller number, e.g.,/24.

Under Distribution, after the general statement, localities are arranged by state and province in the order Alaska, Canada, Contiguous United States, and Mexico. Occasionally the records are separated into western and eastern records for purposes of special discussion. Identifications of adults are by Sabrosky and of larvae and puparia by Bennett unless stated otherwise.

For seasonal range, available data may be summarized, but the information is not considered significant and dates have usually not been given. The larval period corresponds to the nesting period of the bird hosts, and this governs the appearance of adult flies. However, the latter are long lived and apparently overwinter, and thus there are records of occurrence outside of the nesting season. Available records are also complicated by the frequent failure to specify on labels whether the dates given are those of collection of nests or of emergence of adult flies. Any unusual dates are noted.

Many factors make this genus of closely related species a particularly difficult one. There is some variation, often wide, in all the characters of chaetotaxy and proportions (e.g., note the ranges in Tables V and VI) that one must use because of the paucity of effective differences. Some of these characters are further modified by the degree of maturity of individual specimens. Measurements given in this paper are based on examples judged to be mature (i.e., fully extended and hardened). Body color is little affected by slight immaturity, and its correct interpretation is rarely jeopardized. If a specimen is greasy or dirty, or has become wet while in a killing jar or while being

Table V. Ratios of measures of male Protocalliphora

Species	No.	frons: head	frons: oc. span	pfac: frons	pfac: oc. span	pfac: vibr. int.	cheek: eye
aenea	25	.085(.07-.10)	1.32(1.00-1.50)	1.33(1.18-1.67)	1.75(1.50-2.02)	.70(.64-.83)	.40(.37-.42)
asiovora	25	.13(.115-.14)	1.68(1.44-1.88)	1.14(1.00-1.31)	1.93(1.75-2.20)	1.04(.91-1.14)	.47(.44-.52)
avium	25	.16(.15-.18)	1.96(1.68-2.19)	.955(.86-1.06)	1.87(1.56-2.27)	1.16(1.00-1.29)	.45(.42-.48)
beameri	10	.12(.11-.14)		1.06(.92-1.28)	1.70(1.50-1.93)	.82(.75-.92)	.44(.41-.47)
bicolor	25	.065(.05-.08)	1.08(.83-1.40)	1.67(1.29-2.00)	1.81(1.50-2.10)	.71(.615-.81)	.38(.36-.40)
braueri	25	.05(.04-.07)	.73(.60-.93)	1.79(1.31-2.28)	1.31(1.07-1.60)	.64(.56-.75)	.35(.32-.39)
brunneisquama	5	.10(.07-.13)	1.26(1.00-1.375)	1.35(1.00-2.00)	1.70(1.50-2.00)	.88(.82-.90)	.41(40-.43)
chrysorrhoea	25	.115(.095-1.35	1.63(1.29-1.86)	1.22(1.07-1.50)	1.99(1.67-2.29)	1.00(.87-1.07)	.44(.40-.48)
cuprina	25	.06(.05-.07)	.89(.71-1.18)	1.74(1.31-2.00)	1.55(1.38-1.83)	.72(.615-.83)	.38(.34-.42)
deceptor	22	.07(.05-.08)	1.01(.77-1.25)	1.65(1.43-2.20)	1.67(1.38-1.83)	.76(.65-.79)	.35(.31-.40)
fallisi	21	.11(.085-.13)	1.44(1.14-1.67)	1.15(1.04-1.38)	1.66(1.57-1.83)	.86(.77-1.00)	.42(.39-.44)
halli	25	.08(.07-.09)	1.07(.86-1.50)	1.47(1.22-1.75)	1.57(1.36-1.83)	.75(.65-.85)	.41(.38-.43)
hesperia	24	.08(.08-.09)	1.15(.94-1.33)	1.37(1.20-1.57)	1.57(1.33-1.83)	.74(.67-.79)	.41(.37-.44)
hesperioides	25	.06(.05-.07)	.83(.69-1.00)	1.77(1.50-2.00)	1.47(1.29-1.75)	.68(.61-.77)	.37(.31-.41)
hirundo	25	.10(.09-.11)	1.37(1.21-1.58)	1.28(1.17-1.50)	1.76(1.50-1.91)	.89(.82-.96)	.43(.40-.45)
interrupta	25	.09(.075-.11)	1.18(1.00-1.40)	1.31(1.00-1.50)	1.54(1.23-1.80)	.70(.53-.80)	.41(.36-.415)
lata	8	.34(.32-.35)	4.15(3.74-4.60)	.645(.595-.66)	2.67(2.27-3.00)	1.83(1.47-2.00)	.635(.61-.67)
metallica	25	.08(.06-.09)	1.11(.85-1.44)	1.51(1.33-2.00)	1.68(1.43-2.00)	.78(.70-.86)	.36(.33-.39)
parorum	25	.08(.06-.10)	1.13(.91-1.42)	1.41(1.18-1.70)	1.60(1.43-1.83)	.77(.69-.86)	.42(.38-.45)
sapphira	1	.17	1.90	.85	1.60	.79	.46
seminuda	9	.08(.06-.09)	1.16(1.00-1.50)	1.50(1.33-1.83)	1.74(1.36-2.08)	.90(.75-.92)	.38(.36-.41)
shannoni	25	.06(.05-.07)	.985(.83-1.18)	1.78(1.46-2.20)	1.76(1.50-2.20)	.70(.58-.815)	.37(.34-.40)
sialia	25	.09(.08-.11)	1.37(1.13-1.67)	1.43(1.20-1.62)	1.96(1.69-2.18)	1.04(.93-1.13)	.42(.38-.45)
spatulata	25	.12(.10-.13)	1.60(1.33-1.79)	1.19(1.08-1.38)	1.90(1.71-2.29)	.90(.79-1.09)	.47(.42-.49)
spenceri	17	.07(.06.-.08)	1.14(1.00-1.30)	1.48(1.29-1.67)	1.68(1.50-1.90)	.66(.57-.79)	.38(.36-.41)
tundrae	25	.135(.12-.16)	1.43(1.22-1.73)	1.00(.91-1.13)	1.43(1.22-1.62)	.86(.73-1.00)	.46(.43-.48)

Table VI. Ratios of measures of female Protocalliphora

Species	No.	frons: head	frons: oc. span	pfac: frons	pfac: oc. span	pfac: vibr. int.	cheek: eye
aenea	25	.26(.24-.295)	3.41(2.86-3.91)	.48(.45-.55)	1.64(1.43-2.00)	.78(.67-.89)	.40(.37-.48)
asiovora	25	.295(.28-.32)	3.52(3.00-4.29)	.56(.50-.59)	1.96(1.75-2.27)	1.13(1.00-1.24)	.49(.42-.525)
avium	25	.295(.27-.32)	3.45(2.89-3.88)	.54(.50-.59)	1.86(1.56-2.25)	1.15(1.00-1.36)	.47(.42-.50)
beameri	1	.29	4.00	.54	2.14	1.25	.47
bicolor	25	.24(.22-.26)	3.29(3.00-3.83)	.50(.425-.57)	1.65(1.38-2.00)	.77(.65-.86)	.38(.36-.42)
braueri	25	.22(.21-.25)	2.86(2.43-3.38)	.44(.36-.50)	1.25(1.07-1.43)	.67(.57-.82)	.34(.30-.37)
brunneisquama	6	.30(.30-.31)	3.48(3.44-4.12)	.47(.44-.48)	1.65(1.455-1.94)	.92(.84-1.00)	.44(.46-.49)
chrysorrhoea	25	.29(.26-.31)	3.64(3.22-3.88)	.53(.48-.58)	1.93(1.67-2.25)	1.07(.94-1.21)	.47(.42-.54)
cuprina	25	.26(.24-.27)	3.20(2.86-3.67)	.48(.44-.54)	1.54(1.36-1.77)	.79(.69-.92)	.40(.37-.44)
deceptor	25	.23(.20-.25)	3.01(2.56-3.43)	.52(.44-.56)	1.56(1.25-1.82)	.77(.64-.92)	.34(.29-.37)
fallisi	20	.29(.265-.31)	3.54(3.14-4.00)	.50(.46-.58)	1.77(1.60-2.10)	.97(.92-1.04)	.45(.42-.48)
halli	25	.26(.23-.29)	3.24(2.86-3.86)	.50(.44-.55)	1.61(1.50-1.71)	.81(.73-.92)	.44(.40-.47)
hesperia	25	.26(.23-.28)	3.19(2.75-3.57)	.49(.41-.57)	1.57(1.36-1.83)	.81(.73-.92)	.43(.40-.46)
hesperioides	25	.26(.23-.27)	3.08(2.81-3.57)	.45(.40-.50)	1.40(1.25-1.64)	.74(.64-.85)	.40(.37-.45)
hirundo	25	.28(.26-.30)	3.58(3.00-4.33)	.54(.48-.58)	1.92(1.62-2.33)	1.00(.93-1.15)	.44(.41-.47)
interrupta	25	.29(.26-.31)	3.40(3.14-3.92)	.46(.43-.52)	1.58(1.43-1.83)	.78(.69-.85)	.42(.39-.43)
lata	20	.35(.32-.38)	4.63(4.30-5.20)	.61(.57-.66)	2.84(2.60-3.12)	1.90(1.53-2.23)	.62(.60-.66)
metallica	25	.27(.25-.29)	3.58(3.00-4.17)	.45(.40-.53)	1.59(1.43-1.91)	.82(.72-.92)	.36(.32-.39)
parorum	25	.26(.25-.28)	3.33(2.875-3.71)	.47(.43-.52)	1.55(1.25-1.71)	.81(.69-.92)	.43(.40-.47)
sapphira	2	(.30-.31)	1.73	.54	1.73	1.08	(.49-.50)
seminuda	6	.33(.32-.36)	4.00(3.56-4.43)	.41(.39-.45)	1.69(1.50-1.875)	.92(.88-1.00)	.44(.42-.48)
shannoni	25	.25(.23-.27)	3.275(3.00-3.54)	.495(.44-.54)	1.62(1.43-1.78)	.81(.72-.88)	.39(.37-.41)
sialia	25	.265(.24-.29)	3.31(3.06-3.73)	.555(.52-.62)	1.84(1.65-2.00)	1.12(1.00-1.32)	.44(.42-.46)
spatulata	25	.31(.29-.33)	3.56(3.10-4.00)	.485(.42-.54)	1.73(1.40-2.00)	.98(.85-1.10)	.50(.45-.54)
spenceri	24	.25(.24-.275)	3.22(2.86-3.62)	.485(.45-.52)	1.56(1.31-1.69)	.75(.68-.85)	.40(.37-.435)
tundrae	25	.31(.30-.32)	3.16(2.89-3.62)	.50(.44-.54)	1.59(1.29-1.75)	.99(.76-1.04)	.49(.47-.51)

relaxed for extraction of the male terminalia, the color of the calypteres is generally darker than normal, e.g., normally whitish calypteres may appear yellowish, or yellowish calypteres a deeper yellow. All such factors will usually be compensated for in an adequate series reared from the same nest, but any one of them may seriously affect identification of single individuals.

Small size affects other characters. In small adults, the bristles are more slender and thus appear less well developed than in other species, even to the extent that ordinarily small bristles may appear to be absent altogether because they are hairlike and hardly distinguishable from surrounding hairs. For example, distinct accessory notopleural bristles are absent in the tiny specimens of *braueri* that were originally described as *hirudo* var. *parva*, although they are regularly characteristic of *braueri*. Undersized specimens may also yield atypical proportions. Because so few examples are available, two small females of eastern *braueri* are included in the measures for that species. Their ratios are the lowest in five of the six proportions used and obviously affected the ranges. For some proportions the ratios are extremely atypical.

Hall (1948) in his revision of the Nearctic species relied a great deal on measurements and ratios, but we find that his key is almost impossible to use because of the slight differences between species that are closely related and difficult to distinguish. For example, in his couplet 23 (his p. 101), he separated *hirundo* from *sialia* and *metallica* on the ratio of frons to head width, 0.09 vs. 0.10 or more in males, 0.27 vs. 0.28 in females. Moreover he stated that "Proportional measurements in specimens of the same species remain nearly constant." However, after hundreds of measurements, we find that there is a wide range in most ratios (cf. Tables V, VI) and that mean or median ratios can only be used effectively in separation of species where the differences are considerable. Standard deviations are of limited use where there are extensive overlaps of ranges. The range of ratios must obviously be considered in judging the usefulness of the mean ratio for a species, although extremes of ranges often represent one or few far-out specimens that would not seriously affect the identification of a great majority of specimens. The greatest differences in mean ratios often exist between species that are also distinct in some outstanding features that are easier to use, as for example the differences in male terminalia between *avium* and *sialia*.

Hall measured few specimens, five per species, or ten at most, and he included "the largest and smallest individuals, specimens from the extremes of the distribution, and several of average size," a mixture that would have greatly affected his ratios, especially in a statistical sample of few specimens. We differ in not measuring the smallest individuals, because the underfed "runts" in a series may and undoubtedly do show measurements and proportions that are not typical of the species. In our opinion, there is a normal or optimum size for each species, and measurements of such specimens will give a truer basis for comparison of species identities.

A random selection that included teneral specimens would give some distorted measurements. Selection must be confined to mature and measurable specimens, after which selection can be random to assure representation of the geographic range of the species. Even where it was necessary to measure a number of specimens from the same nest, presumably siblings, there was a range of variation and not the "nearly constant" measures postulated by Hall.

Accordingly the descriptions have been confined to a relatively few characters in which the mean ratios show some appreciable differences, and hence promise demonstrable significance as well as practicable usability in dealing with specimens. The best and most useful characters were found to be ratios between certain relatively stable features. These are listed below, with notes on how the measures were made, in order to promote uniform treatment and interpretation (cf. also Figs. 2a, b).

Width of frons: Measured across the narrowest part. In males this is usually anterior to the median ocellus, and in females at the vertex and measured across the posterior ocelli.

Width of head: The maximum over-all width, measured from the extreme convexity of each eye, as viewed directly from above. Special care must be given to the angle of vision, to assure the position for getting the maximum width.

Ocellar span: The width of the ocellar tubercle, measured between the outer margins of the posterior ocelli.

Width of parafacial: Measured opposite the anterior corner of the frontal vitta, at right angles to the eye margin and just below the foremost frontal bristle.

Vibrissal interval: The distance between the vibrissae, as measured between the inner margins of the two vibrissal sockets. This may not be too reliable. The epistoma often appears wrinkled or somewhat folded, or even concave, as if it is one of the last areas to become fully extended and hardened, and this could affect the measure and ratios based upon it. In some measured specimens, the vibrissal interval was not measured because of this suspicion. The vibrissal interval may be characteristic of a species, but the practical usefulness of the character is limited by the degree of maturity of the specimen, often a serious problem with so many reared specimens.

Length of eye: The maximum length that can be observed, usually at an angle somewhat above direct profile. The maximum length is usually greater than its length as seen in direct profile, although the difference is not as great as in measures of the cheek height and does not seriously affect the proportion of cheek height to eye height (length).

Height (breadth) of cheek: The maximum height, measured at a right angle to the surface of the cheek, from a median point on the oral margin, midway between the vibrissa and the back of the oral cavity, to the lowest point on the eye margin. The height of the cheek compared with the length of the eye is a useful ratio, but the angle of measurement makes considerable difference in the cheek measurement and therefore in the ratios based upon it. This has been discussed by Sabrosky (1955) in connection with the tachinid genus *Archytas*. In many muscoid flies, such as *Protocalliphora* and *Archytas*, the cheek slopes or curves ventro-mesad from the eye, and its maximum height, measured at right angles to its surface, is somewhat greater than the height as seen in direct profile.

Features characteristic of one sex only are so specified, except that features found only in males (aedeagus, surstylus) are left unspecified. For other characters, if there is no qualifying note a statement applies to both sexes.

In the description of the male, the width of a parafacial is sometimes compared with the breadth of the 3rd antennal segment, but no figures and mean ratios are given. This is a relationship that can easily be judged by eye, but is not so easily measured, and it is used for gross comparisons, such as "obviously wider than" compared to "equal to."

Chaetotaxy: The chaetotaxy is rather uniform throughout the genus. The most notable variations are absence of reclinate orbital bristles (frontal bristles of Gregor and Povolný, 1959, and Peus, 1960) in females of *braueri*, presence of an accessory notopleural bristle (a smaller bristle between the two strong notopleural bristles) in that same species (Fig. 3a), and presence of two (occasionally three) posterior bristles on the fore tibia of *chrysorrhoea* and predominantly in *tundrae* and *sapphira*. Occasional individuals of other species, especially *hesperia*, show a second posterior bristle on the fore tibia on at least one side, and, less commonly, an accessory notopleural bristle may be seen. A distinct tympanic tuft of setae (Fig. 3b) is characteristic of *chrysorrhoea* and *sapphira*, but a few tympanic hairs occur occasionally in individuals of other species, and sometimes (e.g., *braueri*) in a considerable proportion of the individuals of a species. Variation in these characters is noted elsewhere, in the discussion of the species and in the discussion of the status of *Apaulina* Hall.

Male terminalia (Figs. 5-32): Examination of the male terminalia is usually essential for precise identification, and is always desirable. For some species, it is virtually the only significantly differentiating character, and species that are externally indistinguishable in the male sex may have quite different terminalia. *Protocalliphora deceptor* is the most startling example, the type series having been misidentified originally as a variety of *aenea*, until the terminalia revealed broad surstyli (Fig. 9) like those of *avium*. A few species can be identified without recourse to the terminalia, but even in those cases the terminalia can often confirm the identification. Frequently they will give an immediate clue to the identity, when variation is confusing or a teneral condition discourages measurement. The male terminalia are easily pulled into an extended position for study, even in dried specimens that have been relaxed over moist sand, although very old specimens will be difficult. For a discussion of technique and an explanatory figure see Hall (1948:37-39). In the present revision the "inner forceps" and "outer forceps" of Hall are termed cercus and surstylus. Past revisions of *Protocalliphora*, especially in Europe, have often showed the terminalia in posterior aspect. Unfortunately differences between the cerci of different species are usually slight, if any, and the posterior aspect misses the characteristic shape of the surstyli as seen in profile. Accordingly for most species we have figured a surstylus in side view to show how these differ greatly and how useful they can be in the differentiation of the species (Figs. 7-32). The aedeagus is fairly uniform in the subgenus *Protocalliphora*, differing from that in the subgenus *Trypocalliphora* (Fig. 6a-c).

The female terminalia showed only slight differences, and these chiefly in species that were already amply distinct on external characters.

Size: In species reared from large species of birds, the young of which remain in the nest for a relatively long time, there is usually little difference in the size of the flies. Series reared from birds that develop quickly often show many undersized specimens, or "runts". Where material is adequate, the length is stated as a "normal length," i.e., what appears to be the length normally attained by mature, fully fed individuals, but not including obviously immature or extremely undersized specimens. Most specimens of the genus are within 8-10 mm in length.

The sexes are described separately. The general matters of structure and chaetotaxy, described in detail for the male, apply also to the female unless otherwise noted. The normal differences between the sexes, e.g., the broader frons and the orbital bristles of the female, occur in all species and are not mentioned in the descriptions, except for the statements of ratios.

Most of the terms used are standard for flies (Diptera) and are explained and figured in the *Manual of Nearctic Diptera*, volume 1 (McAlpine et al. 1981), and usually in textbooks of entomology. The terms and measures used in this work are illustrated, somewhat diagrammatically, in Figures 1 - 4. Only a few terms need special mention:

(1) Accessory notopleural bristles (Fig. 3a): On each notopleuron in *P. braueri*, and only occasionally on other species, one to several bristles shorter and weaker than the standard 1 + 1 notopleural bristles or macrochaetae, but still clearly bristlelike compared with ordinary clothing hairs.

(2) Microtomentum (microtomentose): The fine coating or "dusting", called pruinosity in the *Manual of Nearctic Diptera* or pollinosity by some authors, is actually composed of ultramicroscopic outgrowths of the cuticle (Sabrosky 1983). These are cylindrical to more or less flattened and scalelike outgrowths, which grade from minute and fine to distinct and coarse. For descriptive purposes the term microtomentose can be appropriately modified, if desired, as finely, coarsely, thinly, or densely.

(3) Postalar wall (Fig. 3b): The vertical surface below the postalar callus. It may be bare or with few fine hairs, or with a tuft of coarse hairs.

(4) Preocellar area (Figs. 1, 2a): The part of the frontal vitta immediately anterior to the median ocellus. This may be a triangular to acuminate extension of the ocellar tubercle, especially in males, and dull or rugose or shining, or it may be an indefintely bounded polished spot, or it may be entirely dull and microtomentose, without any special spot or ridge or area, especially in females.

(5) Prevertical area (Fig. 2a): The narrow upper part of each parafrontal, just anterior to the vertical bristles.

(6) Surstylus (Figs. 7-32): In males, one of the clasping structures known by various authors as outer forceps, paralobes, or edita.

(7) Tympanic membrane or tympanum (Fig. 3b): An oval membrane, slightly countersunk in a "tympanic pit" behind the base of the wing and at the base of the calypteres, at the anterior end of the suprasquamal ridge. Like the postalar wall, it may be bare or with a few fine hairs, or with a tuft of long hairs (e.g., in *chrysorrhoea*).

Identification of the Adults

Few species can be identified satisfactorily in both sexes. In some species, males are distinct but females are not; in others, the revarse is true. In still other species, the immature stages are more distinctive than the adults. Crosskey (1981) has referred to a similar situation in the taxonomy of the Simuliidae: "Certain species in the better-known faunas can occasionally be quickly and easily identified on some distinctive feature ..., but it is unusual for a taxon to be equally distinct in the larvae, the pupae and both sexes of the adults."

The surest identifications of *Protocalliphora* depend, with few exceptions, on reared series in which both sexes are represented and puparia are associated. Accordingly the main key has been constructed on the basis that all three elements are present. The identity of the host may in some cases be helpful, or at least suggestive, but it is never by itself decisive. Lone specimens are extremely difficult to identify with any assurance, except for a very few unusually characteristic species, such as *braueri* and *lata*, or well-marked sexes such as males of *deceptor*.

The complex of species with curved surstylus could not have been solved without series of both sexes and their puparia, and identifications should be based on the same elements. When all these are available, some species can be spotted easily, such as *brunneisquama* with dark brown calypteres in both sexes, *metallica* and *interrupta* with greatly reduced spines on the puparium, *metallica* and *bicolor* with cupreous 5th tergite in females, and 5th cupreous also in male *bicolor*, and *fallisi* and *tundrae* with the posterior spine bands on the ventral side of the puparium vestigial or lacking. Recognition of these species narrows the problem to where, for the remaining species, ratios of measures can be utilized, variable though they are. However, when one tries to devise a key to males alone, the advantages of characters on the females and puparia fall by the wayside and, except for *bicolor* with its female-like cupreous 5th tergite, separation of species is much more difficult and the keys are less satisfactory. Likewise, for a key to females, once the cupreous 5th tergite has removed *metallica*, *bicolor*, and *cuprina*, identification of the remaining species is difficult or impossible.

We recognize, however, that it may sometime be desirable, for distribution or host records, to try to name isolated specimens, or a series composed of one sex only, or because the occasional mixture of species in a nest--12.1% in Bennett's studies (1957), 7.1% in Whitworth's (1976)--may be misleading and the sexes will require separate attention. (In all cases so far, one of the pairings, either of males or females or immature stages, alerted us to the existence of a mixture of species). Accordingly separate keys to males, females, and puparia have been prepared in which use of various characteristics, together with geographical distribution, will enable the user to eliminate some species and narrow the possibilities to one or a few names. In all three keys, after the especially characteristic species have been keyed out, discrimination between species becomes increasingly difficult because of variation in the few useful taxonomic characters, in the adults in the color of the calypteres, the number of posterior bristles on the fore tibia, and ratios derived from measurements. The key to males is more reliable than that to females because of the extra dimension provided by the male terminalia. Females of a number of species are "look alikes," and beyond couplet 12 that key is particularly weak. Puparia, except for a few relatively characteristic species, are difficult to key out by themselves, although in the main key they can be exceedingly useful in separating certain pairs of similar species, e.g., *avium-asiovora*, or *metallica-cuprina.*

In mixed infestations, the importance of careful association of each adult with the puparium from which it emerged is readily apparent. If puparia are available but not individually associated, they may do no more than reinforce a conclusion of a one-species series or of a mixed infestation, whichever is appropriate, although in some cases they may help to distinguish one or more of the species present, or to narrow the possibilities. In mixed infestations, more than any other, one must always be alert to the fact that not all collectors are careful to associate with an adult *the* very puparium from which that adult emerged.

In the separate keys to males and to females, puparia are sometimes cited for use when a series of reared specimens consists of one sex only, or on the chance that they might be accompanying single specimens, although isolated specimens are most likely to be net-collected "loners."

Emphasis has been placed on characters that are easiest to use as well as most reliable. There was a temptation to begin with the choice of one or two posterior bristles on the fore tibia, the latter being one of the chief features of the species that may be termed a Palearctic or Holarctic element in our fauna: *chrysorrhoea*, *lata*, and *tundrae*. However, a small number of specimens of species with a normal number of one posterior bristle have two on one or both fore tibiae, and some with a normal complement of two have only one, which would confuse anyone trying to identify specimens if that were the opening couplet (see notes on variation in *tundrae* and *hesperia*, for example). In a long series, one can determine the usual and presumably typical number, but this would not necessarily be true in every series. Consequently that character has been avoided as a primary choice, though mentioned as a supplementary character.

Species Groups in the Subgenus *Protocalliphora*

The species of the subgenus *Protocalliphora* can be arbitrarily divided into groups of convenience in various ways, depending on what characters are used, thus white calypteres vs. brownish, wide vs. intermediate vs. narrow frons in males (variable measures), monochromatic vs. dichromatic species, etc. The most useful division, and one that is especially good because the character states are distinct and can be illustrated, is based on the shape of the surstylus as seen in profile.

A. Surstylus broadly rounded (Figs. 7-9): *asiovora*, *avium*, *deceptor*

B. Surstylus digitate (fingerlike or thumblike), parallel sided and straight, or nearly so (Figs. 10-19)

   Calypteres brown or brownish, especially margins: *cuprina*, *halli*, *hesperia*, *hesperioides*, *lata*

   Calypteres white: *aenea*, *chrysorrhoea*, *hirundo*, *shannoni*, *sialia* (yellowish in *aenea*)

C. Surstylus curved, usually slender (Figs. 21-32)
 Calypteres brown or brownish: *brunneisquama*
 Calypteres white:
  Females with 5th tergite cupreous or cupreous green: *bicolor*, *metallica*
  Female with bluish abdomen as in male: *beameri*, *fallisi*, *interrupta*, *parorum*, *sapphira*, *seminuda*, *spatulata*, *spenceri*, *tundrae*

Taxonomy and Taxonomic Characters of the Immature Stages

Chiefly because of the economic significance of the family, probably more attention has been given to the characters of the immature stages of the Calliphoridae than for any other group of the Cyclorrhapha, an effort directed primarily to easily available species. Only relatively recently, when more species were recognized in *Protocalliphora*, has greater attention been given to its immature stages. The 3rd-instar larva and puparium of *Protocalliphora* were first described and figured by Dufour in 1845 for what he called *Lucilia dispar*, now a synonym of *P. azurea*. By modern standards, this description was too generalized and incomplete. Hall (1948), while recognizing ten species of Nearctic *Protocalliphora*, gave descriptions of the immature stages of only two (3rd-instar larva of both, and puparium of one of those). A paper on the 3rd-instar larvae of three European species was published by Rohdendorf (1957), well illustrated with figures of the important characteristics.

For the present study, early observations by D. M. Davies at Algonquin Park, Ontario, in 1949 were continued by the Bennett during 1950-56. Studies on the immature stages of the species in that area were broadened to include others, in order to parallel and complement the studies of the taxonomy of the adults being carried out by Sabrosky. A discussion of this joint study and an exhibit of both adults and immature stages were presented at the Tenth International Congress of Entomology in 1956 (Sabrosky and Bennett 1958). Studies since that time have added materially to the knowledge of the immature stages of the Nearctic species.

The present revision is fortunate in having material of the immature stages, in full or in part, of 23 of the 26 North American species recognized from adults (Table I). Puparia are known for 23 species and 3rd-instar larvae for 16 (plus two characters derived from the puparium for six others). As for the other stages, which are less useful, eggs are known for four species, 1st-instar larvae for three, and 2nd-instar for seven. Long series, especially of puparia, were available for some of the species so that the extent of the variation could be assessed.

Preparation of Material

Treatment of material varied in detail from species to species and in decade to decade, but in general was as follows:

Second- and 3rd-instar larvae were split open from the prothoracic to anal segment along the lateral line (so as not to destroy the important diagnostic characters on the dorsal and ventral surfaces) and heated in a 5 molar potassium hydroxide (KOH) solution in small containers in a water bath. Heating continued until the sclerotized structures appeared to be adequately decolored, which varied from about 15 minutes to two hours. The larvae were then washed in hot water until all the internal structures washed free, leaving the cuticle with its attached spines, stigmatal plates and cephalopharyngeal skeleton. The larval skin was then cleared and dehydrated in beechwood creosote (which leaves the cuticle pliable) and mounted in balsam or "Permount", care being taken to ensure that the stigmatal area was folded to provide an unobstructed view of the stigmatal plates.

The puparia were heated for a variable length of time in 5 molar potash solution until adequately decolored and washed as above. The posterior region, portions of the dorsal and ventral cuticle, and the cephalopharyngeal skeleton (if present) were cut with a sharp scalpel from the pliable puparium, cleared and dehydrated in beechwood creosote and mounted as above. The procedure varied as conditions demanded. Certain puparia, such as those of *P. aenea*, with less color than others, required a shorter period of hydroxide treatment, while puparia such as those of *P. avium* and *P. lata* required much longer treatment. It was also true that old puparia, such as those from long-abandoned nests, required relatively little hydroxide.

All measurements are given as the average, the range (in parentheses) and the total number of measurements, e.g., 24 μ (14-41)/100. For spines, ideally ten measurements were made for spines on each portion of the cuticle for each individual puparium. When material was abundant, puparia were taken at random for measurements of length and breadth.

General Discussion of the Immature Stages

Several of the characters used in the following descriptions of *Protocalliphora* larvae are not in common use in entomological literature, and therefore a discussion of the characters of the immature stages and a Glossary of terms are presented for the sake of brevity and clarity in the descriptions of species.

Glossary

anterior band: band of spines on anterior part of a segment's ventral surface, posterad of a ventral pad.

anterior patch: dense patch of small, slender spines in the anterior band, immediately posterior to a ventral pad.

button (*ecdysial scar*): vestigial remains or scar from the shedding of a stigmatal plate of the previous instar, usually appearing circular or buttonlike at the base of the stigmatal plates.

cephalopharyngeal skeleton: the sclerites of the feeding apparatus of the larvae (mouth hooks, pharyngeal sclerites).

circumstigmatal folds: prominent ridges of cuticle surrounding and outlining the stigmatal area.

cuticular folds: prominent ridges of cuticle in the posterior region.

cuticular plaques: irregular masses of cuticle, heavily sclerotized, scattered throughout the cuticle. They are readily seen in the puparia and in stained preparations of the larvae.

cuticular ridges: ridges of cuticle, smaller than the cuticular folds, on the dorsal and ventral surfaces.

dorsal cuticle: cuticle of the dorsal surface of the larva and the puparium. Spines on the dorsal cuticle are usually directed posteriorly but in some species also anteriorly.

hyperstigmatal area: the area of cuticle immediately dorsal to the stigmatal area and bearing *hyperstigmatal spines*.

hypostigmatal area: the area of cuticle immediately ventral to the stigmatal area and bearing the *hypostigmatal spines*.

medial band: band of spines on medial part of a segment's ventral surface, midway between the two ventral pads. The medial band is usually entire and pronounced, but in some species it can be markedly thinned or broken (interrupted).

mesostigmatal area: the area of cuticle between the stigmatal plates. It may bear *mesostigmatal spines*, *folds* and/or *plaques*.

parastigmate forms: puparia characterized by having the stigmatal plates abnormally close together as well as having reduction in ratios and other features. Parastigmate forms are usually produced by under-fed larvae (see text).

peritreme: dark, sclerotized ring bounding each stigmatal plate.

plaques: see cuticular plaques.

posterior band: band of spines on posterior part of a segment's ventral surface. It is usually a short arc of small spines immediately anterior to each ventral pad. It is sometimes absent in some species, and in others present only on the anterior five to six segments.

posterior region: at the posterior end (12th segment) of larva and puparium; that region bounded by a ring of posterior tubercles (cf. Hall, 1948, pl. 35D, E). The tubercles are fleshy and evident in the larvae, but are usually represented only by scars in puparia (especially in *Protocalliphora*).

prothoracic fringe (Hall 1948, pl. 38D): band of long spines on anterior margin of prothoracic segment of larvae and puparium. On ecloded puparia, it is often missing along with the ventral half of the puparial cap.

stigmatal area: broadly rectangular area surrounding and including the stigmatal plates; usually slightly depressed and frequently outlined by cuticular folds and/or small plaques. The area is usually divided medially by mesostigmatal folds and/or plaques, in some species bisected so sharply as to form two separate areas.

stigmatal plate (*spiracular plate*): each of two heavily sclerotized plates that bear two (2nd instar) or three (3rd instar and puparium) spiracular slits (stigmata); sometimes called spiracular plates.

stigmatal ratio: ratio of the distance between the buttons to the distance across the two stigmatal plates, measured between their outer margins at the level of the buttons.

ventral band ratio: ratio of the summation of the widths of the three ventral spine bands of any segment, measured on the midline, to the total distance between the adjacent margins of the ventral pads bordering that segment.

ventral cuticle: cuticle of the ventral surface of the larva and the puparium.

ventral pads (*creeping welts*): ventral, ovoid, fluid-filled, spine-free protruding areas of the larvae, situated on the midline in the intersegmental grooves, and apparently aiding in locomotion: comparable in position and function to spined, tuberculate, or roughened areas in other cyclorrhaphous larvae. The puparia show ovoid scars in the corresponding areas.

Egg

The eggs are of typical calliphorid shape, rounded at one end and tapering slightly at the other, and lacking an operculum. The cuticle of the egg is variously ornamented with spines and raised ridges. Although eggs of only four species are described, the cuticle pattern offers specific differences, and examination of eggs of other species may also offer useful specific characters.

Larval stages:

First instar: The 1st-instar larva differs, other than in size, from larvae of other instars as follows: (i) the lateral hooks and hypopharyngeal portion of the cephalopharyngeal skeleton appear to be elongated in comparison to those of other instars, but conversely the pharyngeal portion appears to be much shorter and the dorsal and ventral depressions are lacking; (ii) the prothoracic fringe is very short or absent; (iii) the spines of the anterior patch are two- to four-pronged in comparison to the simple spines of the other instars; (iv) mesostigmatal spines are absent; (v) hypostigmatal spines are scarce or absent; (vi) there is neither peritreme nor button.

Second instar. The spine distribution is the same as that of the 3rd instar, as is the general structure of the cephalopharyngeal skeleton.

Third instar. The tapering anterior end terminates in the cephalic segment which can be retracted into the prothoracic segment, on the anterior margin of which is a dense band of spines, the *prothoracic fringe*. These spines are longer than any others on the body. The measurements of the spines of the fringe are based on the longest which are those of the posterior rows. In the formation of the puparium, the cephalic segment is withdrawn into the prothoracic, the most anterior external segment of the puparium. In this formation, the prothoracic fringe is contracted and the measurement of the fringe on the puparium is difficult, especially if the material is not well cleared in potash. Furthermore, in cyclorrhaphous flies, the prothoracic fringe remains on the lower half of the emergence cap of the puparium, and this section is frequently lost. The cuticular spines of the larvae are the same length as those found on the puparia. The cephalic annulus bears the oral opening through which the anterior portion of the cephalopharyngeal skeleton is protruded. This skeleton, always present in the larva, adheres to the lower half of the emergence cap of the puparium and consequently is often lacking, as pointed out for the prothoracic fringe. Hence it is measured as from the larva, although the skeleton from the puparium is used in the absence of sufficient larval material. The skeleton was not found to be of use in the larval taxonomy of *Protocalliphora*, and only the overall length is given.

The key to the 3rd-instar larvae is based almost entirely on chaetotaxy, the lengths of the cuticular spines on various parts of the larva having considerable diagnostic importance. Characters such as the ventral band ratio, so useful in diagnosis of the rigid puparia, are not valid with the soft and pliable larvae, which, dependent on fixation technique, can be either contracted or extended.

Puparium

The color ranges from light brown in a thin-walled form to black brown in thick-walled species; puparia formed from underfed larvae are usually lighter in color. The surface is dull in the majority of the species, but appears shiny in the short-spined forms.

The larva will pupate either shortly after attaining the 3rd instar or after several more blood meals. In both cases, adults are produced, but the size range of both adult and puparium is considerable. Small larvae and puparia were found in nests shortly before the nestlings were fledged, and hence the size of the larva is most probably a function of the age of the nestling at time of attack. This is frequently the case with *P. metallica*, normally associated with ground-nesting sparrows and warblers, in which the nestlings are fledged in 8-12 days. On the contrary, in the nests of crows, hawks and other large birds in which the nestlings remain for 5-6 weeks, the puparia of *P. avium* are larger, on average, than any of the other species except *P. lata*. Length and breadth measurements are of limited use because of the degree of variation within a single species; breadth measurements given in text are taken at the widest point of the puparium.

The cuticle of both larvae and puparia of most species is thickly set with spines, the lengths of which vary in different regions of the body. The spines of any given region show a wide range in length, but the majority are of similar size. As more puparia than larvae were available, the measurements of these spines were based on puparial material. The dorsal surface of the puparium is usually thickly set with spines, the largest in the center of the segment and the smallest towards the intersegmental annulus. The lengths of spines given in text are based on spines from the center of segment and these spines are termed *dorsal cuticular spines*. Shrinkage of the cuticle in the formation of the puparium results in the formation of numerous cuticular ridges, seen clearly on the dorsal and ventral cuticule. The size of these ridges seems to be constant for each species although the number found is variable; they are termed pronounced (*sialia*, Fig. 61d), weak (*asiovora*, Fig. 43b), or absent (*aenea*, Fig. 42d), according to size.

The ventral surface of the puparium presents a number of diagnostic features. A small vesicle, fluid filled in life, occurs on the midline of the intersegmental annuli of the larval stage. These vesicles, described by Guberlet and Hotson (1940), apparently serve the larva as a locomotory apparatus and here are termed the *ventral pads* (see Glossary). In the puparium, the remains of each ventral pad appear as a crescentric or ovoid, spine-free scar between segments (Fig. 43c,d). Between two adjacent pads each segment has, typically, three bands of spines. The degree of development in width of each band varies with the species. In the *anterior band*, immediately posterior to the ventral pad, is a dense patch of small, slender spines, termed the *anterior patch*. The second or *medial band* of spines, composed of stouter spines than those of the anterior patch, may be pronounced (Fig. 43c) or reduced (Fig. 50c). The third or *posterior band* of spines likewise may be pronounced (Fig. 44c), reduced (Fig. 43c), or absent (Fig. 42d). The degree of spination is expressed by the *ventral band ratio*, which is a decimal fraction derived from the sum of the widths of the three bands of spines on any segment (measured on the midline) divided by the total distance between the adjacent margins of the ventral pads bordering that segment. In general, and where possible, three or four ratios per puparium were calculated, based on measurements of the central and most typical segments 5 through 8. In samples with few available puparia, all were used but in large samples this was not the case.

The posterior region of the puparium has a number of useful characters. The sclerotization of the larval cuticle fixes the *stigmatal* (spiracular) *plates*, allowing certain measurements to be made. The diameter of a stigmatal plate is given as a maximal diameter measured from the dorsal portion of the peritreme through the central spiracular slit to the ventral margin of the button. The distance between the two buttons (a) is also given, as is the width across the stigmatal plates at the level of the buttons (b), the measurements combined as a/b to give the *stigmatal ratio*. The peritremes of the puparia, unlike those of the larvae, are always closed. The cuticle between the stigmatal plates is termed the *mesostigmatal area*. Mesostigmatal folds and plaques are present or absent, depending on the species; if present, they bisect the area distinctly (*bicolor*, Fig. 45b), weakly, or not at all (Fig. 43a, b). The cuticle immediately dorsal to the stigmatal plates is termed the *hyperstigmatal area*, and with the exception of the prothoracic fringe, the spines of this region are usually the longest of the puparium. The cuticle immediately ventral to the stigmatal plates, the *hypostigmatal area*, characteristically possesses a number of folds; the number, size and configurations of these folds vary with the species. There are generally few spines in the hypostigmatal area compared to the number on the rest of the cuticle. The *circumstigmatal folds* pass laterally and dorsally from the hypostigmatal area around the stigmatal region; their formation, number and size vary with the species. Four basic patterns occur: irregular ridges of no fixed pattern (Fig. 44a); concentric ridges about the area (Fig. 57a); rectangular folds passing laterally but not surrounding the area (Fig. 56b); absence of folds (Fig. 43b). The basic patterns tend to be constant although the number of folds varies markedly; some species show more variability in these characters than others.

The posterior tubercles in the puparia are small scars surrounding the posterior region. They are not pronounced in most *Protocalliphora* (but more so on puparia formed from underfed larvae) and are more distinct on species with few cuticular folds. They are of little value taxonomically.

Production of small puparia by underfed larvae is frequent. These puparia tend to differ from normal puparia in the following respects: (a) the stigmatal plates are close together, forming a *parastigmate* puparium; (b) mesostigmatal characters are reduced or lost, consequent upon the juxtaposition of the plates; (c) circumstigmatal folds are frequently missing; (d) the ventral cuticle is compressed and the ventral spine bands are elided or otherwise distorted; (e) the cuticular spines are set closer together, giving the illusion of greater numbers and length; (f) cuticular ridges are usually absent; (g) posterior tubercles are usually pronounced.

Key to the Nearctic species of *Protocalliphora* (based on males, females and puparia)

1. Male with unique surstylus and aedeagus (Figs. 6c, 20); female lacking upper (reclinate) orbital bristles; one or more accessory notopleural bristles present (Fig. 3a); puparium appearing smooth, cuticular spines small and scarce (Fig. 46), and prothoracic fringe extremely reduced, appearing absent . . . . *P.* (*Trypocalliphora*) *braueri* (Hendel)

-- Male genitalia not as in *braueri*; female with reclinate orbital bristles (Fig. 1); accessory notopleural bristles rarely present (*Protocalliphora*, typical subgenus); puparium (except in two species) finely to coarsely spined, not appearing smooth, and with obvious prothoracic fringe. 2

2. Abdomen of males and females not concolorous, males blue to bluish green or bluish purple (except in *bicolor*), females with 5th tergite cupreous, or abdomen almost entirely aeneous to cupreous 3

-- Abdomen of males and females concolorous, metallic blue to bluish 6

3. Female: thorax dark blue in ground color, abdomen blue or greenish blue with 5th tergite cupreous or at least cupreous tinted (occasionally 4th also); male: surstylus slender or digitate (Figs. 16,21,29) 4

-- Female: thorax and abdomen almost entirely aeneous to almost cupreous green; male: surstylus broadly digitate (Fig. 12) . . . . *P. aenea* S. & D.

4. Abdomen of male blue, of female blue with 5th tergite cupreous (occasionally 4th also, see 5a, *cuprina* var.); male: surstylus digitate, or only moderately slender (Figs. 61, 21); puparium spined or not 5

-- Both sexes with 5th tergite cupreous (sometimes weakly so if teneral) male: surstylus long and slender, slightly curved (Fig. 29); puparium rather strongly spined (Fig. 45); eastern . . . . *P. bicolor*, n. sp.

5. Calypteres white in both sexes; male: surstylus of moderate width, slightly curved, parallel sided (Fig. 21); puparium with most spines present only as minute tubercles, appearing like fine sandpaper (Fig. 56); chiefly eastern . . . . *P. metallica* (Tns.)

-- Calypteres brownish in male, white in female; male: surstylus digitate or only slightly curved, broader than in *metallica* (Fig. 16); puparium with distinct though short spines (Fig. 48); western . . . . *P. cuprina* (Hall)

6. Parafacial unusually broad, over three times breadth of 3rd antennal segment; calypteres dark brown; male: frons unusually broad, at narrowest equal to or even greater than width of an eye, and with both proclinate and reclinate orbital bristles (Fig. 33); puparium strongly ridged and spined (Fig. 55); western . . . *P. lata* n. sp.

-- Parafacial narrower, at most barely twice breadth of 3rd antennal segment; calypteres unusually whitish, brown in only a few western species; male: frons at narrowest obviously much less than width of an eye, usually much less, always lacking orbital bristles (Figs. 34-38); puparium variable, usually not strongly ridged 7

7. Male: surstylus broadly rounded (Figs. 7-9) 8

-- Male: surstylus digitate to slender (Figs. 10-32) 10

8. Male: frons strongly narrowed, narrowest width averaging 0.07 times width of head and obviously narrower than breadth of 3rd antennal segment, parafrontals almost touching; female: frons narrow, half or less the width of an eye, averaging 0.23 times width of head; puparium with minute spines, the appearance like fine sandpaper or little more (Fig. 49); eastern . . . . . . *P. deceptor* n. sp.

-- Male: frons broad for the genus, range 0.115-0.18 times width of head and obviously much greater (about twice) than breadth of 3rd antennal segment, the parafrontals widely separated (Figs. 34, 35); female: frons broad, obviously over half width of an eye and averaging 0.295 times width of head; puparium well ridged and spined . . . (Figs. 43, 44) 9

9. Male: frons typically wider than in *asiovora*, averaging 0.16 times width of head (Fig. 34); puparium with medial spine band of ventral cuticle pronounced and uninterrupted (Fig. 44); eastern . . . *P. avium* S. & D.

-- Male: frons narrower than in *avium*, averaging 0.13 times width of head (Fig. 35); puparium with medial spine band of ventral cuticle pronounced but distinctly interrupted on midline (Fig. 43); western . . . . *P. asiovora* S. & D.

10. Postalar wall and tympanic membrane each with conspicuous tuft of numerous long, brown to blackish hairs (Fig. 3b); fore tibia typically with two posterior bristles; almost always in nests of bank swallow . . . . *P. chrysorrhoea* (Meigen)

-- Postalar wall and tympanic pit bare or with one to few, usually short and inconspicuous, usually pale hairs; fore tibia with one posterior bristle in most species, commonly two in *sapphira* and *tundrae*; rarely in nests of bank swallow 11

11. Calypteres brownish or brownish yellow in males, same in females (except in *halli*), sometimes paler brown but at least brownish tinted with outer rims and fringe brownish to brownish yellow; western species. 12

-- Calypteres in both sexes white 15

12. Surstylus in male exceptionally long and slender, curved (Fig. 28); parafacial obviously much wider (nearly twice) than breadth of 3rd antennal segment . . . . *P. brunneisquama*, n. sp.

-- Surstylus shorter and/or broader, not or only slightly curved 13

13. Calypteres brownish or brownish yellow in both sexes . 14

-- Calypteres brownish or brownish yellow in male, white in female . . . . *P. halli*, n. sp.

14. Preocellar area in female heavily and evenly brownish microtomentose, in male almost always gray and narrowly triangular, rarely somewhat shining; palpi black-tipped . . . . *P. hesperia* S. & D.

-- Preocellar area in female polished black immediately anterior to median ocellus, in male with narrow polished line or low ridge (as in Fig. 2a); palpi usually entirely yellow . . . . *P. hesperioides*, n. sp.

15. Surstylus digitate, not appreciably curved, and broader than opposing species (Figs. 13-15) 16

-- Surstylus distinctly curved, usually slender on distal half or more, or parallel sided throughout (Figs. 22-32) . 19

16. Frons of male obviously wider than breadth of 3rd antennal segment 17

-- Frons of male obviously narrower than breadth of 3rd antennal segment . . . . *P. shannoni*, n. sp.

17. Male with polished black and narrow triangle or streak anterior to median ocellus (Fig. 37); preocellar area in female shining or gray; prothoracic fringe of larva or puparium long and dense, spines averaging 150 in length (80-200) 18

-- Male with preocellar triangle gray microtomentose (Fig. 38); preocellar area in female entirely gray microtomentose; prothoracic fringe not as conspicuous, spines averaging 57 in length (21-103) . . . . *P. hirundo* S. & D.

18. Preocellar area polished black in both sexes; eastern . . . . *P. sialia* S. & D.

-- Preocellar triangle or streak polished black in male, but preocellar area dull and heavily microtomentose in female; western . . . . *P.* sp. (western *sialia*)

*19. Puparium: hyperstigmatal area and dorsal cuticle with distinct spines 20

-- Puparium: Spines of both hyperstigmatal area and dorsal cuticle minute (fine sandpaper appearance) (Fig. 54) . . . . *P. interrupta*, n. sp.

*N.B. The puparium is presently unknown in *beameri* and *sapphira* but assumed to have distinct spines; if not, in both of these the frons is relatively broad compared with that in *interrupta* (0.12-0.17 times head width compared to 0.09 in *interrupta*). If teneral or incompletely colored *metallica* slipped through couplet 3 and came to this point, its surstylus is broader than in *interrupta* (cf. Figs. 21 and 30).

20. Parafacial in both sexes relatively broad, obviously wider than breadth of 3rd antennal segment 21

-- Parafacial in both sexes narrow, equal to or barely wider than breadth of 3rd antennal segment *P. spenceri*, n. sp.

21. Male with preocellar area broadly triangular, evenly bright gray microtomentose (Fig. 36); fore tibia typically with two posterior bristles; far north, arctic Canada and Greenland . . . . *P. tundrae*, n. sp.

-- Male with preocellar area narrowly triangular to acuminate, often more or less rugose and shining; fore tibia with one posterior bristle in most species; not far northern . . 22

22. Male with occiput predominantly white haired behind and below the row of ostocular setae, at most only a few slender black hairs in a partial row *P. beameri*, n. sp.
-- Male with numerous coarse black hairs in two or more rows behind and below the postocular setae 23

23. Frons of male heavily setose, frontal bristles and parafrontal hairs extending posteriorly at least to level of ocellar bristles; parafrontal relatively wide all the way to vertex . . . . *P. spatulata*, n. sp.
-- Frons of male seldom as setose as in *spatulata*, frontal bristles and parafrontal hairs usually ending anterior to level of median ocellus; parafrontal strongly narrowed dorsally in prevertical area 24

24. Frons of male narrow, averaging 0.08 times head width (combined range of three species 0.06-0.13); parafacial of male wider than frons at its narrowest; fore tibia with one posterior bristle 25
-- Frons of male broader, 0.17 times head width in the one known male; parafacial of male narrower than frons (0.85x) . . . . *P. sapphira* (Hall)

25. Parafrontals of male wide, each wider midway than narrowed frontal vitta: parafrontal and parafacial of male sparsely and finely haired, contributing to an "almost bare" appearance, silvery microtomentose . . . . *P. seminuda*, n. sp.
-- Not as above, frontal vitta of male much wider than a parafrontal 26

26. Puparium ventrally with posterior band of spines well developed, thus three spine bands per segment (Fig. 57d); head dark gray microtomentose; typically parasites of chickadees (*Parus* spp.); western . . *P. parorum*, n. sp.
-- Puparium ventrally with posterior spine band vestigial or lacking on at least posterior five or six segments (Fig. 50c); parafrontal and parafacial brighter, silvery gray microtomentose; typically parasites of marsh-inhabiting birds; northeastern . . . . *P. fallisi*, n. sp.

Key to males of Nearctic *Protocalliphora*

1. Aedeagus unique in genus (Fig. 6c); one or more accessory notopleural bristles (Fig. 3a); frons extremely narrow (Fig. 41), averaging only 0.05 times width of head and less than half breadth of 3rd antennal segment, at narrowest parafrontals almost touching . . . . *P.* (*Trypocalliphora*) *braueri* (Hendel)

-- Aedeagus not as in *braueri* (Fig. 6a,b); usually no accessory notopleural bristles; frons usually broader and parafrontals well separated (Figs. 33-40) (typical subgenus *Protocalliphora*) 2

*2. Calypteres deep yellowish to brown or brownish yellow 3

-- Calypteres white or whitish, at most outer margins pale yellow 9

*N.B. *P. aenea*, usually deep yellowish, is keyed both ways because of possible misinterpretation. It is somewhat intermediate, especially if individuals are immature and not fully colored.

(Calypteres deep yellowish to brown)

3. Frons unusually broad for a male, at narrowest equal to or even broader than width of an eye (Fig. 33); proclinate and recinate orbital bristles present, as in females; parafacial unusually broad, over three times breadth of 3rd antennal segment; calypteres dark brown . . . . *P. lata*, n. sp.

-- Frons obviously much narrower than width of an eye; no orbital bristles; parafacial obviously not as broad . . 4

4. Calypteres yellow to deep yellowish . . . *P. aenea* S. & D.

-- Calypteres definitely brown to brownish, at least rims and fringes 5

5. Surstylus exceptionally long, slender, and curved (Fig. 28) . . . . *P. brunneisquama*, n. sp.

-- Surstylus digitate, not curved, straight or nearly so . . . 6

6. Palpus entirely orange-yellow; preocellar area a narrow polished line or low ridge (Fig. 2) . . . . *P. hesperioides*, n. sp.

-- Palpus black-tipped; preocellar area usually narrowly triangular to acuminate 7

7. Preocellar area usually triangular and evenly gray-microtomentose . . . . *P. halli*, n. sp.
-- Preocellar area narrow and acuminate, more or less rugose and shining or subshining 8

8. Parafacial relatively narrow, equal to or barely wider than 3rd antennal segment . . . . *P. cuprina* (Hall)
-- Parafacial obviously wider than 3rd antennal segment . . . . *P. hesperia* S. & D.

(Calypteres white)

9. Surstylus unusually broad (Figs. 7-9) 10
-- Surstylus slender to digitate (Figs. 11, 13-15, 21-32) . 12

10. Frons strongly narrowed, averaging 0.07 times width of head, obviously narrower than 3rd antennal segment, and at narrowest parafrontals almost touching; cerci in posterior view narrowly acuminate, parallel to acute apices (Fig. 5b) . . . . *P. deceptor*, n. sp.
-- Frons broad, averaging 0.13-0.16 times width of head and obviously much wider than breadth of 3rd antennal segment, parafrontals widely separated; cerci sturdy, divergent, and apices thick and rounded (Fig. 5a) 11

11. Frons typically wider than in *asiovora*, averaging 0.16 times width of head (Fig. 34); surstylus narrower in proportion to its length than in *asiovora* (Fig. 7); eastern . . . . *P. avium* S. & D.
-- Frons narrower than in *avium*, averaging 0.13 times width of head (Fig. 35); surstylus broader in proportion to its length than in *avium* (Fig. 8); western . . . . *P. asiovora* S. & D.

12. Fifth tergite cupreous (sometimes weakly so, if teneral) . . . . *P. bicolor*, n. sp.
-- Fifth tergite blue to bluish purple 13

13. Surstylus digitate, not appreciably curved, and broader than opposing species (Figs. 11-15) 14
-- Surstylus distinctly curved, usually slender on distal half, or parallel sided throughout (Figs. 21-32) 18

14. Postalar wall and tympanic membrane each with conspicuous tuft of long, brown to blackish hairs (Fig. 3b); fore tibia typically with two posterior bristles . . . . *P. chrysorrhoea* (Meigen)

-- Postalar wall and tympanic membrane with one to few, usually short and inconspicuous, usually pale hairs; fore tibia with one posterior bristle in most species . . . 15

15. Frons obviously narrower than breadth of 3rd antennal segment 16

-- Frons obviously wider than breadth of 3rd antennal segment 17

16. Calypteres always at least slightly yellowish, with dark yellow margins (Typical males of *aenea* have brownish yellow calypteres with brown margins) . . . *P. aenea* S. & D.

-- Calypteres white with pale yellow margins . . . . *P. shannoni*, n. sp.

17. Preocellar area with polished black and narrow triangle or low ridge (Fig. 37); surstylus in profile relatively long and digitate (Fig. 15) . . . . *P. sialia* S. & D.

-- Preocellar area with broad and evenly gray-microtomentose triangle (Fig. 38); surstylus in profile appearing stubby, relatively short for its length and broad at base (Fig. 14) . . . . *P. hirundo* S. & D.

18. Preocellar area broadly triangular, evenly bright gray microtomentose (Fig. 36); fore tibia typically with two posterior bristles; far north, arctic Canada and Greenland . . . . *P. tundrae*, n. sp.

-- Preocellar area narrowly triangular to acuminate, often more or less rugose and shining; fore tibia with one posterior bristle in most species; not far northern 19

19. Frons heavily setose, frontal bristles and parafrontal hairs extending posteriorly at least to level of ocellar bristles; parafrontal relatively wide all the way to vertex . 20

-- Frons seldom as setose, frontal bristles and parafrontal hairs usually ending anterior to level of median ocellus; parafrontal strongly narrowed dorsally in prevertical area 21

20. Surstylus strongly narrowed on distal half and shorter than cerci (Fig. 31) . . . . *P. beameri*, n. sp.
-- Surstylus parallel sided for most of its length, usually widening rather abruptly at distal end to appear spatulate (Fig. 27) . . . . *P. spatulata*, n. sp.

21. Parafacial wider than width of frons at narrowest; frons averaging 0.08-0.12 times head width; fore tibia usually with one posterior bristle 22
-- Parafacial narrower than width of frons (0.85x); frons 0.17 times head width in the one available specimen; fore tibia with two posterior bristles (probably variable) . . . . *P. sapphira* (Hall)

22. Parafrontal and parafacial dark gray microtomentose; typically parasites of chickadees (*Parus* spp.); western . . . . *P. parorum*, n. sp.
-- Parafrontal and parafacial bright, silvery gray microtomentose 23

23. Parafrontals relatively wide, each wider midway than narrowed frontal vitta; parafrontal and parafacial sparsely and finely haired, contributing to an "almost bare" appearance, silvery microtomentose *P. seminuda*, n. sp.
-- Not as above, frontal vitta much wider than a parafrontal 24

24. Parafacial opposite lunule broader than 3rd antennal segment; 25
-- Parafacial narrower than breadth of 3rd antennal segment, or barely equal to it . . . . *P spenceri*, n. sp.

25. Western . . . . *P. interrupta*, n. sp.
-- Eastern . . . . *P. fallisi*, n. sp.
-- Chiefly eastern . . . . *P. metallica* (Tns.)

Key to females of Nearctic *Protocalliphora*

1. Upper (reclinate) orbital bristle present; usually no accessory notopleural bristles (*P.*, subgenus *Protocalliphora*) . 2
-- *No upper orbital bristle; accessory notopleural bristles present . . . . *P.*, subgenus *Trypocalliphora*) (one species, *P. braueri* (Hendel))

*N.B. See Variation under *P. spenceri*, in which two females are aberrant in lacking upper orbitals, but accessory notopleurals are also lacking.

2. Thorax and abdomen almost entirely aeneous to almost cupreous green . . . . *P. aenea* S. & D.
-- Thorax and abdomen bluish to bluish purple or greenish blue, 5th tergite cupreous in three species 3

*3. Calypteres brown or brownish (tending to be brownish yellow in immature specimens) 4
-- Calypteres white or whitish, at most outer margins pale yellowish 7
*N.B. *P. hesperia* is keyed both ways because in some series females have pale calypteres, yellowish brown or even somewhat whitish, possibly because of immaturity and lack of full coloration in the reared series.

(Calypteres brown)

4. Parafacial unusually broad, over three times breadth of 3rd antennal segment; fore tibia with two posterior bristles; western . . . . *P. lata*, n. sp.
-- Parafacial narrower, at most about twice breadth of 3rd antennal segment; fore tibia with one posterior bristle in most species 5

5. Parafacial obviously much wider than (nearly twice) breadth of 3rd antennal segment . . . *P. brunneisquama*, n. sp.
-- Parafacial obviously much less than twice breadth of 3rd antennal segment 6

6. Preocellar area polished black, small but distinct (cf. Fig. 2a); palpus entirely orange-yellow . *P. hesperioides*, n. sp.
-- Preocellar area dull, heavily gray to brownish microtomentose; palpus black-tipped . . *P. hesperia* S. & D.

(Calypteres white)

7. Fifth tergite cupreous to cupreous green 8
-- Fifth tergite bluish or greenish blue, concolorous with preceding tergites 10

8. Anterolateral area of scutum almost solidly shining blue-black, as viewed from behind; if puparium present, most spines greatly reduced, minute, the appearance like that of fine sandpaper (Fig. 56) . *P. metallica* (Tns.)
-- Anterolateral area of scutum usually not so, as viewed from behind, the shining area deeply incised by extension of posthumeral gray microtomentum, extreme in *cuprina*, weak to somewhat intermediate in *bicolor*; if puparium present, spines are obvious 9

9. Western 9a
-- Eastern . . . . *P. bicolor*, n. sp.

9a. Only 5th tergite cupreous or cupreous green . . . . 5. *P. cuprina* (Hall)
-- Fourth and 5th tergites both cupreous or cupreous green . . . . *P. cuprina* var. or n. sp.

10. Frons exceptionally narrow, averaging 0.23 times width of head; parafrontal relatively wide and frontal vitta less than twice width of a parafrontal; eastern . . . . *P. deceptor*, n. sp.
-- Frons not so narrow and frontal vitta broader, two to three times width of a parafrontal 11

11. Postalar wall and tympanic membrane each with conspicuous tuft of numerous long black hairs (Fig. 3b); fore tibia typically with two posterior bristles . . . . *P. chrysorrhoea* (Meigen)
-- Postalar wall and tympanic membrane bare or with one to few, usually short and inconspicuous, usually pale hairs; fore tibia with one posterior bristle in most species, commonly two in *sapphira* and *tundrae* 12

12. Scutum heavily microtomentose anterior to transverse suture; median shining stripe absent or indistinct as viewed from behind at a low angle, two narrow dark stripes outside acrostical bristles more distinct than usual, the resulting appearance being that of three broad gray stripes; western . . . . *P. spatulata*, n. sp.
-- Scutum not so heavily microtomentose, usually with a broad median shining stripe flanked by broad gray microtomentose stripes, as viewed from behind at a low angle 13

13. Eastern North America, arctic and subarctic Canada, and Greenland 14
-- Western North America 18

14. Parafacial relatively broad, width at lunule obviously much greater than breadth of 3rd antennal segment . . . 15
-- Parafacial relatively narrow, barely wider than breadth of 3rd antennal segment . . . . *P. shannoni*, n. sp.

15. Arctic Canada and Greenland; typically with two posterior bristles on fore tibia . . . . *P. tundrae*, n. sp.
-- Not far northern; typically with one posterior bristle on fore tibia (two in part of *fallisi*!) 16

16. Parafacial about twice breadth of 3rd antennal segment, usually slightly convex and appearing bulging; robust species, normal length 9.5-10 mm . . *P. avium* S. & D.
-- Parafacial not as broad, and not appearing bulging; smaller species, 7.5-9.5 mm 17

17. Preocellar area a large, irregular, polished black spot . . . . *P. sialia* S. & D.
-- Preocellar area dull, finely and evenly microtomentose, sometimes thinly so and subshining . *P. fallisi*, n. sp.

18. Dark species, head dark gray microtomentose; parafrontal posteriorly tending to be thinly microtomentose, hence subshining, blackish . . . . *P. parorum*, n. sp.
-- Brighter species, bright gray microtomentose, often tinted yellowish or brownish 19

19. Preocellar area partly but irregularly polished black, sometimes thinly microtomentose and subshining; parafacial unusually narrow, equal to or barely wider than breadth of 3rd antennal segment . . *P. spenceri*, n. sp.
-- Preocellar area evenly and usually heavily microtomentose; parafacial broader, usually obviously wider than breadth of 3rd antennal segment 20

20. Fore tibia with two posterior bristles, perhaps typically, but probably variable; parafrontal, parafacial, and preocellar area decidedly brownish yellow . . *P. sapphira* (Hall)

-- Not with that combination of characters, fore tibia typically with one posterior bristle (variable in some species!), and parafrontal, parafacial, and usually preocellar area bright gray or light yellowish-gray microtomentose 21

21. Parafacial relatively narrow, appearing equal to or only slightly wider than breadth of 3rd antennal segment 22

-- Parafacial relatively broad, obviously much wider than breadth of 3rd antennal segment, usually almost twice 23

22. Females of three species cannot be reliably separated
. . . . *P. halli*, n. sp.
. . . . *P. hesperia* S. & D.
. . . . *P. interrupta*, n. sp.

23. Females of five species cannot be reliably separated
. . . . *P. asiovora* S. & D.
. . . . *P. beameri*, n. sp.
. . . . *P. hirundo* S. & D.
. . . . *P. seminuda*, n. sp.
. . . . *P.* 'western *sialia*'

Key to the 3rd-instar larvae of Nearctic *Protocalliphora*

1. Prothoracic fringe much reduced (average less than 10); cuticular spines much reduced in number or absent, at most limited to a single narrow band per segment . . . . *P.* (*Trypocalliphora*) *braueri* (Hendel)

-- Prothoracic fringe pronounced (averaging 25 or more); cuticular spines numerous on all parts of body (*Protocalliphora*, typical subgenus) 2

2. Medial band interrupted on midline 3

-- Medial band entire 4

3. Dorsal cuticular spines small, averaging less than 10 . . . . *P. interrupta*, n. sp.

-- Dorsal cuticular spines larger, averaging more than 15 . . . . *P. asiovora* S. & D.

4. Prothoracic fringe averages more than 100 . . . . *P. sialia* S. & D.

-- Prothoracic fringe averages less than 100 5

5. Hyperstigmatal spines as tubercles, averaging 4-6 μ . . 6
-- Hyperstigmatal spines as spines, larger than above . . . 7

6. Entire dorsal cuticle with spines reduced to minute tubercles; spines of medial band average 6 (2-12) μ . . . . *P. metallica* (Tns.)
-- Dorsal cuticle with definite, although relatively few, dorsal spines; spines of medial band average 12 (2-29) μ . . . . *P. deceptor*, n. sp.

7. Posterior band vestigial or absent on at least posterior 5-6 segments 8
-- Posterior band present on all segments 11

8. Posterior band reduced but present on anterior 5-6 segments . . . . *P. tundrae*, n. sp.
-- Posterior band absent on all segments 9

9. Prothoracic fringe averages more than 70 μ; hyperstigmatal spines 34 (16-52) μ . . . . *P. aenea* S. & D.
-- Prothoracic fringe averages less than 60 μ; hyperstigmatal spines average less than 30 μ 10

10. Hyperstigmatal and dorsal cuticular spines average 29 μ or more . . . . *P. halli*, n. sp.
-- Hyperstigmatal and dorsal cuticular spines average less than 16 μ . . . . *P. fallisi*, n. sp.

11. Prothoracic fringe averages 80 μ or more 12
-- Prothoracic fringe averages 75 μ or less 13

12. Hypostigmatal spines average 6 μ . . . . *P. bicolor*, n. sp.
-- Hypostigmatal spines average 15 μ or more . . . . *P. parorum*, n. sp.

13. Prothoracic fringe averages 60 μ or more 14
-- Prothoracic fringe averages less than 60 μ 17

14. Prothoracic fringe averages more than 70 μ 15
-- Prothoracic fringe averages less than 70 μ 16

15. Spines of dorsal cuticle and medial band average more than 40 μ . . . . *P. lata*, n. sp.
-- Spines of dorsal cuticle and medial band average less than 30 μ . . . . *P. hesperioides*, n. sp.

16. Hyperstigmatal spines average 28 μ; hypostigmatal spines average 18 μ . . . . *P. cuprina* (Hall)
-- Hyperstigmatal spines average 11 μ; hypostigmatal spines as tubercles, averaging 6 μ . . . . *P. shannoni*, n. sp.

17. Prothoracic fringe averages 29 (12-62) μ . . . . *P. chrysorrhoea* (Meigen)
-- Prothoracic fringe averages more than 40 μ 18

18. Spines of anterior patch small, average 9-12 μ . . . .
. . . . *P. *spenceri*, n. sp.
. . . . *P. *seminuda*, n. sp.
*N.B. Third-instar larvae of these species cannot be separated on the basis of larval characters alone.
-- Spines of anterior patch averaging 15 μ or larger 19

19. Spines of dorsal cuticle and medial band average 20 μ . . . . *P. avium* S. & D.
-- Spines of dorsal cuticle and medial band average 25 μ or more 20

*20. Posterior band pronounced; ventral band ratio more than 0.80 . . . . *P. spatulata*, n. sp.
-- Posterior band weak; ventral band less than 0.70 . . . . *P. hirundo*, S. & D.
*N.B. Ventral band ratios in larvae are not reliable. Species in couplet 20 can only be identified with certainty on the basis of puparial or adult characters.

Key to the puparia of Nearctic *Protocalliphora*

1. Dorsal cuticular spines much reduced in number, at most limited to a narrow band per segment; prothoracic fringe vestigial, spines averaging less than 10 μ . . . . *P. (Trypocalliphora) braueri* (Hendel)
-- Dorsal cuticular spines numerous (in two species small or minutely tuberculate); prothoracic fringe well developed, spines averaging more than 25 μ (*Protocalliphora*, typical subgenus) 2

2. Medial band interrupted on the midline 3
-- Medial band entire 4

3. Dorsal cuticular spines large, averaging 25 (12-45) μ; ventral band ratio 0.63; posterior band pronounced . . . . *P. asiovora* S. & D.
-- Dorsal cuticular spines small, many as tubercles, averaging 4-6 μ; ventral band ratio 0.27; posterior band usually absent or at best, vestigial . . . . . *P. interrupta*, n. sp.

4. Hyperstigmatal spines as minute tubercles, 4-6 μ 5
-- Hyperstigmatal spines as spines, more than 10 μ 6

5. Dorsal cuticle with spines reduced to minute tubercles, with a fine sandpaper-like appearance; spines of medial band average 6 (2-12) μ . . . . *P. metallica* (Tns.)
-- Dorsal cuticle with definite spines, not appearing as fine sandpaper; spines of medial band 12 (4-20) μ . . . . *P. deceptor*, n. sp.

6. Posterior bands vestigial or absent on at least posterior 5-6 segments; ventral band ratio normally less than 0.50 7
-- Posterior band present (usually pronounced) on all segments; ventral band ratio normally than 0.50 10

7. Posterior band present, although much reduced, on anterior 5-6 segments; circumstigmatal folds present 8
-- Posterior band absent, or at best vestigial, on all segments; circumstigmatal folds absent 9

8. Posterior band, although reduced, present on anterior 5-6 segments . . . . *P. tundrae*, n. sp.
-- Posterior band absent or vestigial on all segments . . . . *P. halli*, n. sp.

9. Prothoracic fringe averages more than 70 μ ventral band ratio more than 0.40; dorsal cuticular spines large . . . . *P. aenea*, S. & D.
-- Prothoracic fringe averages less than 60 μ; ventral ratio less than 0.40; dorsal cuticular spines few, averaging 16 (4-41) μ . . . . *P. fallisi*, n. sp.

10. Ventral band ratio more than 0.80 11
-- Ventral band ratio less than 0.75 12

11. Prothoracic fringe averages 150 (80-200) μ; circumstigmatal folds irregular . . . . *P. sialia*, n. sp.

-- Prothoracic fringe averages 58 (34-90) μ; circumstigmatal folds concentric . . . . *P. spatulata*, n. sp.

12. Dorsal cuticular ridges pronounced; dorsal cuticular spines both anteriorly and posteriorly directed 13
-- Dorsal cuticular ridges weak, dorsal cuticular spines directed posteriorly only 16

13. Mesostigmatal area with folds and plaques reduced or absent; mesostigmatal area entire 14
-- Mesostigmatal area with pronounced folds and/or plaques bisecting the stigmatal area 15

14. Prothoracic fringe averages 74 (31-130) μ; spines of dorsal cuticle and medial band large, averaging 40 (16-64) μ; puparia large (13 mm) . . . . *P. lata*, n. sp.
-- Prothoracic fringe averages 57 (21-103) μ; spines of dorsal cuticle and medial band smaller than above, averaging less than 40 μ (21-41 and 16-37, resp.; puparia medium-sized (7.2-8.9 mm) . . . . *P. hirundo*, S. & D.

15. Circumstigmatal folds pronounced, irregular . . . . *P. avium*, S. & D.
-- Circumstigmatal folds weak, concentric . . . . *P. chrysorrhoea* (Meigen)

16. Mesostigmatal folds and/or plaques bisecting stigmatal plate area 17
-- Mesostigmatal folds and/or plaques weak or absent, not markedly bisecting stigmatal plate area 19

17. Prothoracic fringe averages 75 (55-110) μ; ventral band ratio 0.75 . . . . *P. hesperioides*, n. sp.
-- Prothoracic fringe averages less than 55 μ; ventral band ratio 0.70 or less 18

18. Circumstigmatal folds pronounced; dorsal cuticular ridges absent . . . . *P. seminuda*, n. sp.
-- Circumstigmatal folds weak or absent; dorsal cuticular ridges weak, but always present . . . . *P. spenceri*, n. sp.

19. Hyperstigmatal spines small, averaging 11 (4-24) μ; hypostigmatal spines as tubercles, appearing as fine sandpaper . . . . *P. shannoni*, n. sp.

-- Hyperstigmatal spines large, averaging over 25 μ; hypostigmatal spines as spines 20

20. Mesostigmatal folds and plaques reduced; posterior band reduced but present; prothoracic fringe averages 60 (20-90) μ . . . . *P. cuprina* (Hall)

-- Mesostigmatal folds and plaques pronounced; posterior band pronounced on all segments; prothoracic fringe averaging more than 75 μ 21

21. Ventral band ratio less than 0.60 . . . . . *P. bicolor*, n. sp.

-- Ventral band ratio more than 0.70 . . . *P. parorum*, n. sp.

Protocalliphora (*P.*) *aenea* Shannon and Dobroscky
(Figs. 12, 42, Map 4)

Protocalliphora splendida var. *aenea* Shannon and Dobroscky, 1924: 250, 251 [N.H., Ont., latter now known to be *braueri*; the N.H. specimens were mentioned in Townsend, 1919: 379 as atypical *metallica*].

P. splendida form *aenea*.--Johnson, 1925a: 216 [N.H. record repeated].

Apaulina aenea.--Hall, 1948: 185 [Description mixed with *hirudo* (i.e., *braueri*); B.C. and Mont. specimens are *braueri*, and description of male is of *braueri*].

Protocalliphora aenea.--James, 1955: 24 [No Calif. specimens seen, though listed by Hall (1948) from central Calif.]; Boyes, 1961: 552, 554, 556, 565 [Cytology]; Boyes and van Brink, 1965: 543 [Cytology]; Hall, 1965: 925 [Catalog; widespread]; Boyes and Shewell, 1975: 446 [Mention of cytology]; Whitworth, 1976: 17, 26, 37 [Utah; parasite of American dipper].

Diagnosis. Conspicuously dichromatic, thorax and abdomen shining dark blue to bluish purple in male, aeneous or dull cupreous green in female; superficially similar to *braueri* but without accessory notopleural bristles, with reclinate orbital bristles in female, and without the distinctive male terminalia of *braueri*.

Male. Head with parafrontal and parafacial dark gray to brownish gray microtomentose, parafrontal darkening toward vertex; occiput with two to three irregular rows of long black hairs behind the postocular setae. Body shining dark blue to bluish purple, scutum and abdomen concolorous, with former appearing slightly darker from the gray to brownish gray microtomentum that demarcates three broad shining stripes. Abdomen highly shining in direct view but with a thin gray "bloom" when viewed from behind at a low angle. Calypteres brownish yellow with brown rims, upper calypter slightly paler.

Frons narrow, narrowing slightly anterior to ocellar tubercle, at narrowest 0.085 times head width (0.07-0.10) and slightly broader (1.32x; 1.00-1.50) than ocellar span; ocellar tubercle ending acutely just anterior to median ocellus, usually by less than length of latter, the short triangular preocellar area polished; parafrontals narrow and well separated throughout, the frontal vitta over twice width of a parafrontal, each parafrontal strongly narrowing above and becoming subshining, the microtomentum thin and the prevertical area partly polished; upper portion of parafacial and the adjacent parafrontal with numerous coarse black hairs, on the parafacial in three to four irregular rows; each parafacial of moderate width, 1.33 times (1.18-1.67) width of frons, 1.75 times (1.50-2.02) ocellar span, and 0.70 times (0.64-0.83/24) vibrissal interval; cheek height 0.40 times (0.37-0.42/24) maximum length of eye.

Thorax without accessory notopleural bristles; tympanic membrane typically bare or with only one or few slender and rather inconspicuous black hairs, never densely beset with coarse hairs; postalar wall typically bare, sometimes with a few short hairs. Fore tibia with one median posterior bristle, rarely a second.

Terminalia not distinctive, of a type common to a number of species: cerci distally acuminate and parallel (as in Fig. 5b); surstylus in profile digitate, slightly broadened toward base but distally with sides subparallel and apex bluntly rounded (Fig. 12); distal portion of aedeagus slender, not cylindrical as in *braueri* (cf. Figs. 6a,c).

Female. Body aeneous or dull cupreous green, rather heavily microtomentose, dorsum of thorax more heavily so than abdomen but scutum with three subshining stripes, the abdomen often quite shining aeneous to cupreous. Calypteres usually yellowish tinted, rims strongly so.

Frons broader than in *braueri*, at vertex over 1/4 width of head (0.26x; 0.24-0.295) and 3.41 times (2.86-3.91) ocellar span; frontal vitta with a few weak and inconspicuous hairs, heavily microtomentose, often with small, polished, irregular to triangular preocellar area; parafrontal relatively broad, with one to two irregular rows of hairs, entirely microtomentose throughout, the prevertical area darker but still evenly microtomentose; parafacial well covered with hairs, width 0.48 times (0.45-0.55) width of frons, 1.64 times (1.43-2.00) ocellar span, and 0.78 times (0.67-0.89/23) vibrissal interval; cheek comparatively narrow, 0.40 times (0.37-0.48) maximum length of eye.

Apparent normal length of adults, 7.5-9 mm.

Egg. Length 1130 μ (1080-1190)/18, breadth 320 μ (295-380)/18; cuticle white in color and ornamented with minute spines.

Second instar. *Cephalopharyngeal skeleton* 510 μ (500-524)/2 in length. *Prothoracic fringe* 35 μ (26-54)/30. *Posterior region:* peritreme closed, stigmatal plates 77 μ (69-86)/4 in diameter; button indistinct; mesostigmatal spines absent; hyperstigmatal spines numerous, 10 μ (4-14)/30; hypostigmatal spines scarce, 6 μ (2-10)/30. *Dorsal cuticle* with numerous spines, 12 μ (6-18)/30. *Ventral cuticle:* spines of anterior patch 5 μ (4-6)/30; medial band pronounced, spines 8 μ (4-12)/30; posterior band present.

Third instar. Peritreme usually closed but sometimes open; mesostigmatal and hypostigmatal spines present. *Prothoracic fringe* 78 μ (50-120)/100. *Cephalopharyngeal skeleton* 805 μ (740-1010)/100 in length.

Puparium. Length 8.1 mm (5.7-9.7)/100, breadth 3.8 mm (2.2-4.7)/100; thin-walled, dull, light brown. *Posterior region* (Fig. 42a,b): stigmatal plates 170 μ (130-220)/192 in diameter; distance between buttons 545 μ (340-690)/96, and across stigmatal plates 1100 μ (780-1310)/96; stigmatal ratio 0.49 (0.43-0.52)/96; stigmatal area weakly outlined in cuticular plaques; mesostigmatal spines present, mesostigmatal folds and plaques absent; hyperstigmatal spines 34 μ (16-52)/100; hypostigmatal area with vestigial cuticular folds and numerous true spines, 15 μ (6-27)/100; circumstigmatal folds absent. *Dorsal cuticle* (Fig. 42d) with numerous spines, 24 μ (14-41)/100; cuticular ridges weak or absent. *Ventral cuticle* (Fig. 42c): ventral band ratio 0.47(0.45-50)/100; spines of anterior patch 13 μ (6-23)/100; medial band present, spines 21 μ (8-33)/100; posterior band vestigial or absent (if present, as a few rows of spines in a short arc about the scar of the succeeding ventral pad).

Type Series. Holotype and one paratype, Franconia, N.H. (Mrs. Slosson), and one paratype, "N.H.," all females [USNM]. The original description (p. 251) was vague on the type series, saying only "Type.--Cat. No. 26860 USNM." Considering the entire text, however, *aenea* is listed in the key as from N.H. and Ont., and on p. 252 the authors spell out the details of three females from Franconia, N.H. (Mrs. Annie T. Slosson) and one, probably female, fragmentary specimen from Ontario (And. Halket). For those who would not accept the "Type No." as a designation of holotype, Hall (1948) gave the data of the specimen in full and called it "Type", and that would serve as lectotype designation under Article 74. The broken specimen from Ontario was not labeled paratype by Shannon and Dobroscky and probably was not considered as such because of its condition; it is after all a female *braueri* (cf. succeeding discussion and that under *braueri*).

Specimens Examined. 202 males, 222 females; major sources, Gothic, Colo., 89 males, 110 females, and Algonquin Park, Ont., 94 males, 79 females.

Distribution. Northern, transcontinental, but not often recorded: Canada and northern United States, south in the mountains to California, Colorado, and Virginia. Recorded by Hall (1948) "from central California," but James (1955) recorded none in his study of "The blowflies of California," and we have seen only a recently collected series from Fresno County, California.

Canada. British Columbia: Bowser; Kleanza Creek (14 mi E Terrace); Shames (16 mi SW Terrace); 32 mi SW Terrace; Whytecliff (W side of West Vancouver). **Ontario**: Algonquin Park (chiefly Lake of Two Rivers) (mostly from nests of barn swallow, also of common grackle, eastern phoebe, American robin, house wren). **Quebec**: Burn Creek near Baie Comeau; Duchesnay (25 mi W of Quebec City); Hull; Sept. Iles; Val d'Espoir; Sully; Duncan Lake near Rupert (phoebe).

Contiguous United States. California: Near Bear Creek in Fresno Co. (nest of American dipper). **Colorado**: Gothic (Gunnison Co., N of Crested Butte, 9500 ft, nest of American dipper; Idaho Springs (6 mi SW, 9000 ft); Weld Co. (nest of Say's phoebe); SW corner Jackson Co., 10 mi N junction of highways Colo. 14 and US 40. **Idaho**: Wallace. **Maine**: Little Black River Rapids (near mouth of Allegash River), published as *P. splendida splendida* by Johnson (1925a:6). **New Hampshire**: Franconia (including type series); Mt. Washington. **New York**: Whiteface Mt. in Essex Co. **Oregon**: Mary's Peak (W of Corvallis; paratype of *cuprina* Hall). **Utah**: Temple Fork in Cache Co. near Logan (nest of American dipper). **Virginia**: Shenandoah National Forest (nest of phoebe; puparia only); White Oak Canyon in Shenandoah National Park (nest of Louisiana waterthrush). **Washington**: Deer

Lake near Chevelak. **West Virginia**: 12 mi SW of Berkeley Springs (nest of phoebe).

Of the hand-collected material, the earliest recorded date is May 14 (Whytecliff, B.C.) and the latest September 8 (Little Black River, Maine).

Basis of description of immature stages (all material Algonquin Park, Ontario): *Egg*: Algonquin Park, Ont., from caged female, July 1955. *Larvae:* **2nd instar** (2 specimens): barn swallow, June, 1951; *3rd instar* (10 specimens): barn swallow, June, 1951 (7) and eastern phoebe, June, 1954 (3). *Puparium* (100 specimens): barn swallow, June 1951 (34); June, 1952 (6); July, 1954 (20); June, 1955 (10); eastern phoebe, July, 1953 (10) and July, 1954 (20).

Hosts. American dipper (formerly water ouzel), common grackle, eastern and Say's phoebes, American robin, barn swallow, Louisiana waterthrush, and house wren.

Ecology and Biology. Apparently rare except in certain limited areas. It has been found frequently in nests at any height in or around man-made structures, the most frequent hosts being barn swallows and phoebes. Especially noteworthy are the three records from the American dipper. The record from Gothic, Colo. is based on an unusually long series of 89 males, 110 females reared July 15 and 18, 1960 (C. L. Remington), from "larvae in current nest of *Cinclus mexicanus*" [Yale]. No other species of *Protocalliphora* is known from that species of bird at the present time. Both ecology and biology of *aenea* were studied extensively by Bennett at Algonquin Park.

Variation. In the series of 25 measured females from which the proportions were derived, there were 18 eastern (chiefly Ont. and Va.) and only 7 western (B.C. to Colo.). Proportions were calculated on these separately to determine if the eastern and western populations showed differences, but differences were slight. Subsequently a good series from Gothic, Colo., was received, and the 10 females were measured and proportions calculated as a test of both the proportions based on the whole series and of those based on the few western specimens alone. The comparative proportions are as follows (Table VII) with their ranges (all based on numbers of females shown, except for the parafacial/vibrissal interval, for which two western individuals could not be measured. While the means differ, in general the ranges overlap and it appears that there is only one widely distributed species.

Table VII. Variation in ratios, female Protocalliphora aenea

Proportions	Total (25)	Eastern (18)	Western (7)	Gothic, Colo. (10)
frons/head width	.26(.24-.295)	.26(.24-.27)	.27(.25-.295)	.27(.25-.28)
frons/ocellar span	3.41(2.86-3.91)	3.37(2.86-3.75)	3.50(3.21-3.91)	3.82(3.50-4.00)
parafacial/frons	.48(.45-.55)	.48(.45-.55)	.475(.46-.52)	.485(.45-.50)
parafacial/ocellar span	1.64(1.43-2.00)	1.63(1.43-2.00)	1.66(1.50-1.83)	1.85(1.67-2.00)
parafacial/vibrissal interval	.78(.67-.89)	.77(.67-.89)	.80(.77-.85)	.81(.73-.85)
cheek/eye	.40(.37-.48)	.40(.38-.43)	.42(.39-.48)	.44(.42-.48)

Remarks. Confusion of the aeneous females of *aenea* and *braueri* has apparently been frequent and has led to erroneous records. The broken specimen of "*aenea*" from Ontario, recorded by Shannon and Dobroscky (1924) as if part of the type series although not labeled a paratype by them, is actually a female *braueri*, as shown by the accessory notopleural bristles and the distinctive puparium, fortunately preserved with the specimen. The endoparasitic habit of the larva also indicated *braueri*. The British Columbia and Montana specimens recorded as *aenea* by Hall (1948: 186) were also found to be *braueri* as noted under that species. Hall's description of the male of *aenea* applies to *braueri*, and was probably based, in part at least, on three males from the Hamilton, Mont. series; the other material cited consists only of females. The proportions given for the female of "*aenea*" were apparently influenced by the series of *braueri*, and cannot be relied on for *aenea*.

Males of *aenea* have no striking feature or distinctive combination of characters and will be difficult to identify by themselves. They have probably often been misidentified in collections. Females are striking because of the entirely aeneous coloration, shared only with the female of *braueri*, from which *aenea* is easily separated by having reclinate orbital bristles and broader front, and no accessory notopleural bristles. Associated puparia also will at once distinguish *braueri* and *aenea*.

Teneral females sometimes lack the distinctive aeneous or cupreous green color, and tend to a greenish color tinted with copper. To the naked eye, however, when a series containing both sexes is examined, the females, even though not fully colored, are so distinctly different from the males that there is no question but what a dichromatic species is involved.

Puparia of *aenea* are readily confused with those of *fallisi* and *tundrae* and its close associates. Those of *aenea* are generally distinguished by (i) the long prothoracic fringe, (ii) absence of mesostigmatal folds, and (iii) occurrence in nests of birds in man-made environments.

Protocalliphora (*P.*) *asiovora* Shannon and Dobroscky (Figs. 8, 35, 43, Map 5)

Protocalliphora avium var. *asiovora* Shannon and Dobroscky, 1924: 249, 250 [Wash.]; Shannon, 1926: 128 [In key].

P. avium.--Jellison and Philip, 1933: 28-30 [Mont.]; Gahan, 1933: 31-33 [New chalcidoid parasite].

New genus and n. sp.--Neff, 1945: 73-76 [Calif.].

Apaulina avium in part.--Hall, 1948: 186 [*asiovora* in synonymy under *avium*].

Apaulina basingeri Hall, 1948: 190 [Calif; n. sp. of Neff 1945.].

Protocalliphora larvae.--Hill and Work, 1947: 74-75 [Calif., on sparrow hawk [i.e., American kestrel] and golden eagle].

P. avium.--Pletsch, 1948: 296-7 [Mont.]; Jellison, 1969: 55 [Mont.].

P. sp.--Lee and Ryckman, 1954: 24 [Ariz.].

P. avium and *basingeri*.--James, 1955: 25 [Calif. records all under *avium* (one from Big Dalton Dam is correctly *hirundo*); *basingeri* probably a synonym of *avium*].

P. avium.--Gill, 1956: 652 [Alaska]; Kamal, 1958: 262 [Idaho].

Calliphora sp. in heads of prairie falcons.--White, 1963: 442-3 [Utah; probably *P. asiovora*].

Protocalliphora asiovora.--Hall, 1965: 926 [B.C. to Mont., s. to Calif., N. Mex., and Colo.; *basingeri* a synonym teste Sabrosky and Bennett].

P. asiovora.--James, 1969: 252 [Mention of biology].

P. sp. on hawks.--Seidensticker and Reynolds, 1971: 411 [Mont.; probably *asiovora*].

P. asiovora.--Whitworth, 1976: 22ff [Utah]; Whitworth, 1977: 4933-B.

Diagnosis. As in *P. avium* but frons of male slightly narrower; western, sister species of *avium*.

Male. Color as described for *avium*. Frons (Fig. 35) moderately broad at vertex, but slightly narrower than in *avium*, 0.13 times head width (0.115-0.14; holotype of *basingeri* 0.14) and much less than twice ocellar span (1.68x; 1.44-1.88); preocellar area often granular or weakly shining; parafacial broad, as in *avium*, slightly wider than frons (1.14x; 1.00-1.31), obviously much wider than ocellar span (1.93x; 1.75-2.20), and usually slightly wider than vibrissal interval (1.04x; 0.91-1.14); cheek broad, its height nearly half maximum length of eye (0.47x; 0.44-0.52).

Thorax without accessory notopleural bristles and hairs on tympanic membrane; postalar wall bare. Fore tibia with one posterior bristle.

Terminalia: cerci sturdy, deeply notched, widely divergent in posterior aspect (cf. Fig. 5); surstylus broadly rounded, similar to *avium* but slightly broader in proportion to its length (Fig. 8); aedeagus similar to the of *avium* (cf. Fig. 6a).

Female. Color as in *avium*; concolorous with male but dorsum of thorax more heavily gray microtomentose. Microtomentum of parafacial and parafrontal brownish yellow, not silvery as in male; prevertical area of parafrontal dull, dark gray microtomentose; frons at vertex moderately broad, 0.295 times head width (0.28-0.32) and 3.52 times ocellar span (3.00-4.29); frontal vitta often with some strong bristlelike setae; preocellar area dull; parafacial broad, its width slightly over half that of frons (0.56x; 0.50-0.59), 1.96 times ocellar span (1.75-2.27), and 1.13 times vibrissal interval (1.00-1.24/24); cheek height nearly half maximum length of eye (0.49x; 0.42-0.525).

Apparent normal length of adults 8.5-9 mm, commonly about 9. A very few specimens, no doubt underfed, are as short as 6.5-7 mm.

Third instar. Peritreme open. *Prothoracic fringe* rather sparse, 32 μ (8-66)/72. *Cephalopharyngeal skeleton* 980 μ (850-1080)/11 in length.

Puparium. Length 8.9 mm (8.6-9.1)/6, breadth 4.1 mm (3.6-4.4)/13; medium-thick walled, dull, dark brown. *Posterior region* (Fig. 43a): stigmatal plates 200 μ (150-220)/26 in diameter; distance between buttons 500 μ (460-590)/13, and across stigmatal plates 1200 μ (1000-1390)/13; stigmatal ratio 0.40 (0.36-0.50)/13; stigmatal area not prominently outlined in cuticular plaques; mesostigmatal spines present; mesostigmatal area without plaques but with folds prominently bisecting stigmatal region; hyperstigmatal spines 24 μ (8-31)/150; hypostigmatal area with numerous, prominent cuticular folds; hypostigmatal spines 15 μ (6-27)/150; circumstigmatal folds present, weak and concentric (in some cases, barely discernible and considered absent). *Dorsal cuticle* (Fig. 43b) with numerous spines, 25 μ (12-45)/150; cuticular ridges weak. *Ventral cuticle* (Fig. 43c, d): ventral band ratio 0.63 (0.50-0.69)/10; spines of anterior patch 14 μ (8-27)/150; medial band pronounced, but interrupted on midline by pronounced break, spines 20 μ (10-31)/150; posterior band pronounced.

Types. Type series of *asiovora*: Holotype male, McElroy Lake, Paha [Adams Co.], Wash. June 30, 1920 (R. C. Shannon), reared from larva found in nest of long-eared owl. Type No. 26858 in the U.S. National Museum of Natural History.

Type series of *basingeri*: Holotype male, San Diego, Calif., June 28, 1917 (J. M. Aldrich). Type No. 54929 in the U.S. National Museum of Natural History. Paratypes (all California): 2 males, same data as holotype [USNM]; 4 males, Temecula,

Riverside Co., June 4, 1936 (A. J. Basinger), one bearing a penciled note "from sunshine on bridge" [USNM]; 2 males, Forest Home, San Bernardino Co., May 18, 1934 (A. J. Basinger) [USNM]: 3 males, 2 females (not 2 males, 3 females as published), Le Grand, Merced Co., June 10, 1938, from nest of mourning dove (J. A. Neff) [males, USNM; females, CAS]: 2 females, Anaheim, Orange Co., July 1936, from nest of house finch (A. J. Basinger) [USNM, CAS]; male (now headless), Corona, Riverside Co., Sept. 20, 1934 (Basinger) [CAS]; 6 males, San Diego, Sargent (Santa Clara Co.), Lockwood (Monterey Co.), and Laguna (San Diego Co.) (R. H. and Jack Beamer)[Kans. U.]. Published paratype not found: female, Riverside, Calif. (Basinger).

Specimens Examined. 315 males, 255 females.

Distribution. British Columbia to Montana, south to Baja California and New Mexico. This species should also occur down the Mexican Plateau.

Canada. British Columbia: 2 males, 2 females, Kamloops (American robin nest), Laggan (old railroad station for Lake Louise), Oliver, and Orofino Mt. (4000 ft) near Oliver.

Contiguous United States. Arizona: 17 males, 8 females, Chiricahua Mts. (8000 and 8200 ft) near Portal, Coconino Co. near Flagstaff (nest of great horned owl), Globe, Huachuca Mts. (Ramsey Canyon), Doney Park near Flagstaff (K. A. Walker; total of 55 larvae removed from "various external portions of a dead nestling" as well as from nest material). **California** (by county): 48 males, 32 females, Alameda, Contra Costa, El Dorado, Kern, Los Angeles, Madera, Merced, Modoc, Mono, Monterey, Orange, Riverside, San Bernardino, San Diego, Santa Clara, Shasta, Tulare, and Tuolumne Counties, from nests of mourning dove (Le Grand in Merced Co.), house finch (Anaheim in Orange Co.), great horned owl (O'Neals in Madera Co.), and barn swallow (Adin Bridge in Modoc Co.). Also larvae from Alameda Co. (Sunol, golden eagle), and Santa Clara Co. (Milpitas, two removed from flesh at bottom of crop of golden eagle, one from under left wing of golden eagle, and 5 from ear of great horned owl). Also puparia, Merced Co. (Le Grand, from nest of loggerhead shrike). **Colorado**: 12 males, 6 females, 16 larvae, Boulder, Colorado Springs (black-billed magpie and horned owl nests), Platte Canyon (near Idlewood). **Idaho**: 6 males, 4 females, Franklin in Franklin Co., Lake Waha, Moscow Mt. Also larvae from Boise, Elmore, and Owyhee Counties (all in golden eagle nests). Also larvae on black-billed magpies, Moscow, published by Kamal (1958), probably *asiovora*. **Montana**: 185 males, 185 females, Connor, Hamilton (European starling nest), Gallatin Co. (Fairy

Lake, 9500 ft), Beaverhead Co. (Dillon, long-eared owl nest), Beaverhead Co. (nests of American crow, black-billed magpie, long-eared owl, and Swainson's hawk), Granite Co. (swallow nest), and Ravalli Co. (American crow and black-billed magpie nests; published Jellison and Philip 1933). **Nebraska**: male, Dawes Co., 8-11 mi SE Crawford [See note at end of Remarks]. **Nevada**: male, Douglas Co. (Dagget Pass). **New Mexico**: 3 males, Bernalillo (Cedro Canyon), Otero (Weed), and Torrance (Tajique) Counties. **Oregon**: 6 males, Dufur, Corvallis (cliff swallow nest). Camp Abbot in Deschutes Co., 7 mi SW Beatty in Klamath Co. **Utah**: 28 males, 12 females, Brigham City (Bear River Refuge, black-billed magpie nests), Box Elder Co. (Malaise trap), Cache Co. (Benson, nest of Brewer's blackbird, chiefly infested by *metallica*), Salt Lake City (black-billed magpie nest), and Wasatch Mts. **Washington**: 5 males, 6 females, Almota (Shannon and Dobroscky 1924), Central Ferry, McElroy Lake at Paha (nest of long-eared owl, holotype of *asiovora*), Okanogon Co. (black-billed magpie nest), Pullman, Wenatchee. **Wyoming**: female, South Fork of Powder River, 1.8 mi W of town of Powder River in Natrona Co.

Mexico: male, Baja Calif., Sierra San Pedro Martir, La Grulla, 6900 ft, June 15, 1953 (P. H. Arnaud, Jr.) [CAS].

Basis of description of immature Stages: *Larva, 3rd instar* (5 specimens): Ravalli Co., Mont., black-billed magpie, May 28, 1931. *Puparium* (13 specimens): Le Grand, Calif., mourning dove, June, 1939 (4) and 1943 (6), loggerhead shrike, 1943 (1); Okanogon Co., Wash., black-billed magpie, June, 1939 (2).

Hosts. Brewer's blackbird, American crow, mourning dove, golden eagle, house finch, several hawks (northern goshawk, Cooper's, ferruginous, red-tailed, and Swainson's), pinyon jay, black-billed magpie, great horned and long-eared owls, common raven, American robin, loggerhead shrike, European starling, and barn and cliff swallows. Probable host: American kestrel (sparrow hawk), published by Hill and Work (1947).

Ecology and Biology. The host records thus far known suggest that *asiovora* generally occurs at intermediate or forest canopy levels and usually infests large birds and often raptors, but it occurs more often in smaller birds than does its sister species in the East, *P. avium*.

Variation. Typically, *asiovora* has only one posterior bristle on the fore tibia, but occasional specimens have two, usually on one side only, rarely two on both sides. In a series of 82 from three nests in Ravalli Co., Mont., seven specimens had extra bristles. Six of these had two on one side and one on the other, and one had two on each fore tibia.

Remarks. *P. asiovora* is a sister species and western equivalent of the eastern *P. avium*. The two are very similar in general appearance and in the broad parafacials and exceptionally broad surstylus of the males. Females have proved impossible to distinguish except by their occurrence with males and within the characteristic ranges for the two species. The character, shared only with *interrupta*, of an interrupted median ventral spine band in the larvae, supported by the consistently narrower frons in the male *asiovora* and the different distributions of the two, have led us to recognize them as distinct species.

Protocalliphora asiovora is one of only three species with broadly rounded surstylus in the male (Figs. 7-9), the others being the sister species, *avium*, and *P. deceptor*. *P. avium* and *asiovora* are very close, and the latter was originally described as a mere variety of *avium*. The sharpest distinction here, and the main reason why we recognize *asiovora* as a distinct species, is that the medial spine band on the ventral surface of mature larvae (and of course, puparia) is distinctly interrupted on the midline (Fig. 43c,d). This spine band is continuous in *avium*. Both *avium* and *asiovora* have the mesonotum in both sexes more heavily microtomentose than in most other species, especially in the males. One finds it more appropriate here to speak of two broad gray stripes that set off median and lateral shining stripes, rather than to speak of three shining stripes.

The third species with broadened surstylus, *P. deceptor*, is quite a different species, with narrow frons in both sexes, as narrow as the frons in *P. braueri*, and narrow parafacial. It is a species of eastern United States, and chiefly parasitizes small passeriform birds. It is also worth noting that the cerci in *deceptor* are narrow, acuminate, and parallel to their apices, quite unlike the sturdy and divergent cerci of *avium* and *asiovora* (cf. Figs. 5a and 5b).

Larvae and puparia of *asiovora* are recognized from all other *Protocalliphora* except *interrupta* by the marked interruption of the medial band (Fig. 43c, d). They are separated from *interrupta* by (i) the larger ventral band ratio (0.63, cf. 0.27 for *interrupta*), (ii) the larger spines on the cuticle which appear as spines, not as small tubercles as in *interrupta*, and (iii) the pronounced mesostigmatal folds in *asiovora* compared to their absence in the latter species.

It is probable that *asiovora* is sometimes involved in myiasis like its close relative, the eastern *P. avium*, for which there are many records of aural myiasis in particular. This may be principally the result of greater attention to nestling raptors in eastern North America. Specimens of *asiovora* from Milpitas, Calif. were taken from golden eagles from flesh near the bottom of the crop, and from under the left wing, also from the ear of a great horned owl. Hill and Work (1947) in California reported *Protocalliphora* larvae in the ears and nostrils of sparrow hawk (i.e., American kestrel) and golden eagle, and from this larval habit we can be reasonably sure that they were *asiovora.* The ear canal of the hawk was swollen and filled with squirming larvae, and the nostrils were deformed. In Utah, Whitworth often observed *asiovora* in aural myiasis.

Two males from the northwestern corner of Nebraska are somewhat intermediate between *asiovora* and *avium.* In the one from Dawes County the frons appears narrow as in *asiovora*, the ratio of frons to head width being 0.14, at the upper end of the range for that species. The surstylus and cerci are like those of *asiovora.* In the one from Sioux County, the ratio of frons to head width is 0.16, which would agree with that of *avium.* The male terminalia are unfortunately lacking in this specimen. It would be interesting to have reared series, with immature stages preserved, to solve questions of identity: Do the two species actually overlap here? Or are these specimens extreme variants of one or the other? For the present, each is recorded under what appears to be its species, but with attention called to the need for further study and the possibility of mistaken identification.

Protocalliphora (*P.*) *avium* Shannon and Dobroscky
(Figs. 5a, 6a, 7, 34, 44, Map 5)

Protocalliphora azurea (Fallén).--Coutant, 1915: 135-150 [N.Y., on crow].

Larvae "possibly" *Compsomyia macellaria* (Fabricius).--Burtch, 1915: 293 [N.Y., in ears of Cooper's hawk; undoubtedly *P. avium*].

Protocalliphora avium Shannon and Dobroscky, 1924: 250 [N.Y., from crows].

P. avium.--Dobroscky, 1925: 275 [N.Y.]; Shannon, 1926: 128 [in key]; Johannsen, 1928: 827 [N.Y., Coutant material].

P. sp., larvae in ears of hawks.--Bent, 1937: 189-190 [Probably N.H.; undoubtedly *P. avium*].

P. splendida sialis [sic].--Sargent, 1938: 82-4 [N.Y., nests of hawks, some larvae in ears; undoubtedly *P. avium*].

P. sp., larvae on peregrine falcon.--Hickey, 1942: 189 [N.Y., probably *P. avium*].

Apaulina avium.--Hall, 1948: 186-190 [N. U.S. and s. Canada; *asiovora* here a synonym].

Protocalliphoria [sic] larvae, ears of hawks.--Hamerstrom and Hamerstrom, 1954: 4-8 [Iowa, Wis., Mich.; undoubtedly *P. avium*].

Apaulina avium.--Meng, 1954: 21-22 [N.Y., biology].

Protocalliphora avium.--Boyes, 1961: 552ff. [Cytology]; Boyes and van Brink, 1965: 543 [Cytology]; Hall, 1965: 926 [Wis., Ont., N.Y., Pa.].

P. sp., larvae in ears of young peregrine falcons.--Hagar, 1969: 127 [Mass.] and Hickey, 1969; pl. 43 [Vt.] [Both undoubtedly *P. avium*].

P. avium.--Boyes and Shewell, 1975: 446 [Mention of cytology]; Bohm, 1978: 297 [Minn., in ears of great horned owls]; Tirrell, 1978: 21-5 [N. Dak., on hawks and owls].

P. avium.--Bortolotti, 1985: 165-8 [Frequency on bald eagles, Sask.].

P. *avium*.--Poole and Bromley, 1988: 36 [On gyrfalcon nestling, Central Canadian Arctic].

Diagnosis. Sexes concolorous, monochromatic, shining metallic dark blue to bluish purple; calypteres white; frons of male of moderate width; parafacial broad in both sexes; surstylus of male broadly rounded (Fig. 7); eastern, sister species of *P. asiovora*.

Male. Head chiefly black in ground color, frontal vitta and genal groove reddish, and parafrontal and parafacial bright gray, silvery microtomentose, that of parafrontal posteriorly thinning to shining black opposite ocellar tubercle, parafacial without changeable reflecting spot. Body shining metallic blue to bluish purple, more or less gray microtomentose, heavily so on scutum and scutellum, especially in two broad stripes (as viewed from behind) that set off a median shining stripe and the lateral shining areas, each of which may be weakly separated into a lateral stripe and a supraalar vittula; abdomen appearing shining in direct view but 2nd to 4th tergites gray microtomentose when viewed from behind at a low angle, 5th tergite always shining and non-microtomentose. Calypteres opaque whitish, outer rims pale yellowish.

Frons moderately broad (Fig. 34), at vertex 0.16 times head width (0.15-0.18), and nearly twice ocellar span (1.96x; 1.68-2.19/25); frontal vitta broad throughout, at narrowest twice or more width of a parafrontal; preocellar area granular and subshining or weakly shining; parafrontal distinct throughout, the narrowed upper half with distinct row of fine hairs outside frontal bristles, the row becoming 2-3 irregular rows anteriorly and merging with the numerous long black parafacial hairs; parafacial notably broadened for *Protocalliphora*, approximately equal to width of frons at vertex (0.955x; 0.86-1.06), obviously much wider than ocellar span (1.87x; range 1.56-2.27), and slightly wider than vibrissal interval (1.16x; range 1.00-1.29); cheek broad, nearly half maximum length of eye (0.45x; range 0.42-0.48).

Thorax without accessory notopleural bristles and without hairs on tympanic membrane; postalar wall usually bare, sometimes with a few black hairs. Fore tibia usually with one posterior bristle.

Terminalia: cerci distally slender and tapering, divergent in posterior view (Fig. 5a), strongly curved in profile (Fig. 7); surstylus broadly rounded, subovate (Fig. 7); aedeagus as usual for the genus (Fig. 6a).

Female. Color as described for male, but both thorax and abdomen more heavily gray microtomentose, especially the former. Microtomentum of parafrontal and parafacial dark gray to brownish yellow, not silvery as in male; preocellar area dull, densely brownish microtomentose; prevertical area of each parafrontal dull, dark gray microtomentose; frons at vertex moderately broad, 0.295 times (0.27-0.32) head width and 3.45 times (2.89-3.88) ocellar span; frontal vitta usually with a few strong black mesoclinate setae in addition to weaker and inconspicuous hairs; parafacial unusually broad for the genus, width slightly over half that of frons at vertex (0.54x; 0.50-0.59), 1.86 times (1.56-2.25) ocellar span, 1.15 times (1.00-1.36/21) vibrissal interval, and about twice width of 3rd antennal segment, usually slightly convex and appearing bulging; cheek nearly half maximum length of eye (0.47x; 0.42-0.50).

Apparent normal length of adults, 9.5-10 mm.

Egg. Length 1040 μ (950-1150)/33; breadth 340 μ (275-405)/33; cuticle of yellowish color; cuticle with spinules.

Second instar. *Cephalopharyngeal skeleton* 630 μ (520-730)/5 in length. *Prothoracic fringe* 15 μ (10-26)/50. *Posterior region*: peritreme closed, stigmal plates 94 μ (86-135)/10 in diameter; button indistinct; mesostigmatal spines scarce or absent; hyperstigmatal spines numerous, 12 μ (6-21)/50; hypostigmatal spines numerous, 6 μ (4-12)/50. *Dorsal cuticle* with numerous spines, 10 μ (6-21)/50. *Ventral cuticle*: spines of anterior patch 4 μ (2-12)/50; medial band pronounced 8 μ (4-14)/50; posterior band present.

Third instar. Peritreme usually open but sometimes weakly closed; spines small and scarce between stigmatal plates, numerous and larger in the hypostigmatal area. *Prothoracic fringe* 45 μ (31-61)/100. *Cephalopharyngeal skeleton* 1080 μ (870-1260)/72 in length.

Puparium. Length 9.5 mm (6.3-10.5)/100, breadth 3.8 mm (2.2-4.7)/100; thick-walled, dull, black-brown. *Posterior region* (Fig. 44a, b): stigmatal plates 215 μ (150-260)/196 in diameter; distance between buttons 530 μ (325-655)/100 and across stigmatal plates 1330 μ (860-1620)/100; stigmatal ratio 0.40 (0.38-0.42)/100; stigmatal area outlined in plaques; mesostigmatal spines scarce or absent, mesostigmatal plaques and folds present; hyperstigmatal spines 20 μ (10-33)/100; hypostigmatal area with few, prominent cuticular folds; hypostigmatal spines numerous, 12 μ (6-23)/100; circumstigmatal folds present, pronounced, irregular. *Dorsal cuticule* (Fig. 44d) with numerous spines, 23 μ (10-33)/100; cuticular ridges prominent, usually with two rows of spines associated with each ridge, one each directed posteriorly and anteriorly. *Ventral cuticle* (Fig. 44c): ventral band ratio 0.75 (0.73-0.86); spines of anterior patch 15 μ (6-21)/100; medial band pronounced, spines 20 μ (10-33)/100; posterior band pronounced, present on all segments.

Type series. Holotype male, Ithaca, N.Y., reared from nestling American crow (I.D. Dobroscky), "Cornell U. Lot 1033, Sub 108A," Type No. 26857 in the U.S. National Museum of Natural History. Allotype, labeled only "Cornell U. Lot 1033 Sub 108." Paratypes: male, female, same data as type but Sublot 108, the male labeled May 29, 1923, the female "raised on robins, June 28, 1923" [USNM]; one male, same data as type, and 2 females, Sublot 108, all labeled March 29, 1923 [Cornell U.]. The original description recorded three males, five females, but one female has not been located. The date March 29 was undoubtedly that of the collection of nest and larvae. The holotype and one male paratype [Cornell U.] were each pinned with puparium associated. The holotype bears only the Lot and Sublot label, and the Type Number label, but the other information was published with the original description. The Type No. system is by some not considered designation and it must be admitted that here the original description did not specify the sex of the type nor which

Sublot was involved. To obviate any further argument, I designate the "holotype" as lectotype.

Specimens Examined. 176 males, 212 females.

Distribution. Northern and northeastern, Alaska and Yukon Territory, Northwest Territories, Saskatchewan, and North Dakota to Quebec and Connecticut, south to Nebraska and Pennsylvania.

Alaska: The material for two identifications of "*avium*," the first by M. T. James and the second probably also, could not be located for verification: Fairbanks (G. Gill), published Gill (1956); Matanuska (W. J. Chamberlin), in Chamberlin's mimeographed report on his Alaska work for 1945.

Canada. Manitoba: 3 females, Bird's Hill Provincial Park (from ear of young red-tailed hawk), Churchill ("visiting Solidago"). **Northwest Territories:** larvae (det. Bennett), Mackenzie District, on mainland S of Cambridge Bay and E of Bathhurst Inlet, 68.1° N and 106° 35' W (Kim G. Poole) (upper breast, ears, nostrils, around anus, etc., of gyrfalcon chicks, and from top of wing, beneath wing, left tarsometatarsus, and around cloaca of nestling peregrine falcons). **Ontario:** 57 males, 117 females, larvae and puparia, Algonquin Park (at or near Lake Sasajewan) (nests of gray catbird, American crow, goshawk, broad-winged hawk, American robin), Burleigh Falls, Dryden, Toronto (American crow nest), Osgoode (American crow nest). **Quebec:** 2 females, near Eardley, and La Verendrye Provincial Park. **Saskatchewan:** male, Besnard Lake at 55° 20' N, 106° W (collected as larva in ear of nestling bald eagle). **Yukon Territory:** male, Dempster Highway, Mile 87, July 8-12, 1973 (G. & D. M. Wood) [CNC].

Contiguous United States: Connecticut: male, Stamford. **Illinois:** male, female, McHenry Co., from ear of red-tailed hawk. **Iowa:** Ruthven, Palo Alto Co., larvae reported by Hamerstroms (1954). **Minnesota:** larvae and adults, Trommald (ghost town near Ironton) (larvae from head and neck region of bald eagle), near Foley (in ears of great horned owl) (Bohm 1978). **Nebraska:** male, Monroe Creek Canyon in Sioux Co. NNW of Harrison [See note at end of Discussion under *asiovora*]. **New York:** 66 males, 55 females, many larvae, Chautauqua Co. (from ears of broad-winged hawk), Ithaca (American crow nests), Candor, Catskill Mts., "New Albany" (presumed labeling error for New Albion in Cattaraugus Co.) (from ear of nestling hawk). **North Dakota:** 34 males, 19 females, 38 larvae, 14 localities in Grand Forks Co. (nests of red-tailed and Swainson's hawks and great horned owl). **Pennsylvania:** 10 males, 11 females,

Coraopolis (from larvae in ear canals of nestling broad-winged hawk), State College, Tionesta in Forest Co. **Wisconsin**: 4 males, 4 females, Monroe in Green Co. (from ear of two-week old red-tailed hawk), Waushara Co. (from ear of red-tailed hawk), "Rainbow Flowage" near town of Lake Tomahawk in Oneida Co. (larvae from ear of osprey).

Basis of description of immature stages: *Egg*: Algonquin Park, Ont., from caged female, July 1955. *Larvae, 2nd instar* (5 specimens): Algonquin Park, Ont., American crow, June, 1954; *3rd instar* (10 specimens): Algonquin Park, Ont., American crow, June, 1955. *Puparium* (100 specimens): Algonquin Park, Ont., American crow, June 1952 (25) and June, 1955 (25), broad-winged hawk, August, 1955 (25); Toronto, Ont., American crow, June, 1955 (25).

Hosts. Gray catbird, American crow, bald eagle, gyrfalcon, peregrine falcon, several hawks (broad-winged, Cooper's, red-shouldered, red-tailed, Swainson's, and northern goshawk and northern harrier), osprey, great horned owl, and American robin. The Cooper's hawk was reported by Burtch (1920) and Meng (1954), the goshawk and the red-shouldered hawk by Meng (1954), and the "marsh hawk" (northern harrier) by the Hamerstroms (1954), all records accepted as probably correct because of the known characteristic habit of the larvae, and the eastern localities well apart from *asiovora*.

The published record of *P. avium* from swallows in eastern Canada by Hearle (1938) is uncertain and possibly incorrect. The wording refers to maggots in cysts on the back of swallows, in which case it probably refers to *braueri*. We have no records of *avium* from swallows, although there are a few records from non-raptorial birds.

Ecology and Biology. *P. avium* is found primarily in the nests of large birds such as crows and raptors, and usually in nests in the forest canopy. The larvae are sometimes found subcutaneously, causing true myiasis, but always in the larger birds as contrasted with the myiasis-producing larvae of *braueri*, chiefly in small birds.

Maggots are often noted in the ears and nostrils of nestling raptors, as indicated by many of the records, but the exact damage or extent of damage is seldom recorded. E. C. Schriver (personal communication) has commented that he had "found these parasites in the ear openings of all our native tree-nesting hawks." Burtch (1920) found maggots, probably of *avium* in the ears of three nestling Cooper's hawks. Bent (1937:189-190) quoted a communication from a correspondent, Lewis O. Shelley, about two young red-shouldered hawks that were taken from a nest when about three weeks old, and mounted: "When the young hawks were being skinned both birds were found to have the ears affected with maggots; 20 were collected from the four ears and preserved; ... These maggots had eaten about the ears not only to disfigure the outer ear cavity, leaving the marks of their attacks so that they are preserved in the mounted specimens, but had completely destroyed the ear drums of both birds. It is to be wondered how long these young hawks would have survived had they lived to reach maturity."

The Hamerstroms (1954) found maggots of *Protocalliphora* in the ears of three species of hawks, and they came to suspect that such maggots "may be a rather common childhood disease of hawks." In Iowa they found maggots in marsh hawks [i.e., northern harriers] "embedded under the wings, near the vent or on the crown, and especially in the ears and within the nares." It is noteworthy that the marsh hawk, a ground-nesting species, represents a radical difference from the usual habit of *avium*, yet the larval habit described by the Hamerstroms is characteristically that of *avium*. On this basis we accept the record as probably correct, although we have not seen specimens. The catbird, not a raptor, is also an unusual host for *avium*, nesting as it does at low levels. The nest was three feet above the ground, and 20 *metallica* and one *avium* were reared from it. Occasional aberrant rearings may occur.

Meng (1954), in an intensive four-year study of Cooper's hawks in central New York, found maggots of *avium* in the ears of many young birds of this and several other species of hawks. "When the hawks are very small the only indications one has that they are infected are the droplets of dried blood around the ear openings. The larvae grow rapidly on their diet of blood, and by the time the eyeases are three weeks old their ear openings may be completely plugged by the protruding ends of the screw-worm fly larvae. The maggots are intermittent feeders and do not remain in the ears continuously. At night while the young hawks are sleeping the hungry larvae find their way into the ears, and the full ones crawl out into the nest debris. In this way a given hawk is host to many more parasites then would be indicated by the dozen or so larvae that are usually found in the ear openings on an infected bird." Meng also commented that "None of the eyeases that had maggots in their ears seemed to be weakened appreciably, and by the fourth week all the maggots had left the ears." By that time, of course, the larvae would have matured and dropped away to pupate.

Tirrell (1978) in North Dakota found maggots of *avium* in the auditory meatus of 34 out of 36 nestling red-tailed hawks, as well as in one nestling great horned owl and two of Swainson's hawks. "Eventually larvae plugged and stretched the auditory meatus. The ear opening and surface of the ear canal were covered with a red-brown exudate, and the area surrounding the ear was inflamed and swollen." Nestlings of the red-tailed hawks were closely observed, and they were infested between one week and four to five weeks of age, by which time the *avium* larvae would have matured and dropped from the ears.

Bohm (1978) reared *avium* from larvae in the ear cavities of nestling great horned owls in Minnesota. Infestations were recorded on 26 of 46 (54%) of the nestlings banded over a two-year period. In the same period 41 of 73 (56%) of nestling red-tailed hawks were also infested, undoubtedly also by *avium*, although none were reared and identified.

On a different host, Bortolotti (1985) made repeated observations (461) individually on 72 bald eagle nestlings in a three-year period in north central Saskatchewan. He found that all eaglets had *avium* larvae "in their ears and on their body at some time during development." They "frequently were so numerous that the entire ear opening was blocked. A conspicuous blackish brown exudate was always present in and around the outside edge of an infested chick's ears. Although this substance became dry, encrusted, and difficult to remove, it did not persist, for the ears eventually became clean and to the naked eye showed no trace of ever being parasitized." Bortolotti reported that "nestlings were repeatedly parasitized as they grew older. It was not uncommon to observe a large number of larvae on a bird during one visit, followed by the absence of larvae on the next visit a few days later, and then another heavy infestation on the third visit." The last sentence is especially important in showing that even the larvae in aural cavities leave the host and drop into the nest material between feedings. In this study no ill effects on the birds were observed.

The ecology and biology of *avium* have been investigated in some detail by Bennett (1957) at Algonquin Park, Ontario.

Remarks. *Protocalliphora avium* was adequately diagnosed by Shannon and Dobroscky and described in considerable detail by Hall. It is one of the largest species of Nearctic *Protocalliphora*, and individual specimens are usually up to the apparent norm for the species, probably because the usual hosts have a long nestling period and there is ample time for optimum feeding and growth by the maggots. Of the known species, broadened surstylus in the males is found only in *avium*, *asiovora*, and *deceptor*, and the last named is a much different species than the others, with narrow frons and parafacial in both sexes.

P. avium (northern and northeastern) and *P. asiovora* (western) are very close and adults could easily be confused. Males are consistently different in the width of the frons and they differ slightly in proportions of the surstylus. The immature stages are much more distinct, *asiovora* having an interrupted median ventral spine band. Females are virtually indistinguishable, except by association with males and distribution in East or West. Females of *avium* and *sialia* might also be confused, but the parafacial of *sialia* is not quite as broad as in *avium* and is usually flat, without the bulging appearance of *avium*.

The male from Yukon Territory seems an extraordinary extension of range for *avium*, but the frons: head ratio is greater (0.17), well beyond the observed ratio for *asiovora* and indeed near the upper limit of the range in *avium*. The possible confusion of these two species makes it imperative that puparia be associated with the adults and be available to assist with the identification. Specimens of *avium* are known from the Northwest Territories and Saskatchewan, so that it may be that *avium* is northern and transcontinental compared with *asiovora*.

The puparium of *avium* can be confused with those of *chrysorrhoea*, *hirundo* and *sialia*. *P. avium* differs from *chrysorrhoea* in having much shorter cuticular spines; from *hirundo* in having more circumstigmatal folds, more cuticular ridges and shorter spines on the dorsal cuticule; and from *sialia* by having shorter cuticular spines and much shorter prothoracic fringe. Adults of all species are readily separated.

Protocalliphora (*P.*) *beameri*, new species
(Fig. 31, Map 4)

Protocalliphora n. sp. B.--Horning and Barr, 1970: 73 [Idaho].

Diagnosis: Sexes concolorous, monochromatic, shining metallic blue to bluish purple; calypteres white; frons of male of moderate width; parafacial broad in both sexes; surstylus of male strongly narrowed.

Male. Head with parafrontal and parafacial evenly bright gray, almost silvery microtomentose; hairs on upper half of occiput almost entirely whitish yellow behind row of postocular setae, with a few black hairs adjacent to vertical bristles, sometimes a vague and somewhat incomplete row of black hairs behind the postocular setae. Body shining, metallic blue to dark bluish purple or bluish green, scutum thinly microtomentose, demarcating three full-length shining stripes (as viewed from behind), the microtomentum more evident anteriorly. Abdomen, viewed from behind and especially at a low angle, whitish microtomentose on 2nd to 4th tergites, the 5th shining. Calypteres opaque whitish, outer rims faintly yellowish tinged.

Frons moderately broad, 0.12 times head width (0.11-0.14/10, holotype 0.115) and 1.60 times (1.43-2.00/10) ocellar span; in profile face prominent, fronto-facial and vibrissal angles projecting, and broad parafacial evident; preocellar area polished, short, equilateral acute to short acuminate; each parafrontal slightly less than half width of the frontal vitta, distinctly microtomentose to approximately opposite median ocellus, the prevertical area chiefly shining black, parafrontal with one to two rows of hairs, one row beginning opposite ocellar bristles, becoming two rows midway and merging into the numerous parafacial hairs, latter in about four to five irregular rows; parafacial of moderate width, slightly wider than moderately broad frons (1.06x; 0.92-1.28)/10), obviously much wider than ocellar span (1.70x; 1.50-1.93/10), and 4/5 vibrissal interval (0.82x; 0.75-0.92/10); cheek broad, over 2/5 maximum length of eye (0.44x; 0.41-0.47/10); occiput predominantly whitish haired behind and below postocular setae, at most only a partial row of a few slender black hairs.

Thorax without accessory notopleural bristles and hairs on tympanic membrane; postalar wall with a few short, fine hairs. Fore tibia with one posterior bristle.

Terminalia: cerci narrowly acuminate, deeply notched in posterior aspect (cf. Fig. 5b); surstylus strongly narrowed on distal half and shorter than cerci (Fig. 31); aedeagus as usual in the genus (cf. Fig. 6a).

Female. Concolorous with male, but mesonotum duller and much more heavily microtomentose between the narrow, shining stripes; abdomen slightly more heavily microtomentose than in male. Parafrontal and parafacial almost silvery microtomentose, with slightly changeable spot on upper part of parafacial; prevertical area dull and evenly, heavily microtomentose; hairs on back of head darker than in male, with two irregular rows of black hairs behind postocular setae. Frons at vertex narrower than an eye (0.82 times) in the one available example, 0.29 times width of head, and four times ocellar span, widening anteriorly;

frontal vitta with a few weak hairs mesad of frontal bristles; preocellar area finely microtomentose; 2 proclinate and one reclinate orbital bristles; parafacial broad, its width slightly over half that of frons (0.54x), 2.14 times ocellar span, and 1.25 times vibrissal interval; cheek nearly half (0.47x) maximum eye length.

Apparent normal length of adults, 9.0-9.5 mm.

Immature stages. Unknown.

Type series. Holotype male, and allotype female, Cave Creek Canyon, Chiricahua Mts., Ariz., July 4-8, 1940 (D. G. Hall). Types in the National Museum of Natural History [USNM].

Paratypes: Arizona: male, Flagstaff, July 8, 1941 (R. H. Beamer) [Kans. U.]; male, Prescott, June 9, 1937 (W. Benedict) [USNM]; male, 6 mi N of Chino Valley, Yavapai Co., July 5, 1951 (R. S. Beal, reared from swallow nest) [U. Calif., Berkeley]; male, Ramsey Canyon, 5200 ft, 15 mi S Sierra Vista, Huachuca Mts., Apr. 10, 1967 (R. F. Sternitzky) [CNC]. **California**: male, Westgaard Pass, Inyo Co., June 18, 1969 (S. R. Leffler) [Wash. State U.]; 2 males, Pine Canyon, Inyo Co., Apr. 2, 1953 (J. W. MacSwain) [CAS]. **Idaho**: 4 males, Craters of the Moon National Monument, July 16 (two), 23, and Aug. 23, 1965 (D. S. Horning, Jr.), the July 16 specimens collected on a window, the others in a Malaise trap [CAS, USNM]. **Oregon**: male, Dixie, July 8, 1931 (R. H. Beamer) [Kans. U.]. **Utah**: male, Box Elder Co., July 9, 1969 (K. J. Capelle, Malaise trap) [Whitworth Colln.]. **Washington**: male, Rattlesnake Grade, probably Benton Co., July 18, 1950 (R. B. Spurrier) [Wash. State U.]; male, Fields' Spring State Park, Asotin Co., July 31, 1971 (W. J. Turner, dry ice Malaise trap) [Wash. State U.]. **Mexico: Baja California Norte**: male, Arroyo Santo Domingo, 5.7 mi E Hamilton Ranch, Apr. 23, 1963 (H. B. Leech and P. H. Arnaud, Jr.), and male, Sierra San Pedro Martir, 3000 ft, 2 mi W Socorro, June 4, 1958 (J. Powell) [CAS].

Non-paratype Material. One male, Craters of the Moon National Monument, Idaho, Aug. 23, 1965 (Horning) [U. Idaho] lacks the abdomen and has not been included in the type series. It is labeled "*Chrysothamnus nauseosus*," probably meaning collected on that plant. Another male, Santa Catalina Mts., Ariz., July 16, 1950 (L. D. Beamer) [Kans. U.] has the typical male genitalia and predominantly white-haired occiput and is therefore identified with assurance as *beameri*, but it is undersized (6.5 mm) and the proportions are somewhat different than those of more normal specimens, as is often the case with small specimens. Most notably, the parafacial is narrower than the frons (0.89 times) and less than 2/3 the vibrissal interval

(0.64). The cheek is also much narrower (0.37) than in the typical form. The other proportions fall within the observed range in typical individuals.

Specimens Examined. 20 males, one female.

Distribution. Arizona, California, Idaho, Oregon, Utah, Washington, and Mexico (Baja California).

Hosts. Only one known, a "swallow" from a nest located beneath a road bridge over a small stream, possibly a barn swallow.

Ecology and Biology. Unknown.

Variation. In the male, occiput chiefly whitish haired, at most one somewhat incomplete row of black hairs on the occiput behind the postocular setae. This seems to be unique among the Nearctic species, males of other species having two to three often dense rows of coarse black hairs behind the postocular setae, as in the females. This has been consistent in the males before us, but not in the one available female, which is associated purely on the basis of label data identical to a male *beameri*. Either this is the wrong female, or there are more black occipital hairs in females, or this is a dark-haired variant. From its general appearance and from the features that can be compared between males and females, such as the breadth of the parafacial, the female does appear to be that of *beameri*. Unfortunately, no whitish occipital hairs can be used for *beameri* in the key for females.

Remarks. Strongly narrowed surstylus is characteristic of five species in the group with curved surstylus: *bicolor*, *beameri*, *brunneisquama*, *interrupta*, *spenceri* (Figs. 28-32). The first species has the 5th abdominal tergite cupreous green in both sexes, and *brunneisquama* has dark brown calypteres in both sexes, while the other species in both sexes have the abdomen entirely bluish to bluish purple and the calypteres white. The proportions of *beameri*, notably of frons and parafacial, are different than in the other four species. In male *beameri*, the wide frons and broad parafacial are actually suggestive of species such as *avium*, *asiovora*, and *chrysorrhoea*, which have very broad or straight and digitate surstylus. Females, if we have associated the right one, have no particularly distinctive features.

The superficial similarity of the bluish males to those of other species is evidenced by the fact that one of the males stood in a collection identified as *sialia* and another as *hirundo*, neither having had the genitalia dissected to reveal their characteristic form that is quite unlike that of either *sialia* or *hirundo*.

Etymology. The specific name, in the genitive case, is dedicated to the memory of the late R. H. Beamer of the University of Kansas, leader of many field trips into the Southwest, and collector extraordinary. The holotype and allotype were collected during a field trip by David G. Hall with Beamer and his students.

Protocalliphora (*P.*) *bicolor*, new species
(Figs. 29, 40, 45, Map 11)

Protocalliphora splendida form *splendida* (Macquart).--Johnson, 1925a: 216, in part.
P. splendida typical.--Johnson, 1927c: 79 [N.H., from black-throated blue warbler]; Johnson, 1929a: 29 [N.H., from chestnut-sided warbler]; Johnson, 1930: 171-2 [N.H., from American redstart and solitary vireo].
Apaulina metallica, in part.--George and Mitchell, 1948: 549-550.

Diagnosis. Sexes concolorous, dichromatic, both with 5th abdominal tergite brightly cupreous, contrasting with dark blue to bluish green body; surstylus of male decidedly narrowed.

Male. Head with parafrontal and parafacial bright gray to silvery microtomentose, the microtomentum thinning posteriorly as parafrontals narrow, the prevertical area chiefly polished black; parafacial without changeable spot; occiput with 2 to 3 irregular rows of long black hairs behind row of postocular setae. Thorax shining, dark blue to bluish green or bluish black, thinly microtomentose, microtomentum of scutum outlining three broad shining stripes, as viewed from behind. Abdomen bluish to bluish green (often the latter but specimens possibly not fully colored), tergites 2 to 4 with thin gray microtomentum that is most evident on 2 and 3, 4 often quite shining with the "bloom" visible only from behind when viewed at a low angle, 5th tergite usually bright cupreous, distinctly contrasting with the preceding tergites, sometimes green or greenish blue tinted with cupreous. Calypteres opaque whitish, outer rims slightly yellowish.

Frons (Fig. 40) narrow and narrowing slightly anteriorly, at narrowest only 0.065 times (0.05-0.08) head width and slightly broader than ocellar span (1.08x; 0.83-1.40); parafrontal strongly narrowed posteriorly, frontal vitta over twice width of a parafrontal, polished preocellar area narrow and acuminate; lower third to half of parafrontal with one row of hairs, or irregularly two rows, parafacial sparsely haired, with 18-20 hairs in two or three irregular rows; parafacial wider than breadth of 3rd antennal segment, much wider than frons (1.67x; 1.29-2.00), 1.81 times (1.50-2.10) ocellar span, and 0.71 times (0.615-0.81) vibrissal interval; breadth of cheek nearly 2/5 greatest length of eye (0.38x; 0.36-0.40).

Thorax without accessory notopleural bristles; tympanic membrane and postalar wall usually bare. Fore tibia with a single posterior bristle.

Terminalia: cerci narrowly acuminate (as in Fig. 5b); surstylus elongate and strongly narrowed, in profile its sides subparallel on most of its length (Fig. 29); aedeagus as usual for the genus (as in Fig. 6a).

Female. Parafrontal and parafacial with microtomentum grayish to weakly brownish yellow, not silvery as in male, and likewise without changeable spots; thorax and abdomen, except 5th tergite, dark bluish to bluish green, quite dull compared with male, heavily gray microtomentose, 5th tergite shining cupreous; scutum, viewed from behind, with three shining stripes well separated by heavily microtomentose areas.

Frons narrowed above, width at vertex narrower than usual in genus, obviously much less than width of eye in dorsal aspect, about 1/4 head width (0.24x; 0.22-0.26), and 3.29 times ocellar span (3.00-3.83); frontal vitta with only fine hairs toward the sides, and twice width of a parafrontal, preocellar area microtomentose, sometimes thinly so and subshining; parafrontal narrowing slightly posteriorly, microtomentose throughout but thinly so and darkly subshining in prevertical area; parafacial moderately setose, in two to three irregular rows, its width half as wide as frons (0.50x; 0.425-0.57), 1.65 times (1.38-2.00) ocellar span, and 0.77 times (0.65-0.86) vibrissal interval; breadth of cheek 0.38 times (0.36-0.42) maximum length of eye.

Apparent normal length of adults, 8.5-9 mm.

Third instar. Peritreme closed; spines numerous between stigmatal plates and in hypostigmatal area. *Prothoracic fringe* 82 μ (31-165)/110. *Cephalopharyngeal skeleton* 845 μ (715-960)/41 in length.

Puparium. Length 6.9 mm (4.5-9.0)/25, breadth 3.2 mm (2.3-3.6)/25; thin walled, dull, light brown. *Posterior region* (Fig. 45a, b): stigmatal plates 170 μ (140-200)/76 in diameter; distance between buttons 505 μ (330-660)/38, and across stigmatal plates 1080 μ (730-1250)/38; stigmatal ratio 0.47 (0.42-0.55)/38; stigmatal area outlined in plaques; mesostigmatal spines scarce; mesostigmatal folds and plaques present, markedly separating the stigmatal plates; hyperstigmatal spines slender, 30 μ (14-58)/215; hypostigmatal area with numerous, weak, irregular folds; hypostigmatal spines numerous, 6 μ (4-31)/215; circumstigmatal folds weak and concentric. *Dorsal cuticle* (Fig. 45d) with numerous spines, 25 μ (12-41)/215; cuticular ridges weak or absent. *Ventral cuticle* (Fig. 45c): ventral band ratio 0.55 (0.50-0.65)/50; spines of anterior patch 15 μ (6-31)/215; medial band pronounced, spines 21 μ (10-47)/215; posterior band pronounced.

Type Series. Holotype male, allotype, and 5 males, 20 females paratypes, Ashland, N.H., reared Aug. 4-9, 1927 by C. W. Johnson, from nest of black-throated blue warbler collected by Mrs. R. B. Harding. Type and allotype at MCZ through the courtesy of Prof. A. G. Humes of Boston University, paratypes at Boston U., USNM, and Field Mus. (1 male). Additional paratypes: **Canada: Ontario**: male, two females, Algonquin Park, July 1954 (G. F. Bennett), from yellow warbler nest; male, same locality, October 1954 (Bennett), from nest of eastern kingbird; 3 males, 5 females, same locality, 1955 (Bennett), from nest of myrtle warbler [all USNM]. **Contiguous United States: New Hampshire**: 2 males, Ashland (Mrs. R. B. Harding), from nest of chestnut-sided warbler [Boston U.]; male, female, Ashland, July 20, 1929 (Mrs. Harding, from nest of black-throated blue warbler [CNC]; male, Ashland, August 1927 (Mrs. Harding), nest of black-throated blue warbler [USNM]; 6 males, 4 females, Holderness, July 18, 1919, from black-throated blue warbler [MCZ]; 3 males, 3 females, Holderness, July 20, host not stated [MCZ]; 5 males, 5 females, Holderness, Aug. 10, 1930, nest of American redstart (Mrs. Harding) [MCZ]; 10 males, 6 females, same locality, host, and collector as preceding, but nest collected June 27, 1930, each specimen with puparium [Boston U.]; 2 males, 2 females, Holderness, Aug. 10(?), 1930 (Mrs. Harding), from nest of solitary vireo [MCZ]; 19 males, 20 females, Holderness, July 15, 1929 (Mrs. Harding), from nest of solitary vireo [MCZ]; male, Rumney, Dec. 22, 1924 (P. J. Darlington), "chopping dead pine" [Boston U.]; 2 males, Grafton Co., Livermore, East Pond, Aug. 23-24, 1980, on *Solidago* (J. F. Burger) [U. of New Hampshire]. **New York**: male, Lake Clear Junction, Franklin Co., Aug. 10, 1946 (J. L. George, R. T. Mitchell), with puparium [USNM], mixed with *Apaulina metallica* of George and Mitchell (1948); male, Whiteface Mt., Essex Co., Sept. 13, 1979 (A. Freidberg) [USNM]. **Wisconsin**: 9 males, 5 females, Madison, with puparia, from nest of alder flycatcher, July 1974 [U. Wis., USNM].

Non-paratype Material: A male and two females from various New Hampshire lots, and 2 males, 2 females, from Algonquin Park, Ont., from nest of eastern kingbird, are teneral and have not been included as paratypes. Puparia identified as *bicolor* were found by Bennett at Algonquin Park in nests of common grackle, American tree swallow, wood thrush, and red-eyed vireo, and at Toronto, Ont., in nest of American goldfinch, but no adults were reared. Two isolated females, from Peterboro, N.H., Aug. 29, 1924 (C. F. Batchelder) and Little Black River Rapids, Maine (near mouth of Allegash River, upper Aroostook Co.), Sept. 8, 1907 (J. A. Cushman) [Boston U.] are probably *bicolor*, having narrow frons, broad parafacials, and long third antennal segment. At least the second of these was recorded by Johnson (1925: 216) as *Protocalliphora splendida* form *splendida*.

Two females, Val d'Or, Que. [CNC], each mounted with puparium, definitely are *bicolor*, but there may have been some confusion in labeling. The abbreviated labels signify "Ex *Estigmene acraea* (Arctiidae), emerged 7-29-1942, Recherches des insectes forestiers," but the association with the lepidopteran is certainly incorrect.

Specimens Examined. 77 males, 83 females.

Distribution. Northeastern: Wisconsin, Ontario, New York, and New Hampshire, and probably Quebec and Maine.

Basis of description of immature stages: *Larva, 3rd instar* (7 specimens): Algonquin Park, Ont., yellow warbler, July, 1954 (6), and myrtle warbler, July, 1955 (1). *Puparium* (38 specimens, partly from type series): Algonquin Park, Ont., yellow warbler, July, 1954 (22) and American goldfinch, August, 1954 (2); Ashland, N.H., black-throated blue warbler (9); Holderness, N. H., solitary vireo (2), and American redstart (3).

Hosts. Alder flycatcher, eastern kingbird, American redstart, solitary vireo, and various warblers (black-throated blue, chestnut-sided, myrtle, and yellow); also, recorded from puparia alone, American goldfinch, common grackle, American tree swallow, wood thrush, and red-eyed vireo.

Ecology and Biology. This species has been taken in only a few localities, chiefly from birds nesting in the intermediate levels of deciduous forests. If *bicolor* occurs characteristically in this habitat it may be more common and more widely distributed than the records indicate because the habitat is less thoroughly explored than others and is more difficult to survey.

Remarks. *Protocalliphora bicolor* is unique among the eastern species in having the surstylus so strongly narrowed. Several western species have equally narrow surstylus, but they are easily distinguished from *bicolor* by the entirely blue abdomen of the males. Females of *bicolor* fall in the group of species with abdomen predominantly bluish but fifth tergite cupreous, the other species being *metallica* and the western *cuprina*. These will be difficult to separate satisfactorily, except when males and/or immature stages are associated with them, and isolated females have usually not been positively identified in this study. In the East, where females of *bicolor* and *metallica* can easily be confused, the puparia are very distinctive. The integument in *bicolor* is well covered with distinct spines, whereas in *metallica* the cuticular spines are fine and minute, and the dorsal plate appears to be finely granular, with a sandpaperlike appearance. Furthermore in females of *bicolor*, the parafacial has no changeable spot, whereas the other two species have an arrangement of microtomentum on the upper part of each parafacial that results in a large dark spot on a light parafacial, viewed from above, changing to a light spot on a dark field when viewed from below.

Puparia of *bicolor* are easily confused with those of *cuprina* and *parorum*. They are separated with difficulty from *cuprina* on the basis of the somewhat longer prothoracic fringe and short tubercle-like spines of the hypostigmatal area and from *parorum* on the basis of the smaller ventral band ratio and shorter tubercle-like spines of the hypostigmatal area. This species has thus far only been recorded in the eastern portion of the continent while the other two are western in distribution.

Teneral males of *bicolor* often have the fifth tergite only slightly cupreous, and these may be confused with teneral males of *metallica* in which the fifth tergite is occasionally somewhat greenish tinted. Male terminalia or puparia will easily settle this question of identity.

The holotype and specimens reared with it were recorded as typical *P. splendida* by Johnson (1927c: 79), who reared a total of 9 males, 26 females from the nest on Aug. 4, 5, and 9, 1927. Johnson remarked that the abdomen of both sexes showed a coppery fifth tergite but he believed that this merely refuted Shannon and Dobroscky's (1924) belief in sexual dichromatism for *splendida*. However, their specimens were correctly interpreted (cf. discussion under *metallica*), and Johnson's specimens represented an undescribed species, here described as *bicolor*. Both series from the American redstart and that from the solitary vireo (called in Johnson the blue-headed vireo) were recorded as typical *splendida* by Johnson (1930), and the Ashland specimens from the chestnut-sided warbler were recorded as the same by Johnson (1929a: 29).

The small series from the myrtle warbler and the 1929 series from the solitary vireo contain many small and certainly undersized specimens, ranging from 5-8 mm in length. In these small specimens, the fifth tergite is usually only slightly tinted with cupreous, not strongly cupreous as in more typical specimens. Some of these are teneral, which would also affect the development of mature color.

Etymology: The specific name is a Latin adjective, referring to the two colors of the abdomen, and invariable.

Protocalliphora (*Trypocalliphora*) *braueri* (Hendel)
(Figs. 6c, 20, 41, 46, Map 3)

Avihospita braueri Hendel, 1901: 30 [Provenance not stated, later known to be Austria].

Protocalliphora chrysorrhaea [sic] (Meigen).--Walton, 1914: 175, in part [N. Mex., later paratypes of *hirudo*].

P. azurea (Fallén).--Arnold, 1919: 147-8 [Colo.; paratypes of *hirudo*].

Phormia metallica Townsend.--Plath, 1919c: 376 [Wash.; became type series of *Protocalliphora hirudo* var. *cuprea*]; Townsend, 1919: 379-80 [Wash., atypical form of *metallica*, also proved to be *hirudo cuprea*].

Protocalliphora azurea.--Cole and Lovett, 1921: 309 [Oreg.].

P. hirudo, *P. hirudo* var. *parva* (subspecies?), and *P. hirudo* var. *cuprea* Shannon and Dobroscky, 1924: 249, 252-3 [*hirudo*, Colo., N. Mex.; *parva*, Kans.; *cuprea*, Wash.].

P. hirudo var. *cuprea*.--Johnson, 1925a: 216 [Mass.]; Johnson, 1925b: 51-3 [Mass.].

P. hirudo and var. *parva*.--Shannon, 1926: 128 [In key].

P. splendida.--Bergtold, 1927: 106-7 [Colo.; larvae in abscess, undoubtedly *braueri*].

P. hirudo cuprea and *P.* sp. on baby chicks.--Hearle, 1938: 60-1 [Undoubtedly *braueri*].

P. sp. larvae in golden eagle nest.--Philip, 1938: 486-7 [Alaska].

P. hesperia.--Guberlet and Hotson, 1940: 65-8 [Wash.].

Apaulina hirudo.--Hall, 1948: 192-5 [Revision; varr. *cuprea* and *parva* possibly synonyms; Kansas as type locality for *parva* corrected to Greeley, Colo.]; George and Mitchell, 1948: 550-2 [N.Y.].

Protocalliphora hirudo.--Munro, 1949: 35-38 [B.C.].

P. sp. or spp.--Wood, 1953: 77 [B.C.; probably *braueri* in part]; Dodge and Seago, 1954: 53-4 [Ga.].
Trypocalliphora hirudo.--Peus, 1960: 225 [Europe; reference to new genus, *hirudo* and *braueri* considered distinct species].
Protocalliphora hirudo.--Hall, 1965: 926 [Catalog; widespread; *parva* and *cuprea* synonyms].
P. hirudo.--Cais, 1965: 183-191 [Poland; good figures, male genitalia and larvae].
P. (*Trypocalliphora*) *hirudo*.--Sabrosky, 1967: 122 [*Trypocalliphora* accepted as subgenus rather than genus].
P. hirudo.--James, 1969: 252 [Mention of biology]; Rausch, 1972: 1-4 [Alaska]; Whitworth, 1976: 16 ff.; Whitworth, 1977: 4933-B [Mention]; Bedard and McNeil, 1979: 111 [Quebec].
Trypocalliphora braueri.--Rognes, 1985: 373 [Europe; detailed description and notes on type series from Austria; *hirudo* a synonym].

Diagnosis.--Conspicuously dichromatic, the thorax and abdomen shining, metallic bluish black in male, aeneous green and only weakly shining in female; accessory notopleural bristle almost always present, rarely absent on both sides; male with extremely narrow frons and uniquely characteristic surstylus and aedeagus (Figs. 6c, 20); female lacking reclinate orbital bristles, and with narrow frons.

Male.--Head with microtomentum of parafrontal and parafacial yellowish to brownish tinted. Thorax shining dark bluish black, or with slight greenish-blue tint, the microtomentum thin and fine, usually evident only on anterior slope of scutum; fringe of mesothoracic spiracle orange-yellow to brownish, not as bright as in *Phormia regina* but brighter than in any other *Protocalliphora*. Abdomen dark blue to bluish black, highly shining, with thin gray "bloom" evident only at a low angle from behind. Calypteres yellow to brownish tinted, perhaps typically light brown in mature specimens.

Frons narrow (Fig. 41), obviously narrowing anterior to ocelli, at narrowest 0.05 times head width (0.04-0.07, with 19 of the sample between 0.04 and 0.06), and narrower than ocellar span (0.73x; 0.60-0.93); parafrontals clearly separated, though narrowly, the frontal vitta visible throughout, subequal to or wider than a parafrontal and widening strongly anterior to the narrowest point; preocellar area narrow, acuminate, rugose, shining and not microtomentose; parafrontal narrow and microtomentose below, with a single row of hairs outside the frontal bristles, above strongly narrowed and thinly microtomentose or bare, shining black; parafacial narrow, densely microtomentose, with two to three irregular rows of black hairs, weakly ridged, width of each parafacial only a little wider than ocellar span (1.31x; 1.07-1.60), wider than narrow frons (1.79x; 1.31-2.28), and not 2/3 vibrissal interval (0.64x; 0.56-0.75/23); frontal bristles rapidly decreasing in size above, the uppermost short and hairlike, rows ending well before median ocellus, and approximately opposite apex of the acuminate preocellar area; cheek comparatively narrow, its maximum height 1/3 maximum length of eye (0.35x; 0.32-0.39).

Thorax: Accessory notopleural bristle almost invariably present, rarely absent and then usually on one side only (cf. Fig. 3a); tympanic membrane either bare or haired, often with strong black hairs (For both sexes, entirely bare in 43 specimens, but hairs on one or both sides in 28, ranging from a single pale weak hair to 13 coarse black hairs); postalar wall bare, rarely with a weak adventitious hair. Fore tibia with one median posterior bristle.

Terminalia distinctive: cerci straight, long and slender (Fig. 20); surstylus with characteristic profile (Fig. 20), almost boot-shaped, broad at base, narrowing strongly, then slightly broadened distally and tending to curve anteriorly; distal portion of aedeagus cylindrical, distal margin flared (Fig. 6c).

Female. Body aeneous green, rather heavily microtomentose and only weakly shining, scutum usually with the three shining stripes narrow but more or less evident, abdomen more shining than thorax but under microscope obviously densely and finely microtomentose; tergite 5 often especially shining and cupreous green. Calypteres whitish yellow to orange yellow (typical), paler than in male, rims dark yellow.

Frons extremely narrow for female *Protocalliphora*, half or less width of an eye, less than 1/4 width of head (0.22x; 0.21-0.25), and less than three times ocellar span (2.86x; 2.43-3.38); parafrontal narrow, dull, gray to brownish gray microtomentose, prevertical area dull and densely microtomentose; frontal vitta with few weak and inconspicuous hairs, preocellar area usually heavily brown-gray microtomentose, contrasting with brown of frontal vitta, occasionally thinly microtomentose and more shining; reclinate orbital bristles absent; parafacial narrow, its width less than half width of frons (0.44x; 0.36-0.50), only 1 1/4 times ocellar span (1.25x; 1.07-1.43), and 2/3 vibrissal interval (0.67x; 0.57-0.82/22); cheek comparatively narrow, 1/3 maximum eye length (0.34x; 0.30-0.37).

Thorax and legs as in male.

Apparent normal length of adults, 7.5-8.5 mm.

Second instar. *Cephalopharyngeal skeleton* 370 μ in length. *Prothoracic fringe* 8 μ (4-15)/15; spines scarce. *Posterior region:* Peritreme open, button indistinct; stigmatal plates 71 μ in diameter; mesostigmatal spines absent; hyperstigmatal spines scarce, 7 μ (2-14)/30; hypostigmatal spines absent. *Dorsal cuticle* with narrow band of spines, one-fourth width of segment, on anterior margin of each segment, passing ventrally as the anterior band, spines 9 μ (4-18)/30. *Ventral cuticle* with single (anterior) band of spines, 9 μ (4-18)/30; spines of anterior patch scarce and small, 5 μ (2-8)/30; medial and posterior bands absent.

Third instar. Peritreme usually open; spines absent from meso- and hypostigmatal areas. *Prothoracic fringe* 7 μ (3-10)/60. *Cephalopharyngeal skeleton* 734 μ (560-860)/23 in length.

Puparium. Length 7.2 mm. (4.3-8.3)/26, breadth 3.2 mm. (1.9-3.9)/32; thin walled, semi-translucent, shiny, light brown. *Posterior region* (Fig. 46a, b, e): Stigmatal plates 188 μ (130-230)/90 in diameter, distance between buttons 422 μ (260-545)/46, and across stigmatal plates 1050 μ (675-1420)/46; stigmatal ratio 0.40 (0.32-0.46)/46; stigmatal area often outlined in plaques; mesostigmatal spines absent; mesostigmatal folds and/or plaques present; hyperstigmatal spines extremely scarce, 6 μ (3-12)/120, sometimes absent; hypostigmatal area with pronounced, irregular folds (western material), or folds weak or absent (eastern material); circumstigmatal folds variable, in western material prominent and irregular or rectangular (Fig. 46a), in eastern material weak and rectangular, or absent (Fig. 46e). *Dorsal cuticle* with spines or scarce, 8 μ (4-12)/130; cuticular ridges variable, pronounced, weak, or absent. *Ventral cuticle* (Fig. 46 d, f): ventral band ratio 0.11 (0.08-0.25); spines of anterior patch not distinguishable from other spines of anterior band, 9 μ (4-17)/200; medial band absent; posterior band sometimes seen, normally absent; each segment usually subdivided by at least one annulus between the scars of the ventral pads.

Type Series. *P. braueri*: The type series, from Austria [Naturhistorisches Museum, Wien], has been reported on in detail by Rognes (1985).

P. hirudo: Holotype male and allotype, Colorado, July 10, 1911 (M. A. Palmer), "ex nesting warbler"; paratypes, male, 2 females, Koehler, N. Mex. (W. R. Walton, Webster No. 7707), and male, 3 females, Colorado Springs, Colo., August 1916 (W. W. Arnold), "from maggot in sparrow a[nd] goldfinch" [last 3 females in MCZ]. One of the female paratypes from Koehler, N. Mex., has not been located. [All USNM except as noted.].

P. hirudo var. or subsp. *parva*: Holotype male and allotype, "Kansas," but correctly Greeley, Colo. [USNM].

P. hirudo var. *cuprea*: Holotype and 2 paratypes, all females, published as Seattle, Wash., labeled as Puget Sound (O. E. Plath), from nest of "western" [= American] robin. [USNM].

Under typical *hirudo* in the USNM collection are two other females with data like the Koehler, N. Mex., paratypes. One of these is also labeled "ex horned lark *Otocoris alpestris*." These Koehler specimens are part of the 30 specimens of "*Protocalliphora chrysorrhaea*" [sic] reported by Walton (1914: 175), which was a mixed infestation (See *P. seminuda* n. sp. for a part of this series).

The type material of *P. hirudo parva* was twice recorded in error, according to information in J. M. Aldrich's card catalogue of Diptera [USNM]. Aldrich, on a card then headed *Protocalliphora chrysorrhoea*, noted that "Williston gave me two pairs that he got from a man in Kansas, who found the larvae under the skin on the back of the head in young birds; I labeled them 'Young birds, Kans., S.W.W.'" Shannon and Dobroscky, having one male and one female of this lot (the former headless and the latter without abdomen!) described *P. hirudo parva* as "reared from fledglings, Kansas, (S.W. Williston through Prof. L. L. Adams)." Aldrich thereupon noted on his card for *parva*: "This should be reversed, it really was Greeley, Colo. (L. L. Adams, through S. W. Williston)," and he placed a label in the collection, "Prof. L. L. Adams, Greeley, Colo. Ex nestlings." Hall (1948: 195) apparently missed the Aldrich notes, and credited collection of the specimens to Williston, although correctly recording the type locality as Greeley, Colo. Incidentally, the specimens of *parva* are undersized individuals, and the male holotype lacks distinct accessory notopleural bristles, as noted by Hall. The allotype has an accessory notopleural on the left side only.

The variety *cuprea* was apparently based on three females, the holotype and a paratype deposited in the U.S. National Museum and a female paratype that came later with the Shannon Collection. A fourth female from the same rearing is teneral, with ptilinum extruded, and was not labeled as a paratype. These are the "four Puget Sound specimens" included by Townsend (1919) in the type series of his *Phormia metallica*.

Specimens Examined. 121 males, 137 females. If a nest is not specified, larvae were taken directly from the bird noted. 102 males, 114 females were from the West, the balance from the East.

Distribution. Canadian and Transition Zones: in the West, Alaska to Saskatchewan, south to Mexico (Durango), New Mexico, and Colorado; in the East, Ontario and Michigan to Labrador and Virginia, also Georgia.

Alaska: Mentasta Lake, Mentasta Road (Wilson's warbler); Rapids, S of Big Delta (golden eagle nest); Steese Highway, mile 105 (American tree sparrow nest).

Canada: **Alberta**: Edmonton (house wren). **British Columbia**: Beaver Lake, SW of Likely (young bird); Bella Coola (junco nest); Kamloops (nests of Brewer's blackbird, cliff swallow, and western kingbird); Lilloet, Mt. McLean (chipping sparrow); Prince George (house sparrow nest); Qualicum (barn swallow); Terrace; 6 mi W Terrace; 32 mi S Terrace; Vancouver (barn swallow nest, house sparrow, western meadowlark, Swainson's thrush); North Vancouver (budgerigar); Vanderhoof, W of Prince George (song sparrow); Vernon; Victoria (barn swallow). **Labrador**: Cartwright (white-crowned sparrow). **Newfoundland**: Codroy Valley, near Port-aux-Basques (American robin nest). **Ontario**: Algonquin Park (great crested and alder flycatchers, vesper sparrow, and Canada and myrtle warblers, and in nests of common grackle, ovenbird, vesper sparrow, barn and bank swallows, and yellow warbler); Ogoki. **Prince Edward Island**: Cavendish Beach, Prince Edward Island National Park. **Quebec**: Isle Verte, 15 km E of Rivière-du-Loup (savannah sparrow); La Verendrye Provincial Park; Val d'Espoir. **Saskatchewan**: Muscow, S of Fort Qu'Appelle (wren).

Contiguous United States, West: **California**: Columbia; Fresno Co. (Bear Creek); 8 mi E Georgetown (mountain chickadee); Tuolumne Co. (black-headed grosbeak). **Colorado**: "Colo." (warbler); Colorado Springs (sparrow and goldfinch); near Fort Collins (loggerhead shrike); Fort Collins; Greeley. **Idaho**: Lake Waha; Lewiston. **Montana**: Flathead Lake, Lake Co. (Brewer's blackbird); Hamilton (sparrow); Ravalli Co. (violet-green swallow, bluebird nest); Sunburst, Toole Co. (sparrow, probably vesper). **Nevada**: Washoe Co. (house wren). **New Mexico**: Abique Dam, Rio Arriba Co. (cliff swallow nest); Koehler (horned lark). **Oregon**: Lents, now in Portland (young bird); Milwaukie (tree swallow). **Utah**: Mantua in Box Elder Co. (house wren, subcutaneous). **Washington**: Olympia (barn swallow); Puyallup (house sparrow); Seattle (yellow warbler, and in nests of American robin and a sparrow); Tacoma (barn swallow). **Wyoming**: Laramie (house sparrow); Moran (pine siskin).

Contiguous United States, East: Georgia: Rabun Co., Rabun Bald at summit, 4714 ft. **Massachusetts**: near Middleboro (sparrow). **Michigan**: Iron Co., Gaastra ("ground sparrow"). **New York**: Franklin Co., Lake Clear Junction (chipping sparrow). **Virginia**: Giles Co. (great crested flycatcher); Roanoke (house wren); larvae only, Augusta Co. (loggerhead shrike).

Mexico: Male, Durango, 11 mi W Durango, June 20, 1964 (J. F. McAlpine) [CNC]; female, Rio Chico, 20 mi W Durango, July 22, 1964 (J. F. McAlpine) [CNC].

Most available specimens were reared June-August. An early record was the rearing of three males, three females at Puyallup, Wash., Apr. 28, 1926, "from larvae on *Passer domesticus*" (Theo. H. Schaffer).

Basis of description of immature stages: *Larvae*, *2nd instar* (1 specimen): Algonquin Park, Ont., ovenbird, June 1954. *Third instar* (9 specimens, plus some measures of prothoracic fringe and cephalopharyngeal skeleton from western puparia): Algonquin Park, Ont., ovenbird, June 1954 (4), and yellow warbler, June 1954 (5). *Puparium*, western (25 specimens): Hamilton, Mont., "sparrow" fledgling, July 1936 (10); Kamloops, B. C., Arkansas (i.e., western) kingbird, August 1945 (4); Vanderhoof, B. C., song sparrow, July 1945 (7); Lilloet, B. C., chipping sparrow, June 1947 (2); Lulu Island, Vancouver, B. C., western meadowlark, June 1932 (2).

Puparium, eastern (21 specimens): Algonquin Park, Ont., ovenbird, June 1954 (10), myrtle warbler, June 1945 (6), and Canada warbler, June 1945 (2); "Ontario" (Halket), "sparrow" (1); Lake Clear, N.Y. (2).

Hosts. Western: Brewer's and yellow-headed blackbirds, a bluebird (Montana, Oregon), mountain chickadee, golden eagle, house finch, American goldfinch, black-headed grosbeak, a junco (probably Oregon junco, from north of Vancouver, B.C., in breeding range of Oregon junco), western kingbird, horned lark, black-billed magpie, western meadowlark, American robin, loggerhead shrike, pine siskin, Townsend's solitaire (probably; Munro 1949), sparrows (chipping, house, song, American tree, and probably vesper), European starling, swallows (barn, cliff, tree, and violet-green), Swainson's thrush, warblers (Wilson's, yellow, and unidentified warbler in Colorado), and house and long-billed marsh wrens, also budgerigar and possibly baby chicks (Hearle 1938).

Eastern: Great-crested and alder flycatchers, common grackle, ovenbird, American robin, loggerhead shrike, sparrows (chipping, savannah, vesper, white-crowned, and "ground"), bank and barn swallows, warblers (Canada, myrtle, yellow), and house wren.

Ecology and Biology. Presumably ubiquitous, judging from the recorded hosts, but more frequent in nests or on birds at ground or shrub level--or young birds are more often observed closely in these situations! Maggots of this species are often found subcutaneously, causing true myiasis.

The myiasis-producing habit of larvae of *P. braueri* was mentioned when its synonym, *hirudo*, was first described, although it was not then realized how characteristic that was because some specimens were reared from puparia found in nests, as is normal for other species of *Protocalliphora*. Shannon and Dobroscky (1924) described both typical *hirudo* and *hirudo parva* from maggots found in nestlings (cf. Type Series), but they did not give details. They also recorded as "*aenea*" a fragmentary specimen, now recognized as *braueri*, reared from a larva "from brain of living fledgling of sparrow kind." However, Walton (1914) and Arnold (1919) had already published similar observations on myiasis. Walton stated that larvae of what he called *P.* "*chrysorrhaea*" [sic] on a fledgling horned lark "were contained in purulent sores on the sides of the body near the legs and on the neck." (N.B. The infestation was a mixed one; see discussion under *seminuda*). Arnold (1919), writing of *P.* "*azurea*" in his article on "Maggot infested birds," stated that a large number of young birds of various species were infested with maggots, and that "the young [larva] immediately burrows beneath the surface, -- is on the head, near or in the corner of the eyes, about the neck and upper part of the wings....The maggot eats a burrow or chimney into the flesh of its victim, remaining stationary and feeding upon the fluids of the helpless victim until death releases it." Two young meadowlarks attacked by a hundred maggots had bodies "presenting the appearance of having been struck by a load of shot."

A further observation of special interest is that of R. C. Miller (1929). Maggots emerged from a young yellow warbler (*Dendroica aestiva*, now *petechia*) barely able to balance on a twig. Dissection after the bird's death showed the following: "Eleven subcutaneous pockets that had contained one or more larvae were found, three on the head, three on the neck, and the remainder scattered over the body, especially under the wings and on the breast. Two of the lesions entered the pleural cavity, and the lungs had been partly devoured." The author concluded that this seemed definite evidence that when larvae were "present in sufficient numbers they may literally devour the young birds alive." In all 23 larvae emerged or were removed from the bird. A few reared adults were identified by J. M. Aldrich as *Protocalliphora splendida* var. *hesperia*, but a male and female preserved in the National Museum of Natural History in Washington show that the species is *braueri*.

After the Shannon and Dobroscky revision (1924), Johnson (1925) recorded the rearing of adults of *hirudo cuprea* [male, female, now in Boston U.] from maggots in the neck of a sparrow, and in 1931 he added that these had been found in "swellings on the neck." Bergtold (1927) recorded *P. "splendida"* as reared from a young house finch at Denver, Colo., from larvae in an abscess in a good-sized swelling below the right eye, and we believe that this record almost certainly refers to *braueri*. Still later, Guberlet and Hotson (1940) gave details of the myiasis observed in a young house sparrow at Seattle, Wash., from which flies identified as "*P. hesperia*" were reared [A male from this rearing has been found to be *braueri*]. After the death of the young bird, 21 live maggots were removed from the skin and subcutaneous tissue on the sides of the body, the neck, the right wing, and above the right eye. They were in slight swellings, which had small pores at which were located the anal ends of the larvae. George and Mitchell (1948), working at Lake Clear Junction, N.Y., in 1946, reared 3 males, 3 females of "*hirudo*" [USNM] from nestling chipping sparrows, and recorded that 5 of the 12 maggots observed were completely embedded in the skin of one nestling, which died. Munro (1949) found *hirudo* larvae abundant, feeding subcutaneously and causing mortality up to 75 %. Rausch (1972) described a case of subcutaneous myiasis in a fledgling Wilson's warbler in Alaska, with descriptions of both macroscopic and microscopic characteristics of the lesions. The larvae remained in the subcutaneous cavities until the death of the bird. A recent note by Bedard and McNeil (1979) recorded detailed observations on the larvae of *braueri* (as *hirudo*) on savannah sparrow nestlings (four in one brood in 1976, three in another brood in 1977): "The seven nestlings each supported upwards of 10 fly larvae and the site[s] of embedment, in decreasing order of importance were (a) the wing tracts (especially between the base of the remiges), (b) the head including the nasal cavity, the ear duct, and the area close to the eyes, (c) the crural and abdominal tracts, and (d) the legs. One nestling had three larvae moving freely between the naris and a breathing pore on the nape. All larvae were embedded subcutaneously and vigorously resisted extraction, suggesting that *P. hirudo* is an obligate subcutaneous parasite... The presence of these parasites retarded feather growth and the plumage around the site of the embedment had a sticky appearance. One of the nestlings lost an eye due to a parasite larva." The incidence of parasitism was low; one brood of 25 examined in 1976, one of 34 in 1977.

Brief notes on the labels of specimens submitted for identification sometimes state the subcutaneous location of larvae: male, three females, Lilloet, B.C., July 24, 1947 (H. C. Coppel), "internal in *Spizella passerina* " [UBC]; two males, three females, Ravalli Co., Mont., July 1967 (Sheri Marino and W. L. Jellison), "reared ex larvae in cysts of *Tachycineta thalassina lepida*" [USNM]. At Algonquin Park, Ont., Bennett found a subcutaneous larva in a young fledgling alder flycatcher, and three larvae of *braueri* in an open lesion above the eye of a young vesper sparrow. He identified as "*hirudo*" larvae causing myiasis in young sparrows in Quebec City (J. McNeil); at Lents, Oregon, two maggots came out of the head of a young bird about to fly, and one began boring into the hand of the collector, H. A. Darnell [two females, USNM and Oregon State U.]; four males, two females, reared from maggots taken June 20, 1970 from a nestling loggerhead shrike near Fort Collins, Colo. (T. J. Cade) [Cornell U.] (These maggots were completely under the skin,, mostly in the wings between areas of developing primary and secondary feathers, but one under an eye, one in the back of the neck, and some in the nares; none of the birds died).

At Roanoke, Va., Aug. 8, 1972, five young house wrens were observed by C. H. Lewis as they emerged from the nesting box. The feathers on the heads did not appear normal and, having caught two of the birds, he found and removed five larvae from under the skin of one bird and two from the other. Two submitted as puparia for identification proved to be *braueri*; presumably the others were also [Va. Polytechnic Inst.]. One adult male with puparium, and two larvae, came from Edmonton, Alta., July 1975 (Bruce Bembridge) [U. of Alberta]; the 11-day old wren nestlings were said to be "afflicted with boils." At Qualicum, B.C., R. A. Cannings recorded larvae in barn swallows, "found under the skin of the nestlings, in the wings and head." [BC Provincial Mus. (now Royal BC Mus.), USNM]. In Giles Co., Va., B. A. Garrison found larvae in cysts about the head, wings, and beak of a nestling great crested flycatcher [Va. Polytechnic Inst., USNM].

Observations more detailed than usual were recorded by H. W. Prescott at Milwaukie, Oregon (personal communication). These apply to *braueri*; adults of this species were reared from the material in the observed nest. In one of his bluebird nest boxes, he found "a tree swallow nest with four swallow nestlings about five days old and still essentially naked. On the top of the head of one of the nestlings I noted a small round wound slightly less than one sixteenth of an inch in diameter. On looking more closely I noted that this wound was actually a hole in the nestling's integument filled flush to the exterior surface by a fly maggot. At that point I inspected the other nestlings closely but found no more lesions of any sort on them. But to my surprise I found something else equally interesting. Each nostril of every one of the four nestlings was plugged completely shut by a fly larva, the truncate end of all larvae coming flush with the exterior of the nostril surfaces. Six days later, I returned to the swallow's nest. The nestlings were about half grown by then and I saw no further signs of maggots in them." So the nest was removed--a substitute nest was provided--and the nest material sifted for puparia from which adults of *braueri* emerged. He also recalled an earlier experience that also certainly involved *braueri* although no flies were reared. Among his numerous bluebird nest boxes--a total of 470!--two "contained nestlings only a few days old and still naked in both instances. One of the boxes contained nestling house sparrows and the other bluebirds. In both cases some of the nestlings had these small round wounds above described. They seemed to occur at random on the nestling's anatomy. Some were on the wings, some on various parts of the torso, and some on the legs exclusive of the tarsae, some on the head."

When flies are reared from larvae and/or puparia found in the nest, or when labels read simply "on...," the exact feeding habits of the larvae cannot be determined. However, in a few cases the details given show that the larvae of *braueri* do not always live subcutaneously, or not continously. At Moran, Wyo., J. D. Tiner reared 8 males, 11 females "from maggots which had anterior ends buried in craters in the skin of young pine siskin just off nest, and spending night on ground. ... all on head, neck and shoulders of bird." The larvae started falling off as soon as the bird was picked up, on the night of July 16, 1951, and adult flies emerged August 3 [USNM]. J. M. Kinsella made similar observations on a nest of Brewer's blackbirds near Flathead Lake, Flathead Co., Mont.: "The birds were out of the nest in the grass, and, when I picked them up, the larvae immediately began actively leaving the birds and dropping to the ground. Most of the larvae I saw came from the neck region. The larvae pupated in 24 hours and the adults hatched about 10 days later."

In some cases, the birds were apparently not seriously injured, in others the young birds were said to be "weak and emaciated" (Walton 1914). At Columbia, Calif., David C. Rentz reared 12 males, 13 females [CAS, USNM] from "subcutaneous larvae on a young black-headed grosbeak," and the bird died shortly after the larvae left the body. Death was attributed to the maggots. Damage done by maggots of this species was described as follows by Dr. Murray L. Johnson, M.D., Tacoma, Wash. (personal communication), who reared a specimen of "*hirudo*" [Johnson Colln.]: "I found the juvenile barn swallow alongside a small pool of water south of Tacoma on July 19, 1947. It was obviously very ill and unable to fly, though the feathers were well developed. There was a large cavity extending cranially from the right naris, containing two large larvae. The upper beak was weakened and movable; the right eye was closed. The larvae were immediately put in a jar of damp earth; the swallow died about 24 hours later." The adult fly emerged on 7 August. Three males, Lulu Island, Vancouver, B.C., June 1932 (R. A. Cumming) [UBC] were reared from nestling meadowlarks "very much cut up." In the case of cutaneous myiasis by "*hirudo*" in a fledgling Wilson's warbler, described by Rausch (1972), the larvae were in subcutaneous cavities in the head just above the beak. The bird refused food and died the day of its capture. Rausch concluded that "Aside from the possible adverse effects of exsanguination, the extent of destruction of tissue was such that the bird probably could not have survived. The lesions were produced largely by the growth of the larvae within a restricted locus." Under such conditions, notably infestation about the head, *Protocalliphora* larvae can certainly cause great damage and even death, or predispose to later death, or contribute to death from a combination of factors. Cais (1965), in Poland, reached an even stronger conclusion: "It seems definite that all observed cases of parasitism by *P. hirudo* bring about the death of the nestling, even if the number of larvae is small."

Remarks (including Variation). *Protocalliphora braueri* differs from all other species of *Protocalliphora* in North America in several characters that are unique in the genus: The form of both surstylus and aedeagus in the males, the absence of reclinate orbital bristles in the females, and accessory notopleural bristles in both sexes. (Such bristles occur only occasionally in other species as aberrations of chaetotaxy). The females are

Table VIII. Mean ratios and ranges of individual ratios, Nearctic and Palearctic *Protocalliphora braueri*

Samples	frons/head	frons/ocellar span	parafacial/ frons	parafacial/ ocellar span	parafacial/ vibrissal interval	Cheek height/ eye length
			MALES			
braueri						
western (25)	.05(.04-.07)	.73(.60-.93)	1.79(1.31-2.28)	1.31(1.07-1.60)	.64(.56-.75)	.35(.32-.39)
eastern (7)	.04(.03-.05)	.57(.46-.75)	2.40(2.00-2.83)	1.37(1.28-1.50)	.575(.48-.65)	.33(.32-.34)
Calif. (12)	.05(.04-.06)	.75(.64-.91)	1.79(1.60-2.00)	1.34(1.17-1.50)	.64(.58-.69)	.35(.33-.375)
Europe (4)	.05(.04-.06)	.74(.64-.83)	1.94(1.60-2.57)	1.44(1.30-1.64)	.61(.54-.68)	.36(.34-.39)
			FEMALES			
braueri						
western (25)	.22(.21-.25)	2.86(2.43-3.38)	.44(.36-.50)	1.25(1.07-1.43)	.67(.57-.82)	.34(.30-.37)
eastern (13)	.185(.16-.22)	2.46(2.17-3.18)	.48(.43-.53)	1.185(1.00-1.54)	.57(.46-.68)	.30(.28-.32)
Calif. (11)	.22(.20-.23)	2.83(2.57-3.15)	.45(.42-.47)	1.26(1.14-1.41)	.69(.64-.75)	.34(.31-.36)
Europe (4)	.20(.19-.21)	2.56(2.29-2.83)	.51(.47-.53)	1.31(1.14-1.50)	.64(.61-.69)	.34(.31-.35)

further distinguished from those of all known species except *deceptor* by the unusually narrow frons, and from all but females of *aenea* by the aeneous to dull cupreous green body color, contrasting with the metallic blue of most females. Unlike most species of *Protocalliphora*, isolated individuals of either sex can be readily identified, and the larvae and puparia are equally distinct. For a discussion of the possible generic or subgeneric significance of these characters, see the discussion on "Status of *Trypocalliphora*."

In an early stage of the study, *braueri* (as *hirudo*) was regarded as a western species, and the eastern population as a subspecies or a narrowly distinguished sister species with slightly narrower frons and slight differences in other proportions. Later, after reviewing the broadly overlapping range of variation, and also realizing the undue influence on the eastern ratios of some undersized specimens, which had been measured because of the relatively small numbers of available measurable examples, we concluded that only one species was present. For possible future use, however, we have maintained separate ratios for the eastern and western populations. In the description, the ratios and ranges refer to western specimens only. For comparison, Table VIII shows the ratios and ranges for the adults of western or typical Nearctic *braueri*, a California series, and eastern *braueri*, plus those for the few available specimens of *braueri* from central Europe. Third-instar larvae and puparia of western and eastern *braueri* differ slightly as shown in figure 46.

Unfortunately, too few eastern adults were available in measurable condition for the ratios to be strictly comparable with those based on longer series of western examples. The influence of size was also realized. Even so, the ranges of eastern and western measures overlap more or less broadly, and many of the mean ratios do not differ greatly. The male genitalia show no differences, and the biology and ecology, with the characteristic subcutaneous habit, are the same. The larval characters also showed virtually no differences, except for the presence of circumstigmatal folds in some--but not all--western specimens, and differences in hypostigmatal folds that are not regarded as significant. Puparia of *P. braueri* are distinguished from most other Nearctic *Protocalliphora* by the virtual absence of cuticular spines, whether as spines or as tubercles. Spines that do occur are minute as tubercles, giving the impression of fine sandpaper. The occurrence of at least one sub-annulus per segment further separates this species from all others. Some time after measurements and calculations had been completed for the 25 males and 25 females from a variety of western localities, we received an excellent series of 12 males, 13 females from Columbia, Tuolumne Co., Calif., (David C. Rentz), reared from subcutaneous larvae on a young black-headed grosebeak. As a test of the ratios being recorded for *braueri*, this series of probable siblings was also measured and ratios calculated. The results have been included in Table VIII. In all males and in all but one female, the individual ratios of the California specimens fell within the ranges of the main western series; the exception was a trifling difference of 0.003 in the minimum for the frons: head ratio. The mean ratios were remarkably close, differing at most by only 0.03. This essential agreement encourages us to believe that the ratios will usually be useful criteria for species recognition in this genus.

Males with brown calypteres, darker than usual or dark because of discoloration, can be confused with those of *P. hesperia* superficially, although the unique characters noted above permit precise differentiation. Guberlet and Hotson (1940) recorded maggots subcutaneously in house sparrows, and reared a few specimens that were identified as "*hesperia.*" One of these, a male (Seattle, Wash., USNM Collection), has been seen, and as surmised from the biology, it proved to be typical *braueri*. Perhaps this confusion is at the root of an error in Hall (1948). Hall's figure G, plate 21, labeled "*hesperia*" in the explanation on p. 414, was apparently intended to be *braueri* (as *hirudo*), from the reference on p. 194. Unfortunately, also, the figure does not emphasize the distinctive apices of both surstylus and aedeagus.

Shannon and Dobroscky (1924) described the females of "*hirudo*" as having the calypteres "darkened" or "smoky." However, the type series shows that the females, except for the allotype, were dirty and discolored and apparently influenced the description. Actually the calypteres are lighter than in the males, varying from whitish yellow, with outer margins yellow, to deep orange-yellow. Peus (1960) was misled by the original description into believing that true *hirudo* was not sexually dichromatic in the color of the calypteres.

Females of *braueri* have been confused with *aenea* in identifications because of the similarity of color. For example, Shannon and Dobroscky (1924) recorded as "*P. splendida* var. *aenea*" a broken specimen, of which only mesonotum and wings remained, reared "from brain of a living fledgling of sparrow kind" [USNM]. The mesonotum of this still extant fragment shows distinct accessory notopleural bristles, the associated puparium is characteristically that of *braueri*, and the myiasis is typical of *braueri*, hence the specimen assuredly is *braueri*, fragmentary though it is. As a further example, Hall (1948) listed specimens of "*aenea*" reared from cliff swallow nests (Kamloops, B.C.) and from a sparrow fledgling (Hamilton, Mont.), but these all proved to be *braueri*. A male from the Montana series was apparently the basis for Hall's description of the male of "*aenea*" (p. 185), and the measurements for the female of "*aenea*" must have been based on specimens from these misidentified series, in whole or in part.

We have concluded that the "Corvallis," Oregon specimen of *braueri* identified by Melander as *P. azurea* and so recorded by Cole and Lovett (1921) [Oregon State U.] is actually a Lents, Oregon specimen like one now in the U.S. National Museum of Natural History. Both bear the Accession No. 1191, under which a file card in the Department of Entomology, Oregon State University, records two maggots sent in by H. A. Darnell from Lents, Oreg., July 15, 1914, from "the head of a young bird about to fly." By July 24, both had pupated, and adults emerged July 27. The USNM specimen, which came from Melander's collection, bears correct data. The Oregon State specimen has the same lot number but is labeled, apparently erroneously, as Corvallis, July 14, and came to be published that way.

The specimens from Puget Sound, Wash. (Plath), which formed the type series of *P. hirudo* var. *cuprea*, were earlier included in the type series of *Phormia metallica* Townsend, although then noted as a different form that might "perhaps prove to be a good subspecies, or it may even be specifically distinct" (Townsend 1919). However, two other specimens of the same data are females near *metallica* or *halli*, probably the latter. The labels do not make clear whether this species and *braueri* came from the same nest, from a mixed infestation, or whether different nests were involved.

Unconfirmed published records that can be accepted as *braueri* with reasonable confidence because of the typical myiasis are those by Bergtold (1927), who reared two adults of "*splendida*" from larvae found in an abscess in a large swelling below the right eye of a house finch at Denver, Colo., and by Hall (1948), who recorded a male and a female of "*hirudo*" from larvae found under the skin of nestling yellow warblers at Fort Collins, Colo.

A European species was recorded as *hirudo* by Heinz (1954) and Zumpt (1956), based on identification by Sabrosky. However, Peus (1960), who compared a pair from Wyoming (det. Sabrosky) with European material, believed that the European species was distinct and new, and also that both belonged to a new genus distinct from *Protocalliphora*. The new taxa were published as *Trypocalliphora* and *T. lindneri*. He also raised doubt whether *hirudo* sensu Sabrosky was the same as typical *hirudo* Shannon and Dobroscky, noting for the former yellow squamae (calypteres) in the females, brown in the males, whereas *hirudo* was described as having "smoky" squamae in both sexes. As already noted, however, the type series of *hirudo* was somewhat discolored: females normally have yellowish or whitish-yellow calypteres. Thus in reality the characters given by Peus for *hirudo* sensu Sabrosky are typical for true *hirudo*. Both *hirudo* and Peus's *lindneri* are now considered synonyms of *P. braueri* (Hendel) [Revision by Rognes 1985].

The characters of proportion given by Peus in his key, e.g., the width of the frons, have a range of variation, and the characters of color not only vary somewhat but are greatly affected by the condition and/or maturity of the individual specimens. From the descriptive notes given by Peus for *braueri*, we are inclined to suggest that the original reared specimens were slightly teneral and hence paler. In reared series of Nearctic *braueri* before us, we find examples from the same bird or nest, and thus presumably siblings, in which calypteres in the males range from whitish yellow (teneral) through deeper shades of yellow and orange-brown to moderately dark brown (mature). The extremely narrow frons recorded for *braueri* may also stem at least in part from immature reared specimens. The frontal stripe is a weak area that is almost the last to reach the mature stage and harden into normal position.

We have seen authentic *lindneri* Peus, having before us eight examples, including a male and two females from Schlorer's rearing at Diersheim in southern Baden, in other words from the same bird and locality as the holotype and probably siblings of it, though not part of the type series. Ratios from measurements of the few available specimens of *lindneri* have been included in Table VIII as "Europe" for comparison with the measurements of Nearctic *braueri*. It is noteworthy that the ranges of the individual ratios fall within the ranges of western (typical) *hirudo* for most proportions. The exceptions are generally trifling. Admittedly the sample of *lindneri* is too small for mean ratios to have any significance--and note that Peus's sample of Nearctic *hirudo* consisted of one male and one female--but bearing in mind the variation observed in our much greater sample of *hirudo*, we remained unable to separate European *lindneri* from

Nearctic *hirudo*. From our data we are happy to agree with Rognes (1985) in synonymizing both under the still older *braueri*.

Protocalliphora (*P.*) *brunneisquama*, new species
(Fig. 28, Map 14)

Protocalliphora hesperia, in part.--James, 1955: 25 [Los Angeles Co., Calif., July 4, 1950].
P. sp.--Horning and Barr, 1970: 73 [Idaho].
P. hesperia, in part.--Whitworth, 1976: 19, 27 [Utah, chipping sparrow].

Diagnosis. Sexes concolorous, monochromatic, shining metallic dark blue to bluish purple; calypteres dark brown to brownish; in male, surstylus long and narrow (Fig. 28) and cerci exceptionally long and narrow.

Male. Head chiefly black in ground color, parafrontal and parafacial dark gray to silvery gray microtomentose, that of parafrontal thinning posteriorly to shining black prevertical area; parafacial without changeable reflecting spot. Body chiefly shining metallic blue to bluish purple, thinly gray microtomentose, anterior slope of scutum gray microtomentose but rest of scutum sparsely so, viewed from behind only weakly demarcating the shining stripes; abdomen more shining than thorax, brighter blue, tergites 2 and 3 gray microtomentose when viewed from behind at a low angle, tergites 4 and 5 shining. Calypteres dark brown to brownish.

Frons of moderate width, sometimes narrowing slightly anterior to median ocellus, at narrowest 0.10 times head width (0.07-0.13/5) and somewhat broader than ocellar span (1.26x; 1.00-1.375/5); parafrontal strongly narrowed posteriorly to polished prevertical area, with one row of black hairs along narrow part of parafrontal, row widening below to adjoin parafacial rows, each parafacial with 3-4 irregular rows of numerous black hairs; parafacial usually wider than frons (1.35x; 1.00-2.00/5), 1.70 times ocellar span (1.50-2.00/5), and slightly narrower than vibrissal interval (0.88x; 0.82-0.90/4); ocellar tubercle continued forward as a shining sublinear acuminate preocellar area; cheek moderately broad, its height 2/5 maximum length of eye (0.41x; 0.40-0.43/5).

Thorax without accessory notopleural bristles; tympanic membrane and postalar wall with a few short and inconspicuous hairs. Fore tibia usually with one posterior bristle (two on left tibia in male from Utah, and two on both fore tibiae in male from New Mexico, but only one on other three males and all five females).

Terminalia: cerci exceptionally long and narrow; surstylus exceptionally long and narrow, curved, distal third narrower than base (Fig. 28); aedeagus as usual for the genus (cf. Fig. 6a).

Female. Body color as in male, but scutum more heavily gray microtomentose, as viewed from behind, and the shining stripes thus more clearly demarcated than in male; parafrontal and parafacial brownish yellow microtomentose, latter with a changeable reflecting spot just below the frontal bristles; preocellar area brownish microtomentose, more or less subshining but not polished.

Frons at vertex 0.30 times head width (0.30-0.31/6) and 3.48 times ocellar span (3.44-4.12/6); frontal vitta with one to three or four coarse bristlelike setae on each side, mesad of frontal bristles; parafrontal broad throughout, though narrowing somewhat dorsally; parafacial moderately broad, twice width of a 3rd antennal segment, nearly half width of frons (0.47x;0.44-0.48/6), 1.65 times ocellar span (1.455-1.94/6), and 0.92 times vibrissal interval (0.84-1/00/6); cheek height 0.44 times maximum length of eye (0.325-0.49/6), or 0.47x (0.46-0.49/4) if two 'runts' are disregarded [The runts made no difference in other ratios, and the undersized females were counted].

Apparent normal length of adults, 9-11 mm. Two females from Utah are obviously undersized, 6 and 6.5 mm.

Immature Stages. Two puparia are known, but these are with the undersized females and the details cannot be relied upon except in a general way, such as the presence of good-sized spines.

Type Series. Holotype male and allotype, Idaho: Craters of the Moon National Monument, July 3 and 4 (male), 1965, Malaise trap (D. S. Horning, Jr.). In the U.S. National Museum of Natural History.

Additional paratypes. **California**: male, Forest Home, San Bernardino Co., Sept. 21, 1936 (A. J. Basinger) [CAS]; female, Monterey Co., 11 km N Lucia, July 8, 1963 (P. H. Arnaud, Jr.) [CAS]; male, Tanbark Flat, Los Angeles Co., July 4, 1950 (W. O. Marshall), published James 1955 as "*hesperia*" [U. Calif., Davis]; female, White Cloud Campground, Nevada Co., May 31, 1964 (P. H. Arnaud, Jr.) [CAS]. **Colorado**: female, North Cheyenne Canyon, Colorado Springs, June 17, 1926 (E. C. Van Dyke) [CAS]. **Montana**: female, Fairy Lake, Gallatin Co., 9500 ft, Aug.

18, 1983 (W. L. Downes, Jr.), on white-flowered umbellifer [Downes Colln.]. **New Mexico**: male, White Mts., North Fork Ruidoso, about 8300 ft, Aug. 12 (C. H. T. Townsend), on flowers of *Solidago trinervata* [Identified by unknown person as *Calliphora viridescens*] [Brit. Mus. (Nat. Hist.)]. **Utah**: male, two females, Mantua, Box Elder Co., July 7, 1971 (T. L. Whitworth, nest of chipping sparrow; 3 females, Box Elder Co., Mantua-Devil's Gate, June-July, 1969 (T. L. Whitworth, Malaise trap) [USNM, Whitworth Colln.]. **Washington**: female, Malaga, Chelan Co., June 11-13, 1970, Malaise trap [Wash. State U.].

Specimens Examined. 5 males, 12 females.

Distribution. Western, known at present only from the states of California, Colorado, Idaho, Montana, New Mexico, Utah, and Washington. June 11-13 (Washington) to Sept. 21 (southern California). The few records suggest that this is a species of montane areas.

Hosts. Known only from chipping sparrow (Utah). Most available specimens were hand-collected.

Ecology and Biology. Unknown. The size of most of the specimens suggests that the species is usually a parasite of larger birds than chipping sparrow.

Variation. The small sample makes it difficult to rely on the ratios stated in the description. Variation in the posterior bristle on the fore tibia was noted in the description.

Remarks. *P. brunneisquama* is especially easy to recognize. It is one of the few species with dark brown to brownish calypteres in both sexes, and it is the only one of these that has narrow surstylus. The broad parafacial is useful for recognizing females among those with brown calypteres. *P. lata* has unusually broad parafacial, and on the other hand *hesperia* and *hesperioides* have obviously narrow parafacials. *P. brunneisquama* is intermediate between those extremes.

Etymology. The specific name, a noun in apposition, is derived from the Middle Latin *brunneus*, brown, plus the Latin *squama*, a scale.

Protocalliphora (*P.*) *chrysorrhoea* (Meigen)
(Figs. 6b, 11, 47, Map 6)

Musca chrysorrhoea Meigen, 1826: 60 [Austria].

Protocalliphora chrysorrhoea.--Hough, 1899: 66 [Placed in new genus].

P. azurea (Fallén).--Engel, 1920: 249-258 [From bank swallow nests, Dachau, Germany; male terminalia figured].

P. chrysorrhoea.--Hennig, 1939: 363-4 [Possibly an ecological race on bank swallow].

Apaulina sapphira Hall.--Sailer and Lienk, 1951: 209 [From "cliff swallow," error for bank swallow].

P. azurea chrysorrhoea.--Zumpt, 1956: 95, 96 [Central Europe; subspecies, restricted to nests of bank swallows].

P. chrysorrhoea.--Gregor and Povolný, 1959: 43-45 [Comparison with *azurea*; colored figures of each species].

P. (*Orneocalliphora*) *chrysorrhoea*.--Peus, 1960: 208-211 [Central Europe; neotype, female, Aachen, Germany, from Meigen material in Munich].

P. chrysorrhoea.--Whitworth, 1976: 19,20ff [Utah]; Whitworth, 1977: 4933-B.

Diagnosis. Sexes concolorous, monochromatic, dark bluish to bluish purple; calypteres white; tympanic membrane and postalar wall with numerous long black setae; fore tibia with two posterior bristles; frons of male of moderate width; surstylus of male digitate in profile.

Male. Parafrontal and parafacial grayish to brownish gray microtomentose, viewed from above the parafacial with a mottled appearance of irregular light and dark crossbands, because of numerous transverse ridges, and with numerous coarse black hairs in about five irregular rows. Body dark bluish to bluish purple, occasionally with dark greenish tint; thorax and abdomen concolorous, shining, abdomen highly shining with thin gray microtomentum visible only from behind at a low angle, scutum and scutellum duller with two broad sublateral stripes of gray microtomentum demarcating three shining stripes, although to the unaided eye the scutum appears almost uniformly dull. Calypteres whitish with light yellow rims.

Frons of moderate width, narrowing slightly anterior to ocellar triangle, at narrowest 0.115 times head width (0.095-0.135) and much broader than ocellar span (1.63x; 1.29-1.86); preocellar area shining but rugose, elongate triangular; parafrontal narrowing above but clearly evident and microtomentose to level of median ocellus; frontal bristles decreasing posterodorsad and becoming mere hairs, both the frontal bristles and a row of parafrontal hairs extending as far as level of median ocellus; parafacial broad, twice as broad as 3rd antennal segment, width 1.22 times that of frons (1.07-1.50), 1.99 times ocellar span (1.67-2.29), and virtually equal to vibrissal interval (1.00x; 0.87-1.07/19); cheek height 0.44 times (0.40-0.48) maximum length of eye.

Thorax with scutum thickly covered with long, erect, fine hairs, those on presutural area half or more as long as presutural acrostical and dorsocentral bristles; each notopleuron sometimes with an accessory notopleural bristle; postalar wall with a conspicuous patch of black hairs; tympanic membrane with numerous long black hairs, which are especially conspicuous. Fore tibia with two posterior bristles.

Terminalia: cerci distally acuminate and parallel (cf. Fig. 5b); surstylus in profile straight and digitate (Fig. 11), approximately parallel sided and of medium width throughout, rounded apically; aedeagus (Fig. 6b).

Female. Color as in male, monochromatic, but scutum notably duller than in male because the stripes of gray microtomentum are wider and more distinct throughout, the median shining stripe especially narrowed. Frons broad, widening anteriorly, at vertex about 3/10 width of head (0.29x; 0.26-0.31) and 3.64 times ocellar span (3.22-3.88); preocellar and prevertical areas heavily brownish gray microtomentose, the former subshining just anterior to median ocellus; frontal vitta with several pairs of coarse, bristlelike, mesoclinate black hairs and a number of short and inconspicuous hairs just mesad of each row of frontal bristles; parafrontal and parafacial as in male, brownish gray microtomentose, with numerous coarse black hairs; parafacial about half width of frons (0.53x; 0.48-0.58), nearly twice ocellar span (1.93x; 1.67-2.25), and slightly greater than vibrissal interval (1.07x; 0.94-1.21/17); cheek height 0.47 times (0.42-0.54) maximum length of eye. Mesonotal hairs and bristles shorter than in male.

Apparent normal length of adults, 9.5-11.5 mm.

Third instar. Peritreme closed; cuticular spines numerous between stigmatal plates and in hypostigmatal area. *Prothoracic fringe* 29 μ (12-62)/ 21. *Cephalopharyngeal skeleton* 950 μ (870-990)/6 in length.

Puparium. Length 8-10 mm., breadth 4.1 mm. (3.8-4.4)/8; medium-thick walled, dull, brown. *Posterior region* (Fig. 47a, b): stigmatal plates 230 μ (220-250)/16 in diameter; distance between buttons 690 μ (620-780)/8 and across stigmatal plates 1410 μ (1180-1485)/8; stigmatal ratio 0.49 (0.45-0.52)/8; stigmatal area not outlined in large plaques; mesostigmatal area with numerous spines and folds; hyperstigmatal spines 41 μ (14-58)/100; hypostigmatal area with numerous, weak concentric folds; hypostigmatal spines numerous, 18 μ (6-31)/100; circumstigmatal folds present, weak and concentric. *Dorsal cuticle* (Fig. 47c) with numerous spines, 43 μ (18-68)/100; cuticular ridges prominent. *Ventral cuticle* (Fig. 47d): ventral band ratio 0.73 (0.65-0.81)/16; spines of anterior patch 23 μ (12-56)/100; medial band pronounced, spines 38 μ (23-56)/100; posterior band pronounced, present on all segments; cuticular ridges on ventral surface either pronounced or somewhat reduced.

Type. The species was described from Austria, apparently based on a single female. It could not be found by Peus (1960), who thereupon designated a neotype from Aachen, Germany, that had been identified by Meigen himself.

Specimens Examined. 83 males, 76 females.

Distribution. Holarctic; Alaska to Quebec, south to Utah and Montana, in bank swallow nests, probably much more widely distributed than the records indicate.

Alaska [all USNM]. 12 males, 9 females, Ladd Field, Fairbanks, Aug. 6 and 16, 1951, six bank swallow nests in colony along Chena River; 3 males, 1 female, 60 mi E of Tok Junction, July 14, 1948, bank swallow nest (R. I. Sailer) (publ. Sailer and Lienk 1951 as *Apaulina sapphira* from "cliff swallow", error for bank swallow, Sailer in litt.); 10 males, 9 females, Willow (N of Anchorage), July 31, 1974, bank swallow nest (G. E. Haas).

Canada [all CNC]. **Alberta**: 3 males, 3 females, Elkwater (S of Medicine Hat), July 15, 1969 (J. A. Shemanchuk), bank swallow nest. **British Columbia**: 2 males, 1 female, 3 puparia, Atlin, 2200 ft, Aug. 6-7, 1955, barn swallow nest (H. J. Huckel); 3 males, Summit Lake, 4200 ft, mile 392 on Alaska Highway, Aug. 19-21, 1959 (E. E. MacDougall); 4 males, 4 females, 6 puparia, Toad River Lodge, 4500 ft, mile 422 on Alaska Highway, Aug. 15, 1959 (R. E. Leech), nest of violet-green swallow in old bank swallow burrow. **Quebec**: male, Gatineau Co., Mesham Twnp., June 21, 1973 (D. M. Wood); male, Duncan Lake near Rupert (30 mi N of Ottawa), July 31-Aug. 4. 1971 (J. F. McAlpine); male, female, 4 mi N of Eardley (above Ottawa), Aug. 25, 1971 (D. M. Wood). **Yukon Territory**: male with puparium, Rampart House, July 17, 1951 (J. E. H. Martin), bank swallow nest.

Contiguous United States. Montana: 3 males, 1 female, Bigfork, nest July 20, 1917, larvae pupated July 21, adults Aug. 5 (C. H. Danforth), "external par[asite] on bank swallow" [USNM]; male, female, Hamilton, Aug. 4, 1951 (L. E. Hughes), bank swallow nest [Wash. State U.]. **Utah**: 2 males, 2 females, Cache Co., Benson, Aug. 4, 1970 (T. L. Whitworth), barn swallow nest [USNM]; 36 males, 44 females, most with puparia, Cache

Co., Hyrum (Whitworth), 10 different bank swallow nests [USNM, USU, Whitworth Colln.].

Two possible records of *chrysorrhoea* can be mentioned, but there are no voucher specimens to check on the species. Lincoln (1931) referred to a report from Crystal Bay, Minn., of bank swallows with an abundance of "fat maggots protruding from their nostrils." A similar record of numbers of large, blood-filled maggots attached to nestling bank swallows, encircling their necks and on top of their heads, near a beach in Oceana Co., Mich., was reported to us by George C. Steyskal. In both cases the bank swallow parasites might well have been *chrysorrhoea.*

Basis of description of immature stages: *Larva, 3rd instar* (6 specimens): Tok, Alaska, bank swallow, July, 1948. *Puparium* (8 specimens): Tok, Alaska, bank swallow, July, 1948 (7); Bigfork, Mont., bank swallow, July, 1917 (1).

Hosts. Chiefly bank swallow, also violet-green swallow (rare), and isolated records from barn swallow. Whitworth (1976) listed two nests of house sparrow and one of barn swallow infested by *chrysorrhoea*, along with 95 infested nests of bank swallow. The barn swallow record was verified (Benson, Utah), but one of the house sparrow infestations proved to be *sialia*, and perhaps the other was also. Uncommon records still do not seriously affect the unquestioned predominance of *chrysorrhoea* in bank swallow nests. European rearing records of *chrysorrhoea* are likewise from bank swallows, there called sand martins (e.g., Engel 1920, Zumpt 1956, Peus 1960, Nuorteva 1960, Nuorteva and Järvinen 1961). The published record of *chrysorrhoea* from "cliff swallow" (Sailer and Lienk 1951) was an error for bank swallow, according to information from the late Dr. Sailer.

Ecology and Biology. Apparently restricted--with rare exceptions--to cavity nests of the bank swallow, both in North America and in Europe.

The record from the violet-green swallow was noteworthy and was investigated further. According to information kindly furnished by the collector, Robin E. Leech, the violet-green swallow nest was in one group of about 15-20 nest holes from 15-30 inches deep in the vertical face of a roadside gravel pit, about 5-7 feet above the floor of the pit and 3-4 feet below the upper edge of the vertical face. The data indicate almost certainly, according to the late Dr. A. Wetmore of the Smithsonian Institution, that the holes were made by bank swallows and subsequently utilized by the violet-green swallow, which normally uses ready-made holes or crevices. This is an interesting confirmation of Bennett's belief in a ecological basis for the distribution of *Protocalliphora.* Parasitism on bank swallows exclusively had seemed an example of host specificity by *chrysorrhoea* and therefore an exception to the ecological theory. The parasitism of a different host suggests that the apparent host specificity is the result of there being, usually, only one host species in the bank swallow niche.

An interesting observation was provided by J. A. Shemanchuk of the Research Station of Agriculture Canada at Lethbridge, Alta. He reared *chrysorrhoea* at Elkwater, Alta., and reported (letter to G. E. Shewell) that "Some of the adults came from pupae found in the nests and some were reared from larvae that were removed from under the skin and breast muscles of nestlings."

Variation. A long series of 28 males, 39 females--runt specimens not checked--from 10 different nests of bank swallow at Hyrum, Utah, provided a fine opportunity for the evaluation of variation. In the critical character of posterior bristles on the fore tibia, all but three tibiae in the 67 specimens showed 2 bristles and these had 3: a right tibia on a female, a right on a male and a left on another male, the two males in the same nest.

Most of the Alaska and Utah specimens are of the normal size for the species, but many of the Hyrum, Utah specimens are undersized, the smallest barely 6.5 mm in length. Otherwise they are typical *chrysorrhoea*, and one can conclude that these were underfed.

An accessory notopleural bristle was present in 55 of 148 specimens available for checking when this character was noted, but usually on one side only. The species is so different from *braueri* that the presence of an accessory notopleural bristle will give no trouble.

The hindmost presutural intraalar bristle is absent in some specimens, or present on one side only.

Remarks. *Protocalliphora chrysorrhoea* has long been recognized as the parasite of bank swallows, sometimes recorded as *P. azurea* (Fallén), sometimes as a subspecies of *azurea*, and recently as a distinct species, no doubt aided by its apparent host specificity.

Among the Nearctic species, *chrysorrhoea* and *sapphira* are easily distinguished by a character that they share with most of the Palearctic species: tympanic pit and postalar wall setose, and front tibia with 2-3 posterior bristles, rarely only one. The frons of *chrysorrhoea* is considerably narrower than that of *sapphira*, and the surstylus is broader and less curved (cf. Figs. 11 and 26).

Puparia of this species are similar to those of *avium*, *hirundo* and *sialia*, but can be distinguished by the shorter prothoracic fringe. Palearctic examples of puparia have also been studied although our description is based entirely on Nearctic specimens.

Detailed comparison was made of Nearctic and Palearctic material to determine whether we were dealing with a Holarctic species on the Holarctic host, or whether the Nearctic population was a new species or subspecies. This comparison yielded no differences that can be considered reliable, in the face of known variability and the range of ratios. The Palearctic series is unfortunately short, but significant in content. Eleven males and ten females from Europe were available for comparison, including four from Finland and 17 from Dachau, Germany (E. O. Engel), identified as "*azurea*" by Engel and reared from the nests of the "Uferschwalbe" (bank swallow). The Dachau specimens were part of what Peus (1960) referred to as the classic series of *chrysorrhoea* and the basis of the study by Engel (1920).

Nearctic males have a slightly narrower frons than the Palearctic, ranging from 0.095-0.135 times the head width, mean 0.115, compared with a range in available Palearctic males of 0.13-0.14, mean 0.137. Many Nearctic males are somewhat teneral, and the frons is slightly compressed, so the frons: head ratio is undoubtedly low. Allometric difference may also be a factor; a small male from Finland shows a frons: head ratio of 0.12, like those of small Nearctic specimens.

No appreciable difference could be found in the male terminalia. Minor chaetotactic differences were not reliable, and apparently represent variation, possibly differing in degree in different populations, but the present samples are much too small to say. European males more often had an accessory bristle between the two notopleurals. Nearctic specimens more often had one or more fine hairs near the bristlelike infravertical (below and behind each inner vertical bristle). In European males the postocular setae were slightly shorter, and only slightly longer towards the midline, whereas in Nearctic specimens the setae appeared slender--a natural tendency in smaller specimens!--and the dorsal six on each side were obviously longer than the others, but there were exceptions on both sides.

In females, Nearctic specimens commonly have several pairs of coarse bristlelike mesoclinate black setae on the frontal vitta, mesad of the frontal bristles, whereas the few European females available show fine hairs, short or of moderate length.

Puparia of Nearctic and Palearctic specimens are virtually indistinguishable, although puparia of the latter have slightly shorter spines, even though the puparia are larger, and slightly fewer cuticular ridges.

Many, although not all, of the Dachau specimens are larger than available Nearctic specimens. Recent European revisions, usually citing the Dachau material, give a length of 13-15 mm for the species, with 14 as the average, whereas the range in Nearctic material, admittedly somewhat teneral, is 9.5-11 mm. A few Dachau specimens are the same size as the Nearctic, as are the Finnish examples. Perhaps northern specimens of the species tend to be smaller. Also the amount of available food or the length of time the larvae had for feeding would affect the size of the reared adults.

Protocalliphora (*P.*) *cuprina* (Hall)
(Fig. 16, 48, Map 7)

Protocalliphora metallica.--Neff, 1945: 73-6 [Calif.].
Apaulina cuprina Hall, 1948: 191 [Calif., Mont.; other specimens partly *aenea*, partly *metallica*].
Protocalliphora cuprina.--James, 1955: 25 [Calif.]; Hall, 1965: 926 [Catalog; original records repeated].
P. sp. II.--Whitworth, 1976: 17ff.

Diagnosis. Dichromatic, body in male shining metallic dark blue to bluish purple, in female heavily gray microtomentose and dull bluish to bluish green, with tergite 5 of abdomen bright cupreous to cupreous green; calypteres brownish yellow in male, especially the rims, white in female; surstylus digitate, approximately straight.

Male. Parafrontal and parafacial gray microtomentose; palpus black tipped. Body shining, metallic dark blue to bluish purple, thorax darker, bluish black, obscurely and thinly dark gray microtomentose, shining stripes only weakly demarcated. Abdomen thinly gray microtomentose on tergites 2-4, when viewed from behind at a low angle. Calypteres brownish to brownish yellow, especially on rims.

Frons narrow, narrowing anterior to ocellar tubercle, at narrowest 0.06 times head width (0.05-0.07) and 0.89 times ocellar span (0.71-1.18); preocellar area narrowly acuminate, more or less rugose and shining or subshining, sometimes only a ridgelike line; parafrontal strongly narrowed posteriorly, prevertical area shining; frontal bristles weak posteriorly, ending well in advance of median ocellus; parafacial with numerous black hairs in 3-4 irregular rows, equal to or only slightly wider than breadth of 3rd antennal segment, 1.74 times width of frons (1.31-2.00), 1.55 times ocellar span (1.38-1.83), and 0.72 times vibrissal interval (0.615-0.83); cheek height 0.38 times maximum length of eye (0.34-0.42).

Thorax without accessory notopleural bristles; tympanic membrane and postalar wall usually bare; fore tibia typically with one posterior bristle.

Terminalia: cerci distally narrowly acuminate and parallel (as in Fig. 5b); with parallel-sided surstylus of moderate width, digitate or only weakly curved (Fig. 16); aedeagus as usual for the genus (as in Fig. 6a).

Female. Dark bluish black thorax, scutum heavily gray microtomentose, two broad gray stripes demarcating a median stripe and two shining lateral areas, the striping evident to unaided eye, unlike the obscurity of male scutum; abdomen with tergites 2-4 shining bluish green in direct view, but gray microtomentose when viewed from behind at a low angle, tergites 2 and 3 especially dull; tergite 5 cupreous green, usually not as decidedly cupreous as in *metallica*; calypteres white.

Frons at vertex about 1/4 head width (0.26x; 0.24-0.27) and 3.20 times ocellar span (2.86-3.67); preocellar area dull, evenly gray microtomentose; parafacial moderately haired, in three irregular rows, a changeable spot along upper margin, width of parafacial 0.48 times width of frons (0.44-0.54), 1.54 times ocellar span (1.36-1.77), and 0.79 times vibrissal interval (0.69-0.92); height of cheek 0.40 times maximum length of eye (0.37-0.44).

Apparent normal length of adults, 7.5-8.5 mm. Specimens as small as five mm have been seen.

Third instar. No specimens examined, characters derived from the puparium. *Prothoracic fringe* 60 μ (20-90)/60. *Cephalopharyngeal skeleton* 890 μ (775-980)/9 in length.

Puparium. Length 6.7 mm (6-7.3)/6; breadth 3.4 mm (2.6-3.8)/18; thin walled, dull, light brown. *Posterior region* (Fig. 48a, b); stigmatal plates 150 μ (130-180)/36 in diameter; distance between buttons 500 μ (400-550)/18 and across stigmatal plates 1000 μ (875-1180)/18; stigmatal ratio 0.50 (0.46-0.54)/18; stigmatal area not markedly outlined in cuticular plaques or folds; mesostigmatal folds vary from weak and virtually absent to moderately distinct; hyperstigmatal spines numerous, 28 μ (15-40)/120; hypostigmatal area with numerous, weak, irregular folds; hypostigmatal spines numerous, 18 μ (8-40)/120; circumstigmatal folds absent (Fig. 48a), or if present, weak and concentric (Fig. 48b). *Dorsal cuticle* (Fig. 48c) with numerous spines posteriorly directed, 25 μ (13-40)/120; cuticular ridges weak or absent. *Ventral cuticle* (Fig. 48d): ventral band ratio 0.63 (0.52-0.68)/14; spines of anterior patch 15 μ (8-24)/70; anterior band pronounced; medial band narrow, spines 20 μ (16-30)/70; posterior band pronounced.

Type Series. The species was described from a series of 23 males and 35 females, type locality Riverside, Calif., from four species of birds. A few paratype males are actually *hesperia*, the female from Mary's Peak, Oregon is *aenea*, and the female from Mt. Hood, Oregon is *metallica*. Holotype in the U.S. National Museum of Natural History, and paratypes in that collection and the California Academy of Sciences.

Specimens Examined. 265 males, 337 females.

Distribution. Western, British Columbia to Montana, south to California, Utah and Wyoming.

Canada: British Columbia (all from nests): 90 males, 92 females, Haney (39 males, 27 females, yellow warbler), Kamloops and Lac du Bois, Kamloops (western kingbird), Okanagan Landing (yellow warbler), Qualicum (barn swallow), Vernon (house finch).

United States (reared material from nests with one exception, noted): **California**: 26 males, 37 females, Fresno Co. (nestling willow flycatcher, yellow warbler), Le Grand (western kingbird, also adults labeled "pupae dug from ground under dove's nest"), Mt. Lowe (Los Angeles Co.), Riverside ("goldfinch", house finch; bulk of type series of *cuprina*), San Bernardino Co. (Camp Baldy), and Nevada Co. (Truckee, 6000 ft). **Idaho**: 10 males, 14 females, all with puparia, Franklin Co. (Emigration Canyon; western flycatcher). **Montana**: 10 males, 4 females, Corvallis (barn swallow), Ravalli Co. ("wild canary"). **Utah**: 129 males, 182 females, almost all with puparia: Box Elder Co. (Mantua, *Empidonax* flycatcher, and Mantua-Devil's Gate, in Malaise trap), Cache Co. (Benson, barn swallow, western kingbird), Franklin Basin (*Empidonax* flycatcher, western wood-pewee), Logan (eastern kingbird, yellow warbler). **Washington**: 7 females, Puget Sound (one labeled paratype of *cuprina* but not published), Seattle ("western" robin, two as paratypes of *cuprina*), Tacoma. **Wyoming**: female, Fremont Co. (Riverton at Wind River at Highway 789).

Basis of description of immature stages: *Puparium* (18 specimens): Ravalli Co., Mont., American goldfinch, June, 1935 (7); Kamloops, B. C., Arkansas (i.e., western) kingbird, (4); Cache Co., Utah, western wood-pewee, August, 1971 (7).

Hosts. A dove, house finch, flycatchers (dusky, western, and willow), American goldfinch, eastern and western kingbirds, western wood-pewee, American robin, barn swallow, yellow warbler. Whitworth (1976) recorded it also from Audubon's [i.e., yellow-rumped] warbler.

Ecology and Biology. Whitworth (1976) noted that *cuprina* was "virtually the only species found in warbler and flycatcher nests."

Variation. In *P. cuprina*, there is typically one posterior bristle on a fore tibia, but occasionally there are two bristles on one side or the other, and sometimes on both sides. Of 575 specimens checked for this character, 433 or 75.3% have only one posterior bristle on each fore tibia, and this is thus considered the typical number. Only 50 or 8.6% had two on each front tibia, but even this is frequent enough to cause difficulty in using the key. When the 92 specimens with two on one side or the other are added in, 142 or 24.7% show two bristles that might confuse the user of the key. One male from Haney, B. C. has two bristles on the left side and three on the right. In all specimens, the lower posterior bristle is regularly present and outstanding; others, if present, are usually smaller, not always outstanding, and often close to the lower rather than well spaced from it as usual in species with a regular number of two bristles.

Two odd variants: A female from LeGrand, Calif., has the reclinate upper orbital bristles short and hairlike, rather than the normal well developed bristles, and a female from Seattle, Wash., has an accessory notopleural bristle on the left side.

Some color variation also needs mention. In small and/or very teneral individuals, the 5th tergite is only slightly yellowish tinted, a faint yellowish green rather than a cupreous green. Finally, in a series from Vernon, B.C., two females out of 16 have brownish yellow calypteres like the male.

Remarks. *P. cuprina* is in a complex of species with digitate surstylus, but with the advantage in the general key of having a female with cupreous green 5th tergite like the northern and eastern *P. metallica*. Unlike the *avium* (eastern)-*asiovora* (western) pairing, however, *cuprina* and *metallica* differ in the male terminalia and fall into different species groups, *metallica* having the surstylus narrow and curved, *cuprina* straight and digitate or only slightly curved. In addition the two differ greatly in the spination of the puparium: Fine sandpaper appearance in *metallica*, distinct spines in *cuprina*. In the key to males, *cuprina* passes to a small group of species with brownish calypteres, and in the key to females the cupreous 5th tergite associates it readily with the eastern *bicolor* and the chiefly eastern *metallica*.

P. metallica of Plath (1919c) from "western" robin was *cuprina*, at least in part, judging from the two females included in the type series of *cuprina*, plus several others of Plath's material from Puget Sound, Wash.

The series from yellow warbler at Okanagan Landing, B.C., was a mixed infestation of *spenceri* and *cuprina*. Luckily both sexes of the two species are easily distinguished, the males by distinct surstyli, the females by the color of the 5th tergites.

Puparia of *cuprina* are separated with difficulty from those of *bicolor* and *spenceri*. They can be separated from those of *bicolor* on the basis of the shorter prothoracic fringe, but from *spenceri* only by a general evaluation of all the characters.

Protocalliphora (*P.*) *cuprina* var. or new species

Female, Koehler, N. Mex., (W. R. Walton) [USNM]; female, McKinley Co., N. Mex., 4 mi S Fort Wingate, 8000 ft, July 10, 1966 (P. H. Arnaud, Jr.) [USNM]; female, Mexico: Morelos, Cuernavaca, July 9, 1961 (R. & K. Dreisbach) [Mich. State U.]; female, Mexico: Durango, 11 mi W Durango, 7000 ft, July 13, 1964 (J. F. McAlpine) [CNC]; male, Mexico: Durango, 3 mi E El Salto, 8200 ft, July 1964 (J. F. McAlpine) [CNC].

The four females are close to *cuprina* and conceivably could be variants of that species. Both tergites 4 and 5 are coppery green and quite shining, although tergite 4 is finely microtomentose when viewed from behind at a low angle. It is also possible that they represent a new species, hitherto unrecognized, but these isolated females, unassociated with puparia and males, cannot be properly diagnosed and should not be described. They are another indication of the paucity of information on the *Protocalliphora* fauna of the Mexican plateau (cf. Map 1). The lone male could be the male of the species, but only rearing of a good series from one nest and host can give that assurance. A female, Utah; Grand Co., La Sal Mts., Warner Camp, 9500 ft, July 21, 1968 (W. J. Hanson), is not in good condition but may belong here [Whitworth Colln.].

Protocalliphora (*P.*) *deceptor*, new species
(Figs. 5b, 9, 49, Map 4)

Protocalliphora sp., probably new.--McAtee, 1929: 110 [Bell, Md.].
Protocalliphora.--Thomas, 1936: 46 [Hocking, Lawrence, and Jackson Counties, Ohio].
Protocalliphora avium.--Brimley, 1938: 371 [Raleigh, N.C.].
Protocalliphora n.sp.--Nolan, 1959: 355 [Bloomington, Ind.].
Protocalliphora n.sp.--Goddard and Lago, 1983: 482 [Starkville, Miss.].

Diagnosis. Sexes concolorous, monochromatic, dark blue to bluish purple in ground color but mesonotum of female heavily gray microtomentose; calypteres white; frons of both sexes narrow; surstylus of male broadly rounded, as in *avium*.

Male. Head with parafrontal and parafacial brilliant silvery microtomentose, latter without changeable spot; back of head chiefly white haired, with one to two irregular rows of black hairs behind row of postocular setae. Body dark blue to bluish purple, or slightly bluish green, gray microtomentose, the microtomentum on scutum demarcating three broad shining stripes, abdomen appearing highly shining but microtomentum especially heavy on tergites 2 to 4 when viewed from behind, the 5th tergite shining. Calypteres white.

Frons narrow, its width 0.07 times width of head (0.05-0.08/22) and barely greater than ocellar span (1.01x; 0.77-1.25/22), narrowing from vertex 2/5 way to antennal bases, then suddenly widening, at narrowest the parafrontals separated by about width of one; preocellar area polished, elongate, narrowly acuminate; parafrontal and parafacial smoothly and finely microtomentose, the microtomentum thinning in prevertical area, a single row of fine hairs beginning about midway of parafrontal, becoming two rows anteriorly and finally merging with the three to four irregular rows on parafacial; latter obviously wider than frons (1.65x; 1.43-2.20/22), 1.67 times ocellar span (1.38-1.83/22), and much narrower than vibrissal interval (0.76x; 0.65-0.79/21); cheek narrow, in profile little over 1/4 eye height, its greatest breadth approximately 1/3 (0.35x; 0.31-0.40/22) greatest length of eye.

Scutal hairs less dense than in some species; no accessory notopleural bristles; tympanic membrane usually bare, rarely with one to a few fine and inconspicuous hairs; postalar wall bare or with few short black hairs. Fore tibia with one posterior bristle.

Terminalia: cerci relatively short, acuminate, inner margins parallel (Fig. 5b); surstylus longer than the cerci and greatly broadened (Fig. 9), resembling that of *avium* and *asiovora* (cf. Figs. 7-9); aedeagus usual for the genus (as in Fig. 6a).

Female. Parafrontal and parafacial bright and shining gray microtomentose, faintly yellowish tinted, not silvery as in male. Mesonotum much more heavily and extensively microtomentose than in male, the microtomentum of scutum demarcating three relatively narrow shining stripes. Abdomen as in male, much more shining than thorax, but tergites 2 to 4 actually gray microtomentose when viewed from behind at a low angle, tergite 5 shining.

Frons like that of *braueri*, exceptionally narrow for a female, obviously half or less width of an eye, 0.23 times width of head (0.20-0.25) and 3.01 times ocellar span (2.56-3.43), widening gradually from vertex to opposite antennal bases; frontal vitta with a few fine hairs toward sides, a small irregular preocellar area often partly polished; parafrontal almost equibroad throughout, about half width of frontal vitta, evenly microtomentose, including prevertical area, one row of parafrontal hairs, becoming two rows anteriorly; parafacial sparsely haired, with two to three irregular rows, its width 0.52 times that of frons (0.44-0.56), 1.56 times ocellar span (1.25-1.82), and 0.77 times vibrissal interval (0.64-0.92/21); cheek unusually narrow, its breadth 0.34 times maximum length of eye (0.29-0.37).

Apparent normal length of adults, 7.5-8.5 mm.

Third instar. Peritreme open. *Prothoracic fringe* 55 μ (25-90)/81. *Cephalopharyngeal skeleton* 773 μ (675-805)/10 in length.

Puparium. Length 7.7 mm (7.2-8.8)/3; breadth 3.4 mm (3.3-3.4)/3; thin-walled, shiny, light brown. *Posterior region* (Fig. 49a-c): stigmatal plates 175 μ (145-195)/15 in diameter; distance between buttons 530 μ (375-610)/8, and across stigmatal plates 1180 μ (1040-1290)/8; stigmatal ratio 0.45 (0.42-0.47)/8; stigmatal area outlined in large plaques; mesostigmatal spines present only as small tubercles; mesostigmatal plaques and/or folds present; hyperstigmatal spines usually as tubercles, 6 μ (2-14)/100; hypostigmatal area normally without folds but if present, weak and irregular; hypostigmatal spines numerous, as tubercles, 4 μ (2-8)/100; circumstigmatal folds absent. *Dorsal cuticle* (Fig. 49d) with definite but relatively few spines, 12 μ (4-20)/100; cuticular ridges weak. *Ventral cuticle* (Fig. 49e,f): ventral band ratio 0.63 (0.55-0.66)/16; spines of anterior patch 11

μ (4-20)/100; medial band pronounced, spines 12 μ (2-29)/100; posterior band pronounced.

Type Series. Holotype male, allotype, and 8 male, 5 female paratypes, Falls Church, Va., reared July 23, 1929, from [house] wren's nest (E. A. Chapin), each mounted with puparium. In the U.S. National Museum of Natural History. Additional paratypes:**Connecticut**: 11 males, 13 females, Stamford, Aug. 1-4, 1936, from nest of eastern phoebe (S. W. Bromley Colln.) [USNM]. **Delaware**: 3 males, 3 females, Newark, July 10, 14 (male), 1967, reared from pupae in wood thrush nest (C. Strehl)[U.Del., USNM]; male, 5 females, Newark, June 6, 1968, from northern cardinal nest collected May 27 (F. M. Stiner, Jr.), 4 with puparia, and 8 males, 6 females, Newark, July 1 (5 males), Aug. 10 (female), and 22 (3 males, 5 females), 1970 (Vivie E. Davis), reared from wood thrush nests [U. Del., USNM]; male, female, Greenwood, emerged Aug. 1, 1968, reared from nest of common grackle (F. M. Stiner, Jr.) [Stiner Colln.]. **Georgia**: female, Atlanta, July 16, 1935 (P. W. Fattig) [USNM]. **Indiana**: 2 males, Bloomington, June 9, 1954, nest of prairie warbler (Val Nolan) [USNM]. **Maryland**: 2 females, Bell, reared from house wren nest collected July 8, 1927 (W. L. McAtee) [USNM] (basis for record of *Protocalliphora* sp. in McAtee 1929: 110); 23 males, 11 females, with puparia, Plum Point, Calvert Co., May 27, 1986, pupae from Carolina chickadee nest in deserted bluebird nestbox (J. H. Fales), flies emerged May 30-June 5 [USNM]. **Mississippi**: 11 males, 2 females, Starkville, Oktibbeha Co., May 11, 1981, ex pupae in nest of Carolina chickadee (G. Hurst) [U. Miss., USNM]. **Missouri**: female, Kirkwood, Sept. 9, 1952 (W. Downes) [Iowa State U.]. **North Carolina**: male, 4 females, Raleigh, July 9, 1937, reared from nest of [eastern] phoebe collected July 7 (C. S. Brimley) [N. C. Dept. Agr.] (basis for record of "*avium*" by Brimley 1938: 371). **Ohio**: [Ohio State U. and USNM, except as noted]: 13 males, 13 females, near Hanover, Licking Co., May 31-June 5, 1945, from American crow nest (S. E. Airesman, E. S. Thomas); 4 males, 3 females, "Neotoma," Hocking Co., June 23, 1934, from larvae in nest of prairie warbler collected June 10, 1934 (E. S. Thomas) (also 2 males, 2 females, same data, at Amer. Mus. Nat. Hist.); female, Washington Township, Lawrence Co., July 17, 1934, from larva in nest of worm-eating warbler, July 8, 1934 (C. F. Long, E. S. Thomas) (also one male at Amer. Mus. Nat. Hist.); male, 2 females, Jackson Co., July 30 and August (female), 1935 (E. S. Thomas), from pupae in nest of Carolina wren (C. F. Long). **South Carolina**: 4 males, 6 females, Savannah River Plant of Atomic Energy Commn., 1 mi from junction of

Savannah River and Upper Three Run's Creek, May 11, 12, 1959, from nest of Carolina chickadee (W. A. Tarpley) [USNM and Tarpley Colln.]. **Tennessee**: 8 males, 4 females, Nashville, May 16, 1940, reared from northern mockingbird nest, including 8 mounted with puparia (Mrs. Amelia R. Laskey) [USNM]; 2 males, 2 females, Weakley Co., 1979 (T. David Pitts), from nest of Carolina chickadee [USNM]. **Virginia** [all USNM]: 15 males, 16 females, Stafford Co., at mouth of Aquia Creek, NE of Fredericksburg, late May 1969, from nest of Carolina chickadee (E. T. McKnight); 7 males, 7 females, Stafford Co., near mouth of Aquia Creek, June 8-17, 1970, from nest of Carolina chickadee collected June 7 (E. T. McKnight); 19 males, 9 females, Stafford Co., Marlborough Point at mouth of Potomac Creek, June 5, 1970, from puparia collected May 30 in nest of Carolina chickadee.

Non-paratype Material: Ten males, 22 females, broken or undersized specimens, some with same data as cited in the type series, have not been considered as paratypes. Those recorded because of additional localities or hosts (all USNM) are as follows: **Maryland**: 3 males, 3 females, Patuxent Refuge, Bowie, July 5, 1944; **Tennessee**: 4 larvae on which the larval description is based, Nashville, June 25, 1947, from nest of great crested flycatcher (Mrs. Amelia R. Laskey); **Virginia**: female, 4 mi W of Luray, Blue Ridge Mts., July 30, 1956, from northern cardinal nest collected July 21 (A. Wetmore).

Specimens Examined: 156 males, 142 females.

Distribution: Eastern United States, Missouri to Connecticut, south to Georgia and Mississippi.

Basis of description of immature stages: *Larva*, *3rd instar* (3 specimens): Nashville, Tenn., great crested flycatcher, June 25, 1947. *Puparium* (9 specimens, all from type series): Nashville, Tenn., northern mockingbird, May, 1940 (2); Falls Church, Va., house wren, July, 1929 (7).

Hosts: Northern cardinal, Carolina chickadee, brown-headed cowbird [probable; one of three fledglings from infested wood thrush nest, Newark, Del. (V. E. Davis)], American crow, great crested flycatcher, common grackle, northern mockingbird, eastern phoebe, wood thrush, prairie and worm-eating warblers, and house and Carolina wrens.

Ecology and Biology. The hosts thus far known appear to indicate ubiquitous occurrence, especially in cavity nests.

Variation: Some aberrations were observed in the orbital bristles. In females of the Newark, Del., series reared from nest of northern cardinal, one female has only one pair of proclinate orbitals and no reclinate orbitals, and one female lacks the reclinate orbital on one side. The first specimen, lacking the reclinate orbitals altogether, would have appeared to be a female of *braueri* but for the body color.

The series from Stafford Co., Va., reared June 8-17, 1970 from a nest of Carolina chickadee, provided the smallest *Protocalliphora* that we have seen. The smallest female measured only 4.5 mm, compared with others of 7.5-8 mm from the same nest.

Remarks. The very narrow frons and brilliantly silvery parafacials of the males suggest most of the 13 species with such a narrow frons, but in its northeastern range only *bicolor*, *braueri*, *metallica*, and *shannoni* are involved. Of these, *bicolor* has tergite 5 cupreous in the males, *braueri* is much different with yellowish to yellowish brown calypteres and accessory notopleural bristles, and *metallica*, even though having a narrow frons, has the frontal vitta much wider than a parafrontal. But even if the differences in the narrow-fronted individuals are not considered very distinct by an inexperienced person or one without well-identified material for comparison, they will be astonished to dissect the male terminalia and find the surstylus as broadly rounded as in *avium* and *asiovora* (cf. Figs. 7-9). The cerci are, however, narrowly acuminate and parallel distally, not at all like those of *avium* and *asiovora*. This combination of narrow frons and broad surstylus is unique in the genus, as far as known at present.

Females are also unusual. Their narrow frons and broad gray parafrontals at once suggest *braueri* over any of the other species, but *braueri* lacks the reclinate orbital bristles and has accessory notopleural bristles. Females of *deceptor* also are concolorous with the males, metallic bluish or bluish purple, whereas *braueri* is a dichromatic species and the females are aeneous green. If immature stages are available, the numerous spines will easily distinguish *deceptor* from *braueri*.

Puparia of *deceptor* are similar to those of *metallica* and *shannoni*. They can be separated from *shannoni* by the presence of tubercles, rather than spines, in the hyperstigmatal area, and from *metallica* by the longer spines of the dorsal and ventral cuticle.

In the May 1969 lot from Stafford Co., Va., from the nest of a Carolina chickadee, 149 puparia were found in the nest, but many were parasitized by *Nasonia vitripennis*. Numerous *Nasonia* also appeared from puparia collected from the nest taken near the mouth of Aquia Creek in June 1970.

Etymology. Both sexes are very deceptive in appearance, hence the name *deceptor*, a noun in apposition, from the Latin *deceptor*, a deceiver.

Protocalliphora fallisi, new species
(Figs. 24, 50)

Diagnosis. Sexes concolorous, monochromatic, shining metallic bluish to bluish purple; calypteres white; frons of male of medium width, preocellar area narrowly acuminate, shining or somewhat gray microtomentose; surstylus narrow, curved.

Male. Head chiefly black in ground color, parafrontal and parafacial silvery gray microtomentose. Body shining, metallic bluish to bluish purple, scutum slightly gray microtomentose on anterior slope and in two narrow stripes that set off a broad shining (as viewed from behind) median stripe. Abdomen shining in direct view, but thinly gray microtomentose when viewed from behind at a low angle. Calypteres white.

Frons of moderate width, 0.11 times head width (0.085-0.13/21) and 1.44 times ocellar span (1.14-1.67/21); ocellar tubercle typically with four small bristles behind ocellar bristles; preocellar area narrowly acuminate, often somewhat shining, partly somewhat gray microtomentose; parafrontal strongly narrowed above to polished prevertical area, sparsely haired, one irregular row on lower part, row ending about midway; parafacial often with changeable spot along upper margin, and with few to numerous hairs in about three irregular rows, width obviously greater than breadth of 3rd antennal segment, 1.15 times width of frons (1.04-1.38/21), 1.66 times ocellar span (1.57-1.83/21), and 0.86 times vibrissal interval (0.77-1.00/21); height of cheek 0.42 times maximum length of eye (0.39-0.44)/21); occiput with many white hairs, only 2-3 irregular rows of black hairs behind and below the row of postocular setae.

Thorax occasionally with accessory notopleural bristles; tympanic membrane and postalar wall with few short, pale, weak and inconspicuous hairs. Fore tibia with one or two posterior bristles on fore tibia.

Terminalia: cerci long and narrow to acute apices (cf. Fig. 5b); surstylus narrow, curved (Fig. 24); aedeagus as usual in the genus (cf. Fig. 6a).

Female. Concolorous with male, but scutum somewhat more gray microtomentose on anterior slope and the two scutal stripes.

Frons 0.29 times width of head (0.265-0.31/20) and 3.54 times ocellar span (3.14-4.00/20); preocellar area dull, finely and evenly dark yellowish gray to brownish gray microtomentose, as are parafrontal and parafacial; parafacial often with changeable spot along upper margin, and sparsely to moderately haired, in 2-3 irregular rows, width obviously greater than breadth of 3rd antennal segment, one half width of frons (0.50x; 0.46-0.58/20), 1.77 times ocellar span (1.60-2.10/20), and nearly equal to vibrissal interval (0.97x; 0.92-1.04/18); height of cheek 0.45 times maximum length of eye (0.42-0.48/20).

Apparent normal length of adults, 7.5-8.5 mm. (One specimen, a net-caught male from Ogoki, Ontario, is 9 mm, and five specimens are unusually small, 5-5.5 mm, undoubtedly underfed).

Third instar. Peritreme closed; spines present between stigmatal plates and in hypostigmatal area. *Prothoracic fringe* 50 μ (25-78)/87. *Cephalopharyngeal skeleton* 765 μ (690-885)/20 in length.

Puparium. Length 8.0 mm (7.2-8.8)/30; breadth 3.7 mm (3.2-4.0)/30; thick-walled, dull, brown. *Posterior region* (Fig. 50a): stigmatal plates 155 μ (120-190)/42 in diameter; distance between buttons 595 μ (430-700)/20, and across stigmatal plates 1225 μ (900-1380)/20; stigmatal ratio 0.46 (0.44-0.52)/20; stigmatal area outlined in plaques; mesostigmatal spines numerous; mesostigmatal folds and plaques absent; hyperstigmatal spines 16 μ (4-41)/100; hypostigmatal spines numerous, 9 μ (4-20)/100; circumstigmatal folds absent. *Dorsal cuticle* (Fig. 50b) with few spines, 16 μ (4-41)/100; cuticular ridges weak or absent. *Ventral cuticle* (Fig. 50c): ventral band ratio 0.32 (0.20-0.38)/40; anterior patch indistinct, spines of anterior patch 10 μ (4-20)/100; medial band reduced, spines 10 μ (4-20)/100; posterior band vestigial.

Type Series. Holotype male, allotype, and 54 paratypes (26 males, 28 females), Ontario, southern Algonquin Park, near Lake Sasajewan, reared from nests of red-winged blackbird and swamp sparrow in 1955; paratype male, Ogoki, Ont., Aug. 12, 1952 (J. B. Wallis) [Last, CNC]. Holotype and allotype in the U.S. National Museum of Natural History, paratypes there and in the Canadian National Collection. The holotype and allotype were reared from the nest of red-winged blackbird 55-174 (G. F. Bennett).

Non-paratype Material. Six specimens (male and five females), including four reared from common yellowthroat nest in 1954, are small or in poor condition and are not included in the type series.

Specimens Examined. 29 males, 34 females.

Distribution: Now known from two localities in Ontario.

Basis of description of immature stages: *Larva, 3rd instar* (6 specimens): Algonquin Park, Ont., white-throated sparrow, July, 1954. *Puparium* (30 specimens, from the type series): Algonquin Park, Ont., red-winged blackbird, June 1955.

Hosts. Red-winged blackbird (predominant), swamp sparrow, common yellowthroat; also, from puparia only (det. Bennett), common grackle, white-throated sparrow, rough-winged swallow, downy woodpecker.

Ecology and Biology. The few specimens thus far reared were from nests taken in marshes.

Variation. Specimens may have either one or two posterior bristles on the fore tibia, with the former slightly predominating in the available material. Of the 64 specimens, 36 had one posterior bristle on each fore tibia, 15 had two on each, and 13 had two on one tibia or the other.

Remarks. *P. fallisi* is nearest to *P. tundrae* in many respects, and at one time the two were believed to be the same species. However, the appearance of the male frons and the hyperstigmatal spines, in particular, as well as some minor and somewhat variable characters of occipital hairs and the bristles on the ocellar tubercle, have led us to believe that the two are distinct species.

Puparia of *fallisi* are separated with difficulty from those of *aenea*. Generally, *fallisi* has a ventral band ratio smaller than 0.40, a shorter prothoracic fringe, and occurs in nests built in swamps, along river banks, etc.

The species was mixed with *metallica* in the swamp sparrow and two red-winged blackbird nests. Females of *metallica* were easily recognized by the cupreous fifth tergite, but males of the two species are much more difficult to separate in such a mixed infestation.

Etymology. The specific name, in the genitive case, is dedicated to Dr. A. Murray Fallis, respected protozoologist and parasitologist, director of the doctorate study by Bennett, and good friend.

Protocalliphora (*P.*) *halli*, new species
(Figs. 19, 51, Map 8)

Protocalliphora sp. IV.--Whitworth, 1976: 19ff. [Utah].

Diagnosis. Sexes concolorous, monochromatic, shining metallic dark blue to bluish purple; calypteres brown in male, white in female; surstylus digitate.

Male. Parafrontal and parafacial dark gray microtomentose, sometimes faintly brownish or brownish yellow tinted; palpus orange-yellow, black tipped. Body shining dark metallic blue to bluish purple, scutum dark gray microtomentose but three shining stripes only obscurely demarcated, as viewed from behind; abdomen shining in direct view, bright gray microtomentose on tergites 2-4, viewed from behind at a low angle. Calypteres brown, paler but still brownish in teneral specimens.

Frons narrow, 0.08 times head width (0.07-0.09) and 1.07 times ocellar span (0.86-1.50); preocellar area typically dull and more or less triangular but occasionally rugose and acuminate, and rarely linear and shining; frontal bristles decreasing in size posterodorsad, the weak ones ending before level of median ocellus; parafrontal narrowed above to shining prevertical area, middle portion with one row of parafrontal hairs, widening to several rows below at merger with parafacial hairs; parafacial with numerous black hairs in about four irregular rows, width obviously greater than breadth of 3rd antennal segment, 1.47 times width of frons (1.22-1.75), 1.57 times ocellar span (1.36-1.83), and 0.75 times vibrissal interval (0.65-0.85); greatest height of cheek 0.41 times maximum length of eye (0.38-0.43).

Thorax without accessory notopleural bristles and usually without hairs on tympanic membrane and postalar wall; fore tibia typically with one posterior bristle (cf. Variation).

Terminalia: cerci distally narrowed to parallel and acute apical portions (cf. Fig. 5b); surstylus digitate and approximately straight, very weakly curved (Fig. 19); aedeagus as usual for the genus (cf. Fig. 6a).

Female. Concolorous with male, but scutum more heavily gray microtomentose, and demarcation of shining stripes usually obvious to unaided eye, viewed from behind; parafacial with changeable spot along upper margin; calypteres white.

Frons 0.26 times head width (0.23-0.29) and 3.24 times ocellar span (2.86-3.86); preocellar area entirely dull, heavily microtomentose; width of parafacial obviously greater than breadth of 3rd antennal segment, 0.50 times width of frons (0.44-0.55), 1.61 times ocellar span (1.50-1.71), and 0.81 times vibrissal interval (0.73-0.92); greatest height of cheek 0.44 times maximum length of eye (0.40-0.47).

Apparent normal length of adults, 7.5-8.5 mm.

Third instar. No specimens examined, characters derived from puparium. *Prothoracic fringe* 55 μ (30-90)/25. *Cephalopharyngeal skeleton* 900 μ (850-925)/4 in length.

Puparium. Length 7.5 mm (6.7-8.2)/5; breadth 3.3 mm (3.2-3.6)/9; medium thin-walled, dull, brown. *Posterior region* (Fig. 51a,b): stigmatal plates 170 μ (150-200)/18 in diameter; distance between buttons 600 μ (450-700)/9, and across stigmatal plates 1215 μ (925-1425)/9; stigmatal ratio 0.49 (0.46-0.53)/9; stigmatal area outlined in cuticular plaques and folds; mesostigmatal area with a few small plaques and weak folds; mesostigmatal spines scarce; hyperstigmatal spines numerous, 29 μ (15-50)/90; hypostigmatal area with weak, irregular folds; hypostigmatal spines numerous, 18 μ (8-30)/90; circumstigmatal folds weak, concentric. *Dorsal cuticle* (Fig. 51c) with numerous spines, 28 μ (15-50)/90; cuticular ridges, if present, weak. *Ventral cuticle* (Fig. 51d): ventral band ratio 0.49 (0.40-0.60)/17; anterior band pronounced, anterior band not distinct, spines 17 μ (8-25)/90; medial band narrow, thinned on the midline, spines 20 μ (8-40)/90; posterior band much reduced on anterior segments, absent on the posterior ones.

Type Series. Holotype, male, allotype, and 29 paratypes (20 males, 9 females), Utah: Cache Co., Mendon, July 11, 1971 (T. L. Whitworth), barn swallow nest 1155. Type in the U.S. National Museum of Natural History, paratypes there and in the Utah State University and Whitworth Collections. Additional paratypes: **British Columbia**: 135 males, 98 females, Heiron Lease, Kamloops, Lac du Bois at Kamloops, Porcupine, and Vancouver, various dates July-Aug., 1934, 1937, 1941, 1946 (G. J. Spencer), nine barn swallow nests [UBC, USNM]; 78 males, 98 females, Qualicum, July 16-24, 1982 (N. K. Dawe and R. A. Cannings), six barn swallow nests [UBC, USNM]. **California**: 20 males, 16 females, Napa Co., Howell Mt., 3 km NNE Angwin, 396 m, June 27-30, 1980 (Hugh B. Leech), nest of black phoebe [CAS, USNM]; 44 males, 46 females, Riverside, June 1936 (A. J. Basinger), phoebe nest 17 [CAS, USNM]. **Utah**: 153 males, 110 females, Cache Co., Benson, Logan, Mendon, Newton, and Smithfield, various dates July 9 and Aug. 4, 1970, July 4, 11, 21, 24, Aug. 7, 12, 14, and Sept. 2, 1971, from 27 different nests of barn swallows; also 2 males, Rich Co., Walton Canyon, July 22, 1970 (T. L. Whitworth), probably nest of pine grosbeak [USNM, USU, and Whitworth Colln.]. **Wyoming**: 20 males, 6 females, Lincoln Co., La Barge, July 3, 1970 (Whitworth), nest of Say's phoebe, and 3 males, Lincoln Sage, Aug. 12, 1971, nest of barn swallows [USNM, USU, and Whitworth Colln.].

Non-paratype Material: Numerous males and females from lots cited under the type series; also 4 males, 4 females, California: Siskiyou Co., Macdoel, July 21, 1948 (W. W. Wirth), swallow nest under bridge [USNM].

Specimens Examined. 557 males, 481 females, a large number with puparia.

Distribution: British Columbia, California, Utah, Wyoming.

Basis of description of immature stages: *Puparium* (9 specimens): Three, La Barge, Lincoln Co., Wyo., Say's phoebe; 6, Benson, Cache Co., Utah, barn swallow.

Hosts. Pine grosbeak (probable), black and Say's phoebes, barn swallow (predominant).

Ecology and Biology. The few hosts suggest that *halli* prefers cup-nesting birds. The overwhelming proportion of records from nests of barn swallow may or may not be significant for this species. The ready accessibility and number of barn swallow nests would undoubtedly tend to swell the numbers, as seen in the work of Whitworth in Utah and Spencer, Dawe and Cannings in British Columbia. For example, Whitworth (1976) recorded that out of 392 nests of that bird species, 121 were infested by *Protocalliphora* and of these 114 were infested by species IV, i.e. *halli* as here described. Barn swallow is of course an excellent host for *Protocalliphora*, and it is attacked by at least ten species of the genus, according to our records, although some of these may be only occasional or adventitious parasites on barn swallow.

Variation. Like the other species in the "*hesperia* complex," *hesperia*, *hesperioides*, and *cuprina*, variation was noted in the posterior bristles on the fore tibia. Of the 703 specimens that could be counted, 628 (89.3%) had only one bristle on each fore tibia, 31 had two on each fore tibia, and 42 had two on one side (right or left) and one on the other. Two specimens lacked a bristle on one side.

A very few specimens showed a slight yellow tint on the 5th tergite, and in isolated specimens this variation would cause confusion with *cuprina*.

A most disturbing variation was in the appearance of the preocellar area in males, which is usually short triangular and gray microtomentose, but which may appear somewhat rugose or granulated and narrowly acuminate, and even rarely a narrow shining line. Yet in the males of the long reared series the combination of brownish calypteres, black-tipped palpus, narrow frons, and narrow parafacial is so consistent that we believe the differences in the preocellar area in this species must be accepted as variation.

The calypteres of most males are not heavily browned, but are usually more brownish yellow, a paleness that we have often noticed in reared series when the specimens have not been allowed to mature fully. Sometimes only the outer rims are really brownish. In any case, they are obviously different from the white calypteres of females, and from white calypteres of males of species so characterized.

Remarks. For a discussion of the difficulties in the "*hesperia* complex," see Remarks under *hesperia*. Briefly, all four species have brownish calypteres in the male, also brownish in females of *hesperia* and *hesperioides*, but *halli* and *cuprina* have white calypteres in the female. Good series of all the species, and especially of the last two, are available to back up the conclusions. Males with brownish calypteres have in the past usually been identified as *hesperia*.

Puparia of *halli* are easily confused with those of *tundrae*, and to a lesser extent, those of *aenea* and *fallisi*. They are separated from the latter two by weak circumstigmatal folds and presence of some mesostigmatal plaques or folds. *Puparia* of *halli* are separated with difficulty from those of *tundrae*, mainly on the basis that the posterior band, although reduced, is present on the anterior 5-6 segments of *tundrae* but absent on *halli*.

Etymology. The specific name, in the genitive case, is dedicated to the late David G. Hall, Jr., whose 1948 book on "The Blowflies of North America" marked for *Protocalliphora* a major advance after the revision by Shannon and Dobroscky (1924).

Protocalliphora (*P.*) *hesperia* Shannon and Dobroscky (Fig. 17, Map 8)

Phormia azurea (Fallén).--Plath, 1919c: 373-8, in part [Puget Sound, Wash.].

Protocalliphora splendida var. *hesperia* Shannon and Dobroscky, 1924: 249, 251-2 [B.C., Wash., the latter part of Plath's "*azurea*"]; Shannon, 1926: 128 [in key].

Apaulina hesperia.--Hall, 1948: 192 [In part].

Protocalliphora hesperia.--Hall, 1965: 926 [sw. Canada, nw. U.S.]; Whitworth, 1976: 27, 36 ff. [Utah, in part].

Diagnosis. Sexes concolorous, monochromatic, shining dark blue to bluish purple; calypteres brownish or brownish yellow in both sexes; surstylus digitate, not curved.

Male. Parafrontal and parafacial silvery microtomentose. Palpus orange-yellow, black at tip. Body metallic dark blue to bluish purple; scutum gray microtomentose, two broad gray stripes demarcating three broad shining stripes, as viewed from behind; abdomen shining in direct view but gray microtomentose when viewed from behind at a low angle, 2nd and 3rd tergites decidedly gray, 4th less so. Calypteres brownish or brownish yellow in both sexes.

Frons narrow, 0.08 times head width (0.08-0.09/24) and 1.15 times ocellar span (0.94-1.33/24); preocellar area almost always gray and narrowly triangular, rarely somewhat shining; frontal bristles decreasing rapidly in length posteriorly, the rows ending well before median ocellus; parafrontal strongly narrowed, prevertical area shining, mostly polished black; parafacial with numerous black hairs in 3-4 irregular rows, width obviously although not greatly wider than breadth of 3rd antennal segment, 1.37 times width of frons (1.20-1.57/24), 1.57 times ocellar span (1.33-1.83/24), and 0.74 times vibrissal interval (0.67-0.79/21); height of cheek 0.41 times maximum length of eye (0.37-0.44/24).

Thorax lacking accessory notopleural bristles and hairs on tympanic membrane and postalar wall, occasionally a few short, pale and inconspicuous hairs on tympanum. Fore tibia typically with one posterior bristle (cf. Variation).

Terminalia: cerci distally acuminate and parallel (as in Fig. 5b); surstylus digitate, very weakly curved, of moderate width (Fig. 17); aedeagus as usual in the genus (as in Fig. 6a).

Female. Concolorous with male, but scutum not as dark, brighter gray microtomentose; parafrontal, parafacial, and preocellar area light yellowish gray microtomentose, parafacials each with changeable spot along upper margin.

Frons about one-fourth head width (0.26x; 0.23-0.28) and 3.19 times ocellar span (2.75-3.57), preocellar area dull, evenly microtomentose; parafacial as in male, width about half that of frons (0.49x; 0.41-0.57), 1.57 times ocellar span (1.36-1.83), and 0.81 times vibrissal interval (0.73-0.92/21); height of cheek 0.43 times maximum length of eye (0.40-0.46).

Apparent normal length of adults, 8.5-9.5 mm.

Immature Stages. A few puparia of *hesperia* are known, but some confusion in records makes it unwise to describe them at this time. The cuticular spines are distinct, and the prothoracic fringe is well developed, although not long.

Type Series. Shannon and Dobroscky described their variety *hesperia* in a brief two-line description, saying only "Type.--Cat.

no. 26860 U.S.N.M." On the following page they stated that "A large series of this variety was reared from a number of species of birds in regions of Seattle, Washington, summer 1918 (O. E. Plath)" and from a male, five females from Ainsworth, B.C., July 11, 1903 (A. N. Caudell). The Seattle and Ainsworth specimens constitute the type series of *hesperia*, with a type number stated, but unfortunately the authors did not specify either the sex or the type locality. Hall (1948) supplied this information, recording the type as the male from Ainsworth, B.C. Arguably Hall's action could be considered lectotype designation if the original publication were considered not to have fixed the type, but the point need not be belabored; under either interpretation the result is the same.

The collection in the U.S. National Museum of Natural History now contains the holotype and the five females from Ainsworth, B.C., one of the females marked allotype, the other females paratypes, and also three males, two females, hand-lettered "Puget Sd./Wash." "OE. Plath," with two of the males and the two females also bearing the USNM red "paratype" labels. These were not mentioned in the Type Catalog but they could have been received later and labeled with the same number. One female with the same labeling, including a red USNM paratype label, is in the E. S. Thomas Collection now at Ohio State University's Department of Entomology, probably an exchange received from Aldrich or Hall.

A larger series, 13 males and 11 females, in the Museum of Comparative Zoology, bears handwritten labels "Puget Sound Wash." (several are also labeled "Plath") but not in the same handwriting as the USNM specimens. They have Shannon's own characteristic identification labels, and 3 males, 2 females also bear red labels "paratype" which were obviously cut out of the USNM label. It seems probable that these were received on loan by Shannon and so labeled by him for return. Almost all of these are not *hesperia* but the closely related species here described as *hesperioides*. No host labels are on the specimens, nor on the Plath material that is true *hesperia*, so it is impossible to sort out the hosts listed by Plath (1919c). A mixture of species in material reared from a variety of hosts is not unexpected.

Specimens Examined. 75 males, 83 females.

Distribution. Western, Alaska, British Columbia to southern California, Utah, and possibly Colorado.

Alaska: 21 males, 26 females, 5.5 km SE Portage (S of Anchorage, at base of Kenai Peninsula), reared June 19-21, 1978, from nest (? gray jay or thrush) collected June 10 (G. E. Haas) [USNM].

Canada: British Columbia: male, 5 females, Ainsworth [type series, USNM]: male, Kamloops, July 31, 1937, emerged Aug. 19 in laboratory (G. J. Spencer) [UBC]; female, Shames, 18 mi SW Terrace, June 23, 1960 (B. Heming), on flowers of *Heracleum lanatum* [CNC]; female, Gagnon Road, 6 mi W Terrace, June 24, 1960 (C. H. Mann) [CNC].

Contiguous United States: California: 14 males, 18 females, Riverside, May 1936 (A. J. Basinger), house finch, nest 19 [CAS, USNM; 3 males were paratypes of *cuprina*]; female, Riverside, May 4, 1934 (A. J. Basinger) [CAS]; male, Lake Co., Clear Lake, July 29, 1948 (W. W. Wirth), swallow nest under bridge [USNM]; 6 males, 9 females, Modoc Co., Adin Bridge, emerged July 22-27, 1978, barn swallow nest (C. Gold) [U. Calif., Albany]; female, Modoc Co., farmhouse by Adin, Aug. 6, 1978, barn swallow nest (C. Gold) [U. Calif., Albany]; 4 males, 2 females, Tulare Co., Farmersville, May 11, 1939 (T. H. G. Aitken), swallow nest [CAS, USNM]. **Oregon**: male, Jackson Co., Ashland Mt. above timberline, Aug. 5, 1952 (H. A. Scullen) [Oregon State U.]. **Utah** (all reared T. L. Whitworth from nests of American robin): female, Cache Co., Malibu Campground, July 17, 1970; 5 males, same locality and date; male, female, same locality and date; two males, one female, Box Elder Co., Mantua-Devil's Gate, June 22, 1969; 3 males, one female, Mantua, July 14, 1970 [USNM, USU, Whitworth Colln.]. **Washington**: 4 males, 3 females, Puget Sound (in part, in type series) (O. E. Plath) [USNM]; 3 males, Puget Sound (O. E. Plath) [MCZ]; 8 males, 10 females, Seattle, July 1918 (O. E. Plath) [CAS]; female, Mt. Rainier, Burroughs Mt., Aug. 21, 1934 (A. L. Melander) [USNM].

A single female, Colorado: Clear Creek Co., Doolittle Ranch, 9800 ft on Mt. Evans, July 18, 1961 (W. R. M. Mason) [CNC] may be *hesperia*, but it is far from the other known distribution.

Hosts: House finch, American robin, barn swallow. The Alaska series was reared from a nest listed as "? gray jay or thrush." Whitworth (1976) recorded *hesperia* from Cassin's finch (puparia only) and chipping sparrow, but the puparia have not been found for checking, and the specimens from chipping sparrow are herein described as *brunneisquama* n. sp.

Variation. One posterior bristle on the fore tibia appears to be typical for *hesperia*, but there is enough variation to cause some trouble. Of the 115 specimens that could be counted, 82 had just one bristle and 19 had two. Fourteen specimens had one on one side and two on the other, either on right or left.

Remarks. Four western species, *hesperia*, *hesperioides*, *halli*, and *cuprina*, may be termed the "*hesperia* complex". All four may occur in the same area, may attack the same or similar host species, have puparia with distinct spines and generally similar habitus, and have males with brownish calypteres, moderate to narrow frons, and digitate surstylus. *P. cuprina* is more distinct than the others because of the female with cupreous 5th tergite, whereas the first three have all blue females, but males of *cuprina* have been confused with the others.

Difficult as it is to separate the first three species, the uniformity in fairly large samples persuades us to recognize three species, but identifications of single specimens or of one sex only will still be hazardous. Good reared series with both sexes represented will be reassuring, especially in associating males with brownish calypteres with a characteristic female. Females of *hesperia* and *hesperioides* have brownish calypteres like the males, but females of *halli* and *cuprina* have white calypteres. (Warning: Very teneral females of *hesperia* and *hesperioides* have pale calypteres, sometimes quite whitish, but still not the real white of species that have truly white calypteres). In addition, of course, *cuprina* has the cupreous 5th tergite that will separate *halli* and *cuprina* easily. Females of *hesperioides* have the preocellar area irregularly polished black, whereas in the other three the preocellar area is dull, evenly and heavily brownish gray microtomentose. Finally, in both sexes, palpi are entirely yellowish to orange-yellow in *hesperioides*, but brownish- to black-tipped in the other three. Males are very similar, but *hesperioides* has a preocellar polished line or low ridge in addition to the yellow palpi.

Because of the similarities, the name *hesperia* has been widely misused. For example, *hesperia* of Guberlet and Hotson (1940) proved to be *braueri*, and *hesperia* of James (1955) and Whitworth (1976) was a mixture of *hesperia*, *hesperioides*, and *halli*.

Protocalliphora (*P.*) *hesperioides*, new species
(Figs. 18, 52, Map 8))

Phormia azurea (Fallén).--Plath, 1919c: 373-8, in part [Puget Sound, Wash.].

Diagnosis. Sexes concolorous, monochromatic, shining dark metallic blue to bluish purple; calypteres brown in both sexes; palpus entirely orange-yellow; frons of male narrow with polished line in preocellar area; surstylus digitate.

Male. Parafrontal and parafacial gray microtomentose, the parafacial tinted with yellow; palpus entirely orange-yellow, not black tipped. Body shining dark metallic blue to bluish purple, scutum dark gray microtomentose, obscurely demarcating three shining stripes, as viewed from behind under the microscope, but appearing entirely dark to the unaided eye; abdomen highly shining in direct view, but tergites 2 and 3 gray microtomentose when viewed from behind at a low angle, 4th tergite slightly so. Calypteres brown in both sexes.

Frons narrow, 0.06 times width of head (0.05-0.07) and 0.83 times ocellar span (0.69-1.00); preocellar area a polished line or low ridge; frontal bristles decreasing rapidly in length posterodorsad, ending just before median ocellus in several short and weak hairs; parafrontal strongly narrowed above to more or less polished prevertical area, on lower half with one row of hairs becoming two or three rows just before merging with parafacial hairs; parafacial rather densely beset with numerous black hairs in about four irregular rows, width of parafacial only slightly greater than breadth of 3rd antennal segment, 1.77 times width of narrow frons (1.50-2.00), 1.47 times ocellar span (1.29-1.75), and 0.68 times vibrissal interval (0.61-0.77); height of cheek 0.37 times maximum length of eye (0.31-0.41); occiput with about two rows of black hairs behind row of postocular setae.

Thorax without accessory notopleural bristles and hairs on tympanic membrane and postalar wall; fore tibia with one posterior bristle; calypteres brown, often quite dark.

Terminalia: cerci distally narrowed to acute, parallel apical sections (cf. Fig. 5b); surstylus digitate and approximately straight, slightly wider toward base (Fig. 18); aedeagus as usual in the genus (cf. Fig. 6a).

Female. Concolorous with male, but scutum more heavily gray microtomentose and more clearly demarcating shining stripes, as viewed from behind under microscope; microtomentum of both parafrontal and parafacial yellowish tinted, parafacial with changeable spot along upper margin. Frons 1/4 head width (0.25x; 0.23-0.27) and 3.08 times ocellar span (2.81-3.57); preocellar area with irregularly shaped, polished black spot; width of parafacial 0.45 times width of frons (0.40-0.50), 1.40 times ocellar span (1.25-1.64), and 0.74 times vibrissal interval (0.64-0.85); height of cheek 0.40 times maximum length of eye (0.37-0.45).

Apparent normal length of adults, 8.5-9.5 mm. One undersized male was only 5.5 mm long.

Third instar. Peritreme closed; numerous spines present between stigmatal plates and on hypostigmatal area. *Prothoracic fringe* (in combination with measurements derived from puparia) 75 μ (55-110)/70. *Cephalopharyngeal skeleton* 920 μ (750-1020)/7 in length.

Puparium. Length 7.3 mm (7.1-7.5)/7; breadth 3.2 mm (2.9-3.3)/7; medium thick-walled, dull, brown. *Posterior region* (Fig. 52a,b): stigmatal plates 180 μ (155-195)/14 in diameter; distance between buttons 460 μ (425-500)/7, and across stigmatal plates 975 μ (900-1050)/7; stigmatal ratio 0.48 (0.40-0.55)/7; stigmatal area outlined in cuticular plaques and pronounced ridges; mesostigmatal spines numerous; mesostigmatal folds and plaques numerous and pronounced, bisecting stigmatal area; hyperstigmatal spines numerous, 30 μ (18-50)/70; hypostigmatal area rugose with numerous irregular folds; hypostigmatal spines numerous, 20 μ (8-45)/70; circumstigmatal folds weak, concentric. *Dorsal cuticle* (Fig. 52c) with numerous spines directed both posteriorly and anteriorly, 27 μ (18-40)/70; cuticular ridges weak. *Ventral cuticle* (Fig. 52d): ventral band ratio 0.75 (0.71-0.83)/14; ventral cuticle with cuticular ridges; anterior band pronounced, spines of anterior patch not markedly differentiated from surrounding spines, 18 μ (8-30)/70; medial band pronounced, spines 25 μ (15-40)/70; posterior band pronounced, present on all segments.

Type Series. Holotype male, allotype, and 44 paratypes (19 males, 25 females), Riverside, Calif., May 1936 (A. J. Basinger), goldfinch nest 18. Type and allotype in the California Academy of Sciences, paratypes there and in the U.S. National Museum of Natural History. Additional paratypes: **British Columbia**: male, Mt. Revelstoke, 5400 ft, July 7, 1952 (G. J. Spencer) [CNC]. **California**: 19 males, 14 females, Riverside, May 1936 (A. J. Basinger), house finch nest 11 [CAS, USNM]; 10 males, 15 females, Riverside, May 1936 (Basinger), house finch nest 5 [CAS, USNM]; 5 males, 3 females, Anaheim, July 1936 (Basinger), house finch nest 23 [CAS]; male, Anaheim, May 12, 1935 (Basinger), bushtit nest [CAS]; 2 males,, 3 females, Blodgett Forest, El Dorado Co., emerged July 23, 1979 (C. Gold), nest of dusky flycatcher [U. Calif., Albany]; 3 males, 4 females, Poison Meadow, Fresno Co., nest of yellow warbler, puparia July 18, emerged July 24, 1988, conifer forest at 6500 ft. (B. E. Valentine, S. Boland) [USNM]; female, Tanbark Flat, Los Angeles Co., July 7, 1950 (P. D. Hurd) [U. Calif., Berkeley; *hesperia* in James 1955]; female, Mill Valley, Marin Co., April 1965 (P. H. Arnaud, Jr.) [CAS]; 3 males, one female, Woodside, San Mateo Co., Aug.

1, 1954 (E. Hunter), reared but host not stated [CAS]; male, San Mateo Co., Corte de Madera Creek, vicinity of Portola, July 18, 1964 (P. H. Arnaud, Jr.) [CAS].

Non-paratype Material (undersized, teneral, or damaged): 5 males, Anaheim, and one female, Woodside, duplicates of type material; 8 males, 12 females, Monterey Co., Hastings Natural History Reservation, Calif., 1969 (N. A. M. Verbeek), nest of western flycatcher [CAS]; 3 males, 2 females, Berkeley, Calif. [MCZ]; female, Santa Paula, Calif., July 23 (Dean Burk) [U. Calif., Berkeley]; 10 males, 11 females, Puget Sound, Wash. (Plath) [MCZ; three bear "paratype" labels under Shannon's identification label as *P. splendida hesperia*].

Specimens Examined. 91 males, 95 females.

Distribution. British Columbia, California, Washington.

Basis of description of immature stages: *Larva, 3rd instar* (4 specimens): Woodside, San Mateo Co., Calif., host unknown, July, 1954. *Puparium* (7 specimens): Monterey, Calif., western flycatcher, 1969.

Hosts. Bushtit, house finch, goldfinch, dusky and western flycatchers, yellow warbler.

Ecology and Biology. Nothing is known beyond the names of a few hosts.

Variation. Although almost all specimens have entirely orange-yellow palpi, two males of the small El Dorado County series and one male of the Riverside Nest 5 series have black-tipped palpi. One female of Nest 5 lacked the typical polished black preocellar area. Both of these variations appear to be the occasional aberrant occurrences that we have come to expect. Another specimen of Nest 5 has a broad parafacial, and this may be a stray *sialia*; its width seems too great to be mere variation.

As with other western species, the posterior bristles on the fore tibia show some variation, in a character used in keys to some extent and even considered a generic character by Hall, for *Apaulina*. Of 123 specimens that could be counted, 92 had only one posterior bristle on both fore tibiae, but 14 specimens had two. Seventeen had two on right or left tibia but not on both.

Remarks. The four species of the so-called *hesperia* complex, *hesperia*, *hesperioides*, *cuprina*, and *halli*, all have brown calypteres in the male, male genitalia essentially the same, with digitate surstylus, and distinctly spined puparia. Males have commonly been identified as *hesperia*. *P. hesperioides* is distinguished from the other three by the entirely orange-yellow palpi, not black tipped except in the three aberrant specimens already noted. Males are further recognized by the polished black line in the preocellar area. Females have brown calypteres like *hesperia*, but unlike *hesperia* the preocellar area has an irregularly shaped, polished black spot. Females of the other two species of the complex, *cuprina* and *halli*, have white calypteres and dull preocellar area, in addition to which *cuprina* has the 5th tergite cupreous in the female.

Puparia of *hesperioides* can be confused with those of *seminuda* but the longer prothoracic fringe and larger ventral band ratio should separate the two species.

Etymology. The specific name is compounded of *hesperia*, with which the new species has been confused, and *-oides*, like, from the Greek *-eides*, like.

Protocalliphora (*P.*) *hirundo* Shannon and Dobroscky (Figs. 14, 38, 53, Map 9)

Protocalliphora splendida hirundo (subsp.) Shannon and Dobroscky, 1924: 249 (as *hirunda*), 251, 252 [Wash.].
P. sp.--Stoner, 1926: 207-8 [Iowa, bank swallow].
P. splendida hirundo.--Shannon, 1926: 128 [In key].
P. splendida.--Woods, 1926: 12-13 [Mass.].
P. hirundo.--Hall, 1948: 196 [Wash., B. C.]; James, 1955: 25 [Calif.]; Boyes, 1961: 557, pl. I [Cytology]; Boyes and van Brink 1965: 543 [Cytology]; Hall, 1965: 926 [Widespread]; Boyes and Shewell, 1975: 442, 446 [Cytology]; Whitworth, 1976: 20ff.[Utah]; Whitworth, 1977: 4933-B.

Diagnosis. Sexes concolorous, monochromatic, shining dark metallic blue to bluish purple; calypteres white; preocellar area in both sexes dull, heavily microtomentose; surstylus short, broad on basal portion.

Male. Head black in ground color; parafrontal and parafacial strikingly bright, brilliant silvery microtomentose, the parafrontal distinctly microtomentose posteriorly at least to level of median ocellus. Thorax shining, metallic dark blue to bluish purple, dorsum somewhat gray microtomentose, the microtomentum weakly demarcating three broad shining stripes but dorsum still appearing shining and unstriped to the unaided eye. Abdomen concolorous with thorax, but usually brighter blue and more shining, tergites 2 to 4 gray microtomentose when viewed from behind at a low angle. Calypteres opaque whitish, outer rims tinged with pale yellow.

Frons (Fig. 38) slightly less than medium width, only slightly narrowed anterior to ocellar tubercle, at narrowest 0.10 times head width (0.09-0.11) and obviously wider than ocellar span (1.37x; 1.21-1.58); frontal vitta about 1.5 times the width of a parafrontal; preocellar area dull, heavily gray microtomentose; parafrontal distinct throughout, with two irregular rows of fine black hairs on lower part, decreasing above to one row and then disappearing before end of frontal bristles; parafacial of medium width, about twice as wide as 3rd antennal segment, wider than frons (1.28x; 1.17-1.50), wider than ocellar span (1.76x; 1.50-1.91), and 0.89 times vibrissal interval (0.82-0.96), with 3-4 irregular rows of fine black hairs; cheek moderately broad, not half maximum length of eye (0.43x; 0.40-0.45).

Thorax without accessory notopleural bristles and without hairs on tympanic membrane, and with only an occasional seta or two on postalar wall. Fore tibia with one posterior bristle.

Terminalia: cerci distally acuminate, parallel (as in Fig. 5b); surstylus relatively short and broad, and characteristically broader on basal two-fifths (Fig. 14); aedeagus as usual in the genus (cf. Fig. 6a).

Female. Concolorous with male, but microtomentum of frontal vitta, parafrontal, and parafacial dark gray, especially the last, with light brownish yellow tinge, upper part of parafacial, opposite lunule, usually with microtomentum arranged so as to produce a dark brown spot viewed from above, changing to a pale spot against dark background when viewed from below; dorsum of thorax duller than in male, heavily gray microtomentose, the three mesoscutal stripes more distinctly demarcated than in male. Abdomen as in male.

Frons at vertex 0.28 times width of head (0.26-0.30) and 3.58 times ocellar span (3.00-4.33), widening slightly anteriorly; frontal vitta toward the sides with a few short weak hairs; preocellar area dull, densely and evenly microtomentose; parafrontal broad throughout, evenly microtomentose, including prevertical area; reclinate orbital bristles present; parafacial broad, slightly over twice width of 3rd antennal segment, 0.54 times width of frons (0.48-0.58), 1.92 times ocellar span (1.62-2.33), and approximately equal to vibrissal interval (1.00x; 0.93-1.15/24); cheek 0.44 times maximum length of eye (0.41-0.47).

Apparent normal length of adults, 8.0-8.5 mm.

Second instar. *Cephalopharyngeal skeleton* 540 μ /2 in length. *Prothoracic fringe* 56 μ (37-75)/25 in length, spines numerous. *Posterior region:* peritreme closed, button distinct; stigmatal plates 70 μ /4 in diameter; hyperstigmatal spines numerous, 13 μ (6-21)/25; hypostigmatal spines numerous, 5 μ (2-8)/25. *Dorsal cuticle* with numerous spines, 11 μ (8-18)/25. *Ventral cuticle:* spines of anterior patch either single or two- or three-pronged, 11 μ (6-14)/25; medial band pronounced, spines 12 μ (8-19)/25; posterior band present.

Third instar. Peritreme closed, spines present between stigmatal plates and in hypostigmatal area. *Prothoracic fringe* 57 μ (21-103)/48. *Cephalopharyngeal skeleton* 850 μ (750-1000)/18 in length.

Puparium. Length 8 mm (7.2-8.9)/15; breadth 3.5 mm (3.1-3.9)/15; medium thick-walled, dull, brown. *Posterior region* (Fig. 53a, b): stigmatal plates 180 μ (150-200)/39 in diameter; distance between buttons 550 μ (430-600)/20, and across stigmatal plates 1230 μ (960-1410)/20; stigmatal ratio 0.46(0.42-0.48)/20; stigmatal area not outlined in plaques; numerous mesostigmatal spines present; mesostigmatal plaques absent, mesostigmatal folds absent or present, if present, only two or three; hyperstigmatal spines numerous, 31 μ (21-41)/100; hypostigmatal area with few, pronounced irregular folds; hypostigmatal spines numerous, 14 μ (6-33)/100; circumstigmatal folds pronounced, irregular, showing little variation. *Dorsal cuticle* (Fig. 53c) with numerous spines, directed both posteriorly and anteriorly, 31 μ (21-41)/100; cuticular ridges pronounced, irregular, few. *Ventral cuticle* (Fig. 53d): ventral band ratio 0.53 (0.40-0.74)/20; spines of anterior patch 17 μ (10-29)/100; medial band pronounced, spines 27 μ (16-37)/100; posterior band present on all segments, but weak.

Type Series. Holotype male and allotype, Stratford, Wash., July 4, 1920 (R. C. Shannon), from cliff swallow nests. Type No. 26859 in the U. S. National Museum of Natural History. Paratypes: 6 males, 2 females, same data as holotype [USNM], one female, same data as holotype [MCZ]. The original description listed the series as 35 males, 19 females, but the above are the only specimens labeled as paratypes that have been found. One male [MCZ], male, 4 females [USNM], one male, one female from Belkin Collection [now in USNM], and 7 males, 3 females from Melander Collection [now USNM] bear the same data but are not labeled as paratypes, possibly an oversight. The ten from the Melander Collection had been labeled *P. chrysorrhoea*, possibly by Melander.

Specimens Examined. 270 males, 253 females.

Distribution. Northern transcontinental and western. Alaska to Quebec, south to California, New Mexico, Iowa, and Massachusetts. April 11 (Weld Co., Colo.) to Dec. 11 (San Luis Obispo Co., Calif.).

Alaska. Male, female, Alaska Highway bridge over Tok River, nest of cliff swallow; female, Chicken, under Taylor Highway bridge, Chicken Creek, nest of cliff swallow; 4 males, 3 females, Tok Junction, nests of various Hirundinidae.

Canada. British Columbia: 47 males, 58 females, numerous puparia, Atlin, Cariboo, Deadman's Creek (near Walhachin), Kamloops, Lac du Bois (near Kamloops), Minnie Lake Ranch House (Nicola Ranges), Mission City, 108 Mile, Toad River Lodge (Mile 422, Alaska Hgwy), and Vernon; best series, 23 males, 28 females, Kamloops, July 12-20, 1934 (G. J. Spencer), from "eave" [i.e., cliff] swallow. **Ontario**: 56 males, 39 females, numerous puparia, Algonquin Park, 1952-1956, chiefly barn and bank swallow nests; 3 males, 4 females, Port Stanley. **Quebec**: female, Knowlton. **Yukon Territory**: male with puparium, Rampart House, nest bank swallow; male, Km 155 on Dempster Highway.

Contiguous United States. California: 87 males, 75 females, numerous puparia, localities in Del Norte, Kern, Los Angeles, Modoc, Riverside, Sacramento, San Joaquin, San Luis Obispo, Santa Barbara, Solano, Ventura, and Yolo Counties; best series, 28 males, 29 females, many puparia, Point Mugu, Ventura Co., June 23, 1949 (J. N. Belkin), from cliff swallow nest. **Colorado**: male, 2 females, Larimer Co., cliff swallow nest; 8 males, 7 females, Weld Co., barn and cliff swallow nests; male, 2 females, Weld Co., nest of yellow-headed blackbird. **Idaho**: 3 males, 2 females, Bonner's Ferry, nest of small bluebirds; 2 males, 2 females, SW of Dickey, Custer Co., swallow nest. **Iowa**: 4 males, 3 females, 8 puparia, 40 larvae, Bluffton, Winneshiek Co., on nestlings and in nest of cliff swallows. **Massachusetts**: Huntington, male from barn swallow (Woods 1926 as *P. splendida*), female from bluebird. **Michigan**: male, 5 females, Midland. **Montana**: male, Granite Co.; 6 males, 11 females, Hamilton, part from bank swallow nest; 6 males, 6 females, Ravalli Co., part from swallow nests. **New Mexico**: 5 males, 5 females, 7 puparia, Rio Arriba Co., Cañones Creek, 2.5 km SW Abique Dam, 6260 ft, Aug. 23, 1971 (N. Wilson), cliff swallow nests [USNM, U. Northern Iowa]. **New York**: male, Syracuse, bank swallow nest (*P. splendida* by Stoner 1936); 3 males, 2 females, Cranberry Lake, St. Lawrence Co., barn swallow.

Oregon: male with puparium, Nestucca River, Tillamook Co., cliff swallow; 2 males, 1 female, Corvallis; 2 males, 3 females, Chiloquin, Klamath Co., purple martin nest. **South Dakota**: 7 puparia, Redig, Harding Co., cliff swallow nest. **Utah** (Whitworth): 4 males, 4 females, with puparia, Randolph, Rich Co., cliff swallow nest; 9 males, 7 females, Franklin Basin, Cache Co., tree swallow nest; 4 males, 6 females, Tony Grove, Cache Co., purple martin nest. **Washington**: 2 females, Almota (syntypes of *sialia*); 4 males, Stratford, cliff swallow nest. **Wyoming**: male, Rock River.

Basis of description of immature stages: *Larvae, 2nd instar* (2 specimens): Algonquin Park, Ont., bank swallow, July 1955. *Third instar* (5 specimens): same data. *Puparium* (20 specimens): Algonquin Park, Ont., barn swallow, August, 1955 (12); Hamilton, Mont., bank swallow, date unknown (3); Ravalli Co., Mont., bank swallow, date unknown (3); Ventura, Calif., "swallow", June, 1949 (2).

Hosts: Yellow-headed blackbird, eastern bluebird, "a bluebird" (Idaho), common grackle, purple martin, eastern phoebe, European starling, and various swallows (bank, barn, cliff, and American tree). Not seen, but recorded by Whitworth (1976) from violet-green swallow. Commonest in swallow nests.

Ecology and Biology. The species appears to be commonest in the West in cliff swallow nests. In Algonquin Park, Ont., it was common in nests (e.g., barn swallow) in man-made structures, and in cavity nests, particularly those in marshes. It is one of several species, in addition to *chrysorrhoea*, that will invade bank swallow nests.

Variation. In some *hirundo* the postocellar bristles (often called postverticals) are well developed, being strong and bristlelike, approximately equal in length and strength to the ocellar bristles, and two to three times the length of the postocular setae. In *Protocalliphora* in general, the postocellars are not at all conspicuous, being less than or subequal to the postocular row of setae. When this character was tabulated for *hirundo*, we found that all western specimens had the short postocellars typical of the genus. In Algonquin Park material, however, over 2/3 of the specimens (both sexes) had strong postocellars. Material from four nests had uniformly long bristles, including one series of 50 specimens from a barn swallow nest. Six nests (total 20 specimens) were uniform for short postocellars. Three nests were mixed: 6 long, 4 short; 1 long, 1 short; and 3 long, 6 short, respectively. The exact nature of this character is unknown, but it is interpreted from present evidence as probably a Mendelian variant, perhaps a mutation that has become or is becoming established in one part of the range of *hirundo*.

Remarks. *P. hirundo* is one of a large group of species that are usually somewhat difficult to distinguish. In males the short broad surstylus (Fig. 14) is quite distinctive, and the combination of silvery parafrontal and parafacial with the evenly gray-microtomentose preocellar area makes the males of this species among the easiest to recognize. Compared with *sialia*, *shannoni*, and others near it, the habitus of the head is outstanding. In the width of the frons and the silvery parafacial, *hirundo* resembles *sialia* but that species has a polished, acuminate preocellar area, as well as longer digitate parallel-sided surstylus. Females of *hirundo* are less distinctive, but the broad parafacial and entirely gray microtomentose preocellar area will usually separate it readily from other species in the *sialia* complex.

Puparia of *hirundo* are similar to those of *avium*, *sialia* and *chrysorrhoea*. They can be generally separated by the fewer circumstigmatal folds (which are quite characteristic of this species) and their irregular pattern.

Protocalliphora (*P.*) *interrupta*, new species
(Figs. 30, 54, Map 7)

Protocalliphora sp. V.--Whitworth, 1976: 15ff. [Utah].

Diagnosis. Sexes concolorous, monochromatic, shining metallic bluish to bluish purple; calypteres white; surstylus long and slender, strongly curved; puparium with minute spines, appearance like fine sandpaper.

Male. Body shining metallic dark bluish to bluish purple, scutum with gray microtomentum in two broad stripes that demarcate three broad shining areas, as viewed from behind; abdomen brighter than thorax, shining in direct view but tergites 2 and 3 distinctly and tergite 4 thinly gray microtomentose when viewed from behind at a low angle. Calypteres white.

Frons narrow, at narrowest averaging 0.09 times head width (0.075-0.11) and 1.18 times ocellar span (1.00-1.40); preocellar area narrowly triangular, usually rugose and shining, rarely dull gray microtomentose; parafrontal weakly haired on lower 1/3 to 2/5, narrowed above, narrowly microtomentose to or above level of median ocellus; parafacial silvery gray microtomentose, sparsely haired in 2-3 irregular rows, width slightly greater than breadth of 3rd antennal segment, 1.31 times width of frons (1.00-1.50), 1.54 times ocellar span (1.23-1.80), and 0.70 times vibrissal interval (0.53-0.80/23); height of cheek 0.41 times maximum length of eye (0.36-0.415); occiput chiefly white haired, only 1-2 irregular rows of black hairs behind and below postocular setae.

Thorax without accessory notopleural bristles and with none to few weak hairs on tympanic membrane and postalar wall. Fore tibia usually with one posterior bristle.

Terminalia: cerci distally long and narrow to acute apices (cf. Fig. 5b); surstylus long, exceptionally slender on distal three-fifths, strongly curved (Fig. 30); aedeagus as usual in the genus (as in Fig. 6a).

Female. Concolorous with male, but scutum somewhat more heavily gray microtomentose, and shining stripes consequently more clearly demarcated.

Frons of moderate width, at vertex 0.29 times head width (0.26-0.31) and 3.40 times ocellar span (3.14-3.92); preocellar area entirely dull, heavily microtomentose; microtomentum of parafrontal and parafacial brownish gray or yellowish brown tinted; parafrontal broad and gray up to vertex; parafacial sparsely beset with scattered hairs, definite changeable spot usually present along upper margin, width greater than breadth of 3rd antennal segment, 0.46 times width of frons (0.43-0.52), 1.58 times ocellar span (1.43-1.83) and 0.78 times vibrissal interval (0.69-0.85); cheek height 0.42 times maximum length of eye (0.39-0.43).

Apparent normal length of adults, 7.5-8 mm.

Third instar. No specimens examined, characters derived from the puparium. *Prothoracic fringe* 45 (24-80)/50. *Cephalopharyngeal skeleton* 810 (750-925)/7 in length.

Puparium. Length 6.8 mm (6.8-6.8)/5; breadth 2.9 mm (2.8-3.1)/7; medium thick-walled, shiny, light brown. *Posterior region* (Fig. 54a): stigmatal plates 150 (150-150)/14 in diameter; distance between buttons 560 (500-600)/7, and across the stigmatal plates 1060 (950-1200)/7; stigmatal ratio 0.53 (0.46-0.57)/7; stigmatal plate faintly outlined in tubercles; mesostigmatal spines few, represented primarily by small tubercles; mesostigmatal folds absent, a few plaques present; hypostigmatal spines scarce, small and many appearing as small tubercles, 4 (3-5)/70; hypostigmatal area with a few plaques but folds absent; circumstigmatal folds absent. *Dorsal cuticle* (Fig. 54b) with few spines, 6 (4-8)/70; cuticular ridges weak or absent. *Ventral cuticle* (Fig. 54c, d): ventral band ratio 0.27 (0.16-0.38)/14; anterior band represented only by anterior patch, which bears the largest spines of the puparium (except the prothoracic fringe), 10 (8-12)/70; medial band present but reduced, thinned on the midline and frequently interrupted; spines small and usually represented as spine bases only, 4 (2-6)/70; posterior band vestigial or, more usually, absent.

Type Series. Holotype male, allotype, and 27 paratypes (10 males, 17 females), *Utah*: Cache Co., Benson, June 15, 1971 (T. L. Whitworth), Nest 830, yellow-headed blackbird. Type, allotype, and some paratypes in the U.S. National Museum of Natural History, paratypes in Utah State University and Whitworth Colln. Additional paratypes: **Canada, British Columbia**: 8 males, 1 female, New Westminster, June 4, 1958 (Cox), nest of song sparrow; flies reared at Vancouver, June 4-20, 1958 by G. J. Spencer [UBC]. **U.S.: Utah** (all collected by T. L. Whitworth, nests of yellow-headed blackbird except as noted) [USNM, USU, and Whitworth Colln.]: 86 males, 70 females, Benson, June 18, 1970 and June 15 and July 4, 1971; 65 males, 65 females, Mendon, June 16, 1970 and July 11, 1971; 9 males, 7 females, Smithfield, June 5, 1971; 11 males, 12 females, Wellsville, June 16, 1970 (1 male and 2 females from nest of red-winged blackbird).

Non-paratype Material: Five males, five females from Utah and a male from British Columbia have not been included because of poor condition or unusually small size, but the data on these are duplicated in the type seres. There is also a series of 2 males, 25 females, Bear River Refuge, Brigham City, Utah, various dates April 29-June 26, 1965-1968, in trap (K. J. Capelle) [Wash. State U.], which are undoubtedly *interrupta*, but they are not reared and thus lack associated puparia.

A male from California: San Mateo Co., Corte de Madera Creek near Portola, June 17, 1954 (P. H. Arnaud, Jr.) [CAS] appears to be *interrupta*, but as a lone specimen, not reared, it is not included in the type series. Three males, one female, Ontario: Ogoki, Aug. 18, 1952 (J. B. Wallis), and one male, Quebec: La Verendrye Provincial Park, Mi 61, Route 58, Aug. 20, 1965 (D. M. Wood) [CNC] appear to be *interrupta*, although with some suggestion of *spenceri*. For either species, this would be a great eastward extension of known range. Rearing will be necessary to discover the puparial characters that would be decisive.

Specimens Examined. 199 males, 203 females.

Distribution: Utah (many specimens), British Columbia, probably California, and possibly Ontario and Quebec.

Basis of description of immature stages: *Puparium* (7 specimens): Mendon, Utah, yellow-headed blackbird. All from type series.

Hosts. Chiefly yellow-headed blackbird; also Brewer's and red-winged blackbirds, song sparrow, marsh wren (Specimens from Brewer's blackbird and the wren from Whitworth identifications).

Ecology and Biology. Whitworth (1976) noted that this species "was only found in marsh habitats where it commonly infested nests of yellow-headed blackbirds." He reared it also, but in small numbers, from Brewer's and red-winged blackbirds and the long-billed marsh wren. On the other hand, Spencer at Vancouver, B.C., reared it from a song sparrow nest, so it may also parasitize low-nesting birds that are not marsh-nesting.

Variation. Ten males and 2 females from Utah have two posterior bristles on the fore tibia, and 6 males, 4 females have 2 bristles on one side, either left or right, but one on the other.

Remarks. The long series available makes it possible to say with some assurance that this is a small species compared with some others to which it comes close in the keys because of various character states. Even in this small species, there are occasional really small underfed specimens, the smallest only 5 and 5.5 mm in length.

The most remarkable feature is the minutely spined puparium, which resembles that of *metallica*, a dichromatic and chiefly eastern species. In the general key, the fact that males and females of *interrupta* are concolorous takes it far from *metallica*, but in the key to males the similarity of male surstyli brings the two closer together. In some respects, males of *interrupta* key out near *parorum*, but that species is consistently much larger and differs in other respects as well, besides being a parasite of chickadees. Females of *interrupta* are not distinctive and end up with several other species near the end of the key.

The puparium differs from that of *metallica* in having the hindmost of the usual three bands of spines lacking on the ventral surface of the anterior end. Thus the absence of the third band results in a strongly banded appearance of alternating spine bands and bare areas. This is especially noticeable on about three segments posterior to the cap. These segments have the full three bands of spines in *metallica*. The medial band is present in *interrupta* but thin and often interrupted on the midline.

Puparia of *interrupta* can be confused with those of *aenea*, *metallica*, *spenceri*, and *tundrae*. However, the small spines, lack of cuticular ridges and the interrupted medial band serve to readily distinguish it. The "sandpaper-like" puparium also helps to distinguish *interrupta* from other species, such as *fallisi*, *parorum*, and *spenceri*, with narrow curved surstylus and virtually indistinguishable females.

Etymology. The specific name comes directly from the Latin participle *interrupta*, referring to the break in the usual bands.

Protocalliphora (*P.*) *lata*, new species
(Figs. 10, 33, 55, Map 10)

Protocalliphora "*gigantia*" Whitworth, 1976: 17, 18, 27, 35, 37 (as "*gigantica*") [Nomina nuda; Utah, especially on Falconiformes].

Diagnosis. Sexes concolorous, monochromatic, shining dark metallic blue to bluish purple; calypteres dark brown in both sexes; both sexes dichoptic, frons as broad as an eye; surstylus digitate but relatively short and therefore appearing rather broad (Fig. 10); unusually large species, normal length 17-19 mm, the largest Nearctic species of the genus.

Male. Head black in ground color; parafrontal and parafacial yellowish to brownish yellow microtomentose. Body shining, dark metallic blue to bluish purple; scutum thinly dark gray microtomentose, barely interrupting the shine, the usual mesoscutal stripes indistinct. Abdomen more highly shining than thorax, with only a thin grayish "bloom" (microtomentum) visible from behind viewed at a low angle. Calypteres dark brown to brownish black.

Head unusual in genus in having broad frons (Fig. 33), as in female; frons approximately as wide as an eye, 1/3 width of head (0.34x; 0.32-0.35/8) and over four times as wide as ocellar span (4.15x; 3.74-4.60/8); frontal vitta broad, with some bristlelike setae along each side, just mesad of frontal rows of bristles; large, polished black preocellar area; parafrontal broad, anteriorly as broad as parafacial, narrowing strongly posteriorly, microtomentose anteriorly but narrowed prevertical area shining black; one or two pairs of proclinate orbital bristles and a pair of reclinate to lateroclinate upper orbitals; parafacial extraordinarily broad, width over three times that of 3rd antennal segment, nearly 2/3 as broad as frons (0.645x; 0.595-0.66/8), 2.67 times ocellar span (2.27-3.00/8), and 1.83 times vibrissal interval (1.47-2.00/8); cheek broad, 0.635 times maximum length of eye (0.61-0.67/8).

Thorax without accessory notopleural bristles; tympanic membrane nearly bare, with only two or three weak hairs; postalar wall centrally with only two or three fine hairs, occasionally up to six. Fore tibia with two posterior bristles; claws strong, abruptly curved at right angles about midway.

Terminalia: cerci short, stout, straight (as in Fig. 5b); surstylus moderately broad, wider at base and broadly rounded distally (Fig. 10), the breadth making it seem intermediate between the long digitate surstylus of *sialia* and the very broad surstylus of *avium*; aedeagus as usual for the genus (as in Fig. 6a).

Female. Color as in male. Frons broad, at vertex 0.35 times width of head (0.32-0.38/20), and 4.63 times ocellar span (4.30-5.20/20), strongly widening anteriorly and projecting well anterior to eyes; preocellar area large, polished black; frontal vitta with a few fine hairs and some bristlelike setae on each side mesad of frontal bristles; parafrontal broad, with numerous coarse black hairs in from four to five irregular rows opposite the lowest frontal bristles up to two rows at the reclinate orbital bristles; prevertical area chiefly polished black; two proclinate and one reclinate orbitals; parafacial very broad, with numerous black hairs in about six irregular rows, its width 0.61 times width of frons (0.57-0.66/20), 2.84 times ocellar span (2.60-3.12/20) and 1.90 times vibrissal interval (1.53-2.23/20), the interval appearing unusually narrow for *Protocalliphora* because of the wide parafacial; cheek broad, its height 0.62 times maximum length of eye (0.60-0.66/20).

Thorax as in male. Fore tibia likewise with two posterior bristles.

Apparent normal length of adults, 11-12 mm.

Third instar. Peritreme closed, spines present between stigmatal plates, numerous in hyperstigmatal area. *Prothoracic fringe* 74 μ (31-130)/30. *Cephalopharyngeal skeleton* 1085 μ (880-1275)/11 (some measurements derived from puparia).

Puparium. Length 13 mm (on one); breadth 5.1 mm (4.7-5.5)/10; thick-walled, dull, black-brown. *Posterior region* (Fig. 55a, b): stigmatal plates 285 μ (250-325)/24; distance between buttons 745 μ (425-1075)/12, and distance across stigmatal plates 1700 μ (1250-2125)/12; stigmatal ratio 0.44 (0.34-0.51)/12; stigmatal area outlined in pronounced cuticular plaques; mesostigmatal spines absent or very scarce; mesostigmatal folds absent; hyperstigmatal spines pronounced, 37 μ (24-64)/120; hypostigmatal area with few, weak spines, 24 μ (16-48)/120; hypostigmatal folds absent or at best, weak; circumstigmatal folds pronounced, irregular. *Dorsal cuticle* (Fig. 55c) with numerous large spines, 45 μ (24-64)/120; cuticular ridges prominent, usually with two rows of spines associated with each ridge, one directed anteriorly and the other posteriorly. *Ventral cuticle* (Fig. 55d): ventral band ratio 0.63 (0.55-0.73)/25; spines of anterior patch slender, 24 μ (8-40)/120; medial band present, spines 40 μ (16-64)/120; posterior band reduced to one to three rows of small spines.

Type Series. Holotype male, allotype, and 6 male, 7 female paratypes, **Utah**: Smithfield, Cache Co., June 28, 1970 (T. L. Whitworth; nest 208), nest of American kestrel [sparrow hawk]. Type in the U.S. National Museum of Natural History. Additional paratypes: **Utah** [USNM, USU, Whitworth Colln.]: 2 males, 11 females, Mendon, Cache Co., July 20, 1970 (T. L. Whitworth, nest 380), nest of American kestrel; male, Bunch Grass Creek, Cache Co., July 6, 1970 (T. L. Whitworth, nest 340), nest of red-tailed hawk. **British Columbia**: male, Vavenby, Mar. 26, 1934 (V. K. Moilliet) [Brit. Mus. (Nat. Hist.)]. **Oregon**: female, Mt. Hood, 3000-6000 ft, June 26, 1925 (E. C. Van Dyke) [CAS]. **California**: female, Summit Camp, Lassen Co., June 28, 1949 (E. L. Atkinson) [U. Calif., Berkeley]. Almost all the Utah specimens were accompanied by puparia.

Non-paratype Material. 4 males, 7 females, Idaho: Emigration Canyon, Franklin Co., June 28, 1970, nest of red-shafted flicker [USU]. These specimens seemed so much smaller than the "giant" ones that the possibility of still another species had to be considered. The same fundamental features--male terminalia, broad frons and parafacial in both sexes, male with orbital bristles--were present although some proportions differed, as often occurs in smaller specimens. We finally concluded that these were conspecific with the type series, merely undersized specimens, probably underfed, a not uncommon occurrence in *Protocalliphora*. Also a headless male, Logan, Utah, July 7, 1951 (collector unknown) [USU]. Also four empty puparia, so large and heavily ridged that they are certainly the puparia of *lata*: Ravalli Co., Mont., Aug. 14, 1966 (Wm. L. Jellison), from old nest of American kestrel [USNM].

Specimens Examined. 16 males, 28 females.

Distribution. Western, known at present from British Columbia, California, Idaho, Oregon, and Utah, and probably Montana (puparia only).

Basis of description of immature stages: *Larva, 3rd instar* (3 specimens): Pasadena, Calif., brown [California] towhee, June, 1952. *Puparium* (10 specimens): Smithfield, Cache Co., Utah, American kestrel (4); Mendon, Cache Co., Utah, same host (6).

Hosts. Red-shafted flicker, red-tailed hawk, American kestrel [sparrow hawk], brown towhee. It has also been recorded from the northern goshawk by Whitworth (1976).

Ecology and Biology. Unknown, other than the names of a few hosts, which suggest--as does its size-- that the species primarily attacks large birds. Whitworth (1976) suggested that it attacked especially Falconiformes, which seems likely although at this stage the records are few.

Variation. The size difference has already been noted, but this is of common occurrence in *Protocalliphora.* Whitworth reported (personal communication) that the small specimens "were removed from a flicker nest as first and second instars and reared on a young Brewer's blackbird." Most specimens of this series are 7-7.5 mm. in length, far below the normal length of the species. Two males were 9 mm., still smaller than the normal length.

Proclinate orbital bristles are present on one or both sides in all males, but the numbers vary: 2 left and 2 right in seven males, 2-1 or 1-2 in three males, 1-1 in three, and 1-0 in one. Females show the full two pairs in 27 of the 28 available, but only one pair in the female from Mt. Hood, Oregon. Two posterior bristles on each fore tibia is typical of the species, but there is only one bristle on either left or right tibia in 2 of 15 males and 5 of 28 females.

Remarks. The headless male from Logan, Utah, was the first one seen, and it was recognized at once as the largest Nearctic *Protocalliphora* seen up to that time, with male terminalia that also marked it as a distinct species. At the same time, two slightly shorter females from Oregon and California were being recognized as a distinct species because of their unusually wide parafacials. Later, when Whitworth reared the type series of *lata* in Utah, the headless male and the two females were at once recognized as *lata*, and happily the species could be recorded from a more extended range.

The structure of the head is unique among the known Nearctic species, with the unusually broad parafacials, frons broad in both sexes, and males with proclinate and reclinate orbital bristles so that they appear at first glance to be females. The dark brown calypteres in both sexes and the two posterior bristles on the fore tibia, the latter coupled with the virtual absence of hairs on the tympanic membrane and postalar wall, also distinguish it from other species.

The specimens in the type series are by far the largest of the known Nearctic species of *Protocalliphora*. The unusual size and general appearance remind one at once of the large northern blow fly *Cynomya mortuorum* (L.), which is a calliphorine and not a phormiine calliphorid.

Puparia of *lata* are separated from those of *avium*, *hirundo*, and *sialia*, which they broadly resemble, by the generally larger size (*lata* is the largest of the Nearctic *Protocalliphora*). This species is separated from both *avium* and *hirundo* by the generally larger cuticular spines and the longer prothoracic fringe. It is separated from *sialia*, which *lata* closely resembles in most respects, by the fact that the prothoracic fringe of *sialia* is nearly twice as long. All four species have marked cuticular ridges with two rows of spines associated with each ridge, one row pointed posteriorly, the other anteriorly.

Etymology. The specific name is derived from the Latin *latus*, broad, referring to the unusually broad frons of the male and the broad parafacial of both sexes.

Protocalliphora (*P.*) *metallica* (Townsend)
(Figs. 21, 39, 56, Map 11)

Protocalliphora splendida (Macquart).--Harbeck, 1907: 216 [N. J., not rechecked; apparently first reference of *splendida* to *Protocalliphora;* probably *metallica* but *bicolor* is possible].

Protocalliphora azurea (Fallén).--Gibson, 1917: 156 [Ottawa, Ont.].*Phormia metallica* Townsend, 1919: 379 [N.H., Va., Ill., Wis., Wash.; females only in series; type locality, N.H.].

Phormia metallica.--Plath, 1919c: 373-8 [Wash.; ?possibly *cuprina*, only females available].

Protocalliphora splendida Macquart.--Shannon and Dobroscky, 1924: 250 [Va. (part from type series of *metallica*); first recognition of dichromatism].

P. splendida.--Johnson, 1925a: 216, in part [New England]; Dobroscky, 1925: 274-5 [N.Y.]; Shannon, 1926: 128 [In key]; Johannsen, 1928: 827 [N.Y.]; Johnson, 1932: 26 [Mass.]; Petch and Maltais, 1932: 79 [Que.].

P. sp.--Thomas, 1936: 46, in part [Ohio].

P. splendida.--Judd, 1944: 146 [Que.]; Mason, 1944: 232-247 [Mass., mixture of *metallica*, *sialia*, *shannoni*].

Apaulina metallica.--Hall, 1948: 196-200 [Revision]; George and Mitchell, 1948: 550, in part [N.Y].
Protocalliphora metallica.--Nolan, 1955: 55-60 [Ind.]; Stabler, 1959: 48 [Colo.]; Nolan, 1959: 355 [Ind.]; Boyes, 1961: 552ff. [Cytology]; Boyes and van Brink, 1965: 543 [Cytology]; Hall, 1965: 926 [Catalog, widespread]; Stiner, 1969b: 1205-6 [Del.]; Boyes and Shewell, 1975: 446 [Mention of cytology].
P. sp. III.--Whitworth, 1976: 16ff.

Diagnosis. Dichromatic, the thorax and abdomen in male shining metallic dark blue to bluish black, in female heavily gray microtomentose and dull bluish to bluish green, with tergite 5 of abdomen brightly cupreous to cupreous green; surstylus of male long, curved and parallel-sided (Fig. 21).

Male. Parafacial and parafrontal silvery gray, often with a very slight brownish tint. Body highly shining, dark blue to bluish black, the thorax and abdomen concolorous, thorax darker than abdomen and with three broad blackish blue stripes weakly demarcated by the thin gray microtomentum, as viewed from behind. Abdomen with thin gray microtomentum when viewed from behind at a low angle. Calypteres opaque whitish, outer rims at most tinted with yellow.

Frons narrow (Fig. 39), narrowing slightly anterior to ocellar tubercle, at its narrowest 0.08 times head width (0.06-0.09) and slightly broader than ocellar span (1.11x; 0.85-1.44); parafrontal strongly narrowed posteriorly, the prevertical area polished; one row of parafrontal hairs that merge into the 2-3 irregular rows of fine black parafacial hairs; parafacial obviously much wider than width of 3rd antennal segment, 1 1/2 times width of frons (1.51x; 1.33-2.00), 1.68 times ocellar span (1.43-2.00), and about 3/4 vibrissal interval (0.78x; 0.70-0.86); ocellar tubercle continued forward as a sublinear acuminate polished area for a short distance anterior to median ocellus, at least to level of uppermost frontal bristles; cheek comparatively narrow, its height slightly over 1/3 greatest length of eye (0.36x; 0.33-0.39).

Thorax without accessory notopleural bristles; tympanic membrane and postalar wall usually bare. Fore tibia typically with one posterior bristle, seldom with two and then usually only on left or right, rarely on both.

Terminalia characterized by relatively long, moderately narrow, curved but parallel-sided surstylus (Fig. 21); cerci distally acuminate and parallel (as in Fig. 5b); aedeagus as usual for the genus (as in Fig. 6a).

Female. Thorax and abdomen except 5th tergite blue to bluish green, dull to weakly shining, heavily gray microtomentose, 5th tergite bright shining cupreous or cupreous green; scutum viewed from behind with three narrow shining stripes strongly demarcated by gray microtomentum; abdomen with tergites 2 to 4 heavily gray microtomentose, 5th tergite very thinly so and highly shining, thus adding to the contrast in color; in extremely teneral specimens the 5th tergite shining greenish with slightly coppery reflection; parafrontal and parafacial brownish gray microtomentose, upper half of latter, viewed from directly above, appears to have a subtriangular dark brown spot, the spot changing to bright gray or yellowish gray when viewed from below; preocellar area dull or only subshining, heavily microtomentose.

Frons at vertex over 1/4 width of head (0.27x; 0.25-0.29) and 3.58 times ocellar span (3.00-4.17); frontal vitta with few and inconspicuous hairs, usually a narrow polished area adjoining anterior margin of median ocellus, occasionally more extensively polished; parafrontal moderately wide throughout though narrowing slightly above, with one row of fine black hairs, merging below with 2-3 irregular rows on the parafacial; parafacial width almost twice width of 3rd antennal segment, 0.45 times width of frons (0.40-0.53), 1.59 times ocellar span (1.43-1.91), and 0.82 times vibrissal interval (0.72-0.92); cheek comparatively narrow, its height 0.36 times maximum length of eye (0.32-0.39).

Apparent normal length of adults, 8-9 mm. Specimens as small as 5.5 mm have been seen.

Egg. Length 1000 μ (956-1020)/14; breadth 310 μ (275-380)/14; cuticle white in color, with small ridges.

First instar. Length up to three millimeters. *Cephalopharyngeal skeleton* 215 μ in length. *Prothoracic fringe* absent. *Posterior region:* stigmatal plates 18 μ (16-20)/4 in diameter; peritreme absent; mesostigmatal spines absent; hyperstigmatal spines 3 μ (1-5)/20; anal lobes small. *Dorsal cuticle* with numerous spines 6 μ (2-10)/10. *Ventral cuticle:* spines of anterior patch typically two-pronged, 2 μ (1-6)/20; medial band present, spines 3 μ (2-6)/20; posterior band present.

Second instar. *Cephalopharyngeal skeleton* 428 μ (400-456)/4 in length. *Prothoracic fringe* 8 μ (4-12)/50. *Posterior region:* peritreme closed; stigmatal plates 77 μ (69-86)/9 in diameter; button distinct; mesostigmatal spines present; both hyperstigmatal and hypostigmatal spines numerous, as tubercles, 3 μ (2-5)/50. *Dorsal cuticle* with numerous spines,

14 μ (4-10)/50. *Ventral cuticle:* spines of anterior patch 7 μ (6-8)/50; medial band pronounced, spines 9 μ (2-10)/50; posterior band present.

Third instar. Peritreme variable, usually weakly closed; cuticular spines as tubercles, numerous between stigmatal plates and in hypostigmatal area. *Prothoracic fringe* 37 μ (25-49)/100. *Cephalopharyngeal skeleton* 730 μ (610-845)/78 in length.

Puparium. Length 7.3 mm (4.8-9.2)/90; breadth 3.3 mm (2.0-4.2)/90; thin-walled, shiny, semi-translucent, brown. *Posterior region* (Fig. 56a,b): stigmatal plates 160 μ (130-195)/192 in diameter; distance between buttons 560 μ (300-755)/96, and across stigmatal plates 1185 μ (690-1470)/96; stigmatal ratio 0.47 (0.41-0.48)/96; stigmatal area outlined in large plaques; mesostigmatal spines, as tubercles, scarce, but present; mesostigmatal folds absent; mesostigmatal plaques present; hyperstigmatal spines as small tubercles, 4 μ (1-7)/100; hypostigmatal area highly variable, ranging from folds absent (Fig. 56a) to folds numerous and prominent (Fig. 56b); hypostigmatal spines numerous, as tubercles, 3 μ (1-7)/100; circumstigmatal folds highly variable, ranging from concentric (Fig. 56a) to rectangular (Fig. 56b) and from pronounced to virtually absent. *Dorsal cuticle* (Fig. 56c) with numerous minute tubercles, 6 μ (4-14)/100; cuticular ridges weak (in eastern forms), to virtually absent (in western forms). *Ventral cuticle* (Fig. 56d): ventral band ratio 0.62(0.42-0.66)/100; spines of anterior patch 7 μ (2-12)/100; medial band pronounced, spines 6 μ (2-12)/100; posterior band pronounced.

Type Series. Holotype female, Franconia, N.H. (Mrs. Slosson) [USNM], identified by Coquillett as *P. splendida*. This is the only specimen of the type series that is marked in the collection, although the species was based on a series of 28 females, "eight pinned from New Hampshire, Virginia, Illinois, Wisconsin," and twenty (10 in alcohol) from Puget Sound, Wash., reared by O. E. Plath. Undoubtedly the old examples of *metallica* from Milwaukee, Wis., Carlinville, Ill., and Rosslyn, Va., were part of the original series, and they have been labeled as probable paratypes. The three New Hampshire and three of the four Puget Sound specimens cited by Townsend as a color form were used by Shannon and Dobroscky (1924) as the type series of *aenea* (N.H.) and *hirudo cuprea* (Wash.), respectively. One Puget Sound specimen [USNM] and one labeled "ex nest Western Robin" "Seattle, Wn., O. E. Plath"/*Protocalliphora splendida* Macq." [det. Shannon] [Cornell U.] are *metallica* but there is no indication whether these were part of the type series.

Specimens Examined. 469 males, 417 females.

Distribution. Widespread, British Columbia to Quebec and Maine, south to California, Colorado, Iowa, and Virgina. April 19 (Conn.) and 21 (Iowa) to Oct. 10 (Ottawa, Ont.), dates of hand-collected specimens.

Canada: Alberta (Banff), **British Columbia** (Abbotsford, Kamloops, Oliver), **Manitoba** (Ninette near Brandon), **Ontario** (Algonquin Park, Dundas, Low Bush (Abitibi), North Bay, Osgoode, Ottawa, Toronto), **Quebec** (Armstrong, Aylmer, Kirk's Ferry, La Trappe, La Verendrye Provincial Park, Montreal, Parkins Mills, Poltimore).

Contiguous United States: Alabama (Auburn, larvae and puparia only), **California** (Devil's Basin at 8200 ft, Mt. Shasta, Yosemite Park, and Mariposa, Plumas, San Bernardino, and Trinity Counties), **Colorado** (Colorado Springs), **Connecticut** (Colebrook, Milford, New Haven, Stafford), Delaware (Newark), **Idaho** (Moscow, Pocatello Valley in Oneida Co.), **Illinois** (Carlinville, Downer's Grove in DuPage Co., St. Joseph in Champaign Co., Rockford), **Indiana** (Bloomington, Lafayette), **Iowa** (Ames, Des Moines, Lodges State Park in Boone Co.), **Maine** (Stetson in Penobscot Co., Southwest Harbor), **Maryland** (College Park, Cumberland), **Massachusetts** (Bristol Co., Brookline, Forest Hills, Groton, Horse Neck Beach, Martha's Vineyard, North Andover, Swampscott), **Michigan** (Bath, East Lansing, Midland, Owosso, and Pentwater, and Bay, Hillsdale, and Livingston Counties), **Minnesota** (Ft. Snelling, Grand Rapids, and Hennepin, Ramsey, and St. Louis Counties), **Nebraska** (Omaha), **New Hampshire** (Coos Co., Durham, Franconia, Peterboro), **New Jersey** (Clementon, Mountain Lakes, West Orange), **New York** (Ithaca, and Erie, Essex, Franklin, Genesee, and Tompkins Counties, also Babylon, Roslyn, and Mastic on Long Island and Watchogue on Staten Island), **Ohio** (Columbus, Lakewood, Sandusky, and Wadsworth, and Franklin and Union Counties), **Oregon** (Coos Bay, Marshfield, Mt. Hood), **Pennsylvania** (Butler, Pittsburgh, State College, Wind Gap), **South Dakota** (no locality), **Utah** (Benson in Cache Co., Brighton, Box Elder Co.), **Vermont** (Manchester), **Washington** (Asotin Co., Mt. Rainier, Puget Sound, Spokane), **West Virginia** (Cranberry Glades), and **Wisconsin** (Milwaukee).

Basis of description of immature stages: *Egg* (14 specimens): Algonquin Park, Ont., from caged female, July 1955. *Larvae, 1st instar* (2 specimens): Algonquin Park, Ont., red-winged blackbird, June, 1954. *Second instar* (5 specimens): same data. *Third instar* (20 specimens): Algonquin Park, Ont., red-winged blackbird, June, 1955 (7); yellow warbler, June, 1955 (3); song sparrow, July, 1955 (3); slate-colored junco, July, 1954 (7). *Puparium* (96 specimens): Algonquin Park, Ont., red-winged blackbird, June, 1955 (27); common grackle, June, 1955 (26); yellow warbler, July, 1955 (10); song sparrow, August, 1955 (10); gray catbird, July, 1955 (13); swamp sparrow, July, 1953 (10).

Hosts (Eastern). Red-winged blackbird, indigo bunting, northern cardinal, gray catbird, yellow-breasted chat, alder flycatcher, common grackle, slate-colored junco, American robin, a number of sparrows (field, grasshopper, savannah, song, swamp, vesper, and white-throated), American tree swallow, brown thrasher, rufous-sided towhee, prairie and yellow warblers; also, from puparia alone, black-billed cuckoo, American goldfinch, eastern kingbird, chipping sparrow, bank and northern rough-winged swallows, scarlet tanager, Nashville warbler, northern waterthrush, a wren (Alabama), and common yellowthroat.

Hosts (Western). Brewer's and red-winged blackbirds, lazuli bunting, blue grosbeak, fox sparrow, and an unidentified sparrow (Utah).

For the scarlet tanager record, no specimens were seen but the words of Dobroscky (1925) assure us that the species was *metallica.* She described the spines around the stigmatal area of puparia as "so minute that they seemed like mere stipples." Incidentally this is the only record known to us of *Protocalliphora* from that host.

The statement by Van Tyne and Berger (1976) that "54 species of birds are known to be parasitized by *P. metallica* is erroneous, apparently stemming from confusion with the list of 54 species and subspecies cited by Hall (1948) as known hosts at that time of *all* species of *Protocalliphora* in North America. At present we have records of 35 species of hosts of *metallica.*

Ecology and Biology. *P. metallica* primarily infests nests close to the ground or at shrub level, particularly in open fields or marshes, and including cavity nests close to the ground. The biology of this common species has been investigated in some detail by Bennett at Algonquin Park, Ont. (Bennett 1957).

This is the species most often collected by hand, whereas most species have been taken for the most part by rearing from larvae or puparia found in birds' nests. It is also most often collected at flowers, especially at goldenrod in the autumn and fall. Both of these points may rest on the facts that it is an extremely common species and frequents low vegetation levels where it would more often be noticed.

Variation. A few specimens showed one to four fine hairs in the tympanic pit (three on each side in the holotype itself), and a few short black hairs on the postalar wall.

Four males, one female from Groton, Mass., provide a warning note on color variation. Johnson (1932), who called these *splendida*, noted that in the males "the terminal segment of the abdomen is a bright green, not blue." When the original series was found in the C. W. Johnson Collection now at Boston University it proved to be *metallica.* The characteristic "granular-type" puparia were associated with the specimens. The "green" color is more of a greenish blue, exaggerated by the more shining appearance of the 5th tergite compared with the gray-microtomentose preceding segments.

In small or immature females of *metallica*, the distal segment of the abdomen may be shining green with only a faint coppery tint. This variation is clearly shown in ten females reared from a northern cardinal nest at East Falls Church, Va., in which only the most mature specimens show the characteristically copper-colored 5th tergite, sharply contrasted with the preceding segments.

Variation in the parafacial hairs will cause some difficulty in identifying isolated males. Small specimens seem sparsely haired with relatively few hairs in two irregular rows, or occasionally a partial third row, whereas large specimens have more hairs in three or even almost four irregular rows. Inasmuch as *metallica* is especially a parasite of small birds and many specimens are undersized, this characteristic of relatively few hairs will serve for most specimens.

As a possibly interesting check on variation in ratios, 25 males reared by Bennett from six different nests at Algonquin Park, Ont., were measured and ratios calculated for comparison with the 25 males on which were based the ratios used in description. These 25 were a sample from the wide range of the species, coming from 15 localities in 9 states and Ontario, including only five specimens from five nests in Algonquin Park. The following comparison reveals only slight differences:

	General series	Algonquin Park
frons: head	0.08 (.06-.09)	0.08 (.07-.10)
frons: ocellar span	1.11 (.85-1.44)	1.13 (1.00-1.36)
parafacial: frons	1.51 (1.33-2.00)	1.45 (1.17-1.83)
parafacial: ocellar span	1.68 (1.43-2.00)	1.65 (1.43-1.92)
parafacial: vibrissal interval	0.76 (.70-.86)	0.76 (.625-.88)
cheek: eye	0.36 (.33-.39)	0.35 (.32-.375)

Remarks. In older literature, *metallica* was synonymized with *splendida* Macquart, but we concur with Hall (1948) in regarding the latter as a species dubia. For a discussion of this, see the heading *Calliphora splendida* at the end of the species descriptions.

Records of or specimens identified as "*splendida*" cannot be assumed to be *metallica*--although they often are--because there is another northeastern species (*bicolor*) in which females have tergite 5 coppery tinted, and because males identified as *splendida* by older authors often prove to be *sialia*.

Identifications of *metallica* can also be suspect. The possible confusion with *bicolor* has just been noted. In Plath (1919c), "*metallica*" proved in part to be *braueri*, and in Neff (1945) "*metallica*" proved to be *cuprina*. "*P. metallica*" of James (1955), based on a male from Yosemite National Park, Calif., is *spatulata*.

Females of *metallica* have long been regarded as the most distinct in the genus because of the shining cupreous 5th tergite contrasting with the blue or bluish purple of the 2nd to fourth tergites. Unfortunately this can now lead to misidentifications because two other species, *bicolor* in the East and *cuprina* in the West, have the same feature and the females of the three species are very similar and will be difficult to distinguish. The key to females utilizes a minor feature of the microtomentum of the scutum which hopefully will separate females of *metallica* from the others.

Males of *metallica* have no outstanding or peculiar characteristic or combination of characteristics. The surstylus is curved, placing the species in that group (See outline of groups, preceding the keys, and the key to males).

Puparia of *metallica* are distinctive because of the "fine sandpaper" appearance imparted by the short tuberculose spines. Puparia of *metallica* are similar to those of *deceptor*, *interrupta*, and *shannoni*, but separated from *shannoni* by the tubercle-like spines of the hyperstigmatal area and from *deceptor* by the shorter tubercle-like spines of the dorsal and ventral cuticle.

Although similar in many respects to *interrupta*, the latter species may be separated by the interruption of the medial ventral band and the markedly smaller ventral band ratio. Puparia of *metallica* may be separated from *braueri* by the presence of the three distinct ventral spine bands.

Protocalliphora (*P.*) *parorum*, new species
(Figs. 22, 57, Map 10)

Protocalliphora splendida var. near *hirundo*.--Storer, 1927: 227 [From nest of mountain chickadee, San Bernardino Mts., Calif.; material lost, possibly *parorum*, considering the host].

P. sp. VI.--Whitworth, 1976: 16ff. [Utah].

P. n. sp. no. 1.--Gold and Dahlsten, 1984: 560-572 [Calif.; effect on nestlings of chestnut-backed and mountain chickadees].

Diagnosis. Sexes concolorous, monochromatic, shining metallic bluish to bluish purple; parafrontal and parafacial dark gray microtomentose; calypteres white; surstylus narrow and curved.

Male. Head darker than usual in *Protocalliphora*, black in ground color with parafrontal and parafacial dark gray microtomentose. Body shining metallic bluish to bluish purple, scutum partly gray microtomentose, especially anterior to transverse suture, with two broad stripes of microtomentum demarcating three broad shining stripes, as viewed from behind. Abdomen shining in direct view, tergites 2-4 thinly gray microtomentose when viewed from behind at a low angle. Calypteres white.

Frons relatively narrow, at narrowest, anterior to median ocellus, 0.08 times width of head (0.06-0.10) and 1.13 times ocellar span (0.91-1.42); ocellar tubercle with none to few bristles or bristlelike hairs except for ocellar bristles; preocellar area narrowly triangular to acuminate, dark gray microtomentose, slightly shining, sometimes rugose; frontal bristles and parafrontal hairs usually ending well in advance of level of median ocellus, uppermost frontals weak and hairlike; parafrontal with one to two irregular rows of hairs on lower half, strongly narrowed dorsally, its microtomentum narrowed to pinched out, and prevertical area usually polished; parafacial with numerous coarse black hairs in 3-4 irregular rows, no changeable spot along upper margin; parafacial 1.41 times width of frons (1.18-1.70), 1.60 times ocellar span (1.43-1.83), and 0.77 times vibrissal interval (0.69-0.86/21); height of cheek 0.42 times maximum length of eye (0.38-0.45); occiput with about three irregular rows of black hairs behind postocular setae.

Thorax lacking accessory notopleural bristles; tympanic membrane and postalar wall with none to few short, pale and inconspicuous hairs. Fore tibia typically with one posterior bristle.

Terminalia: cerci long and narrow to acute apices (cf. Fig. 5b); surstylus narrow and curved (Fig. 22); aedeagus as usual in the genus (cf. Fig. 6a).

Female. Concolorous with male, but scutum much more gray microtomentose, especially posterior to transverse suture; head dark gray microtomentose to slightly brownish tinted. Frons 0.26 times head width (0.25-0.28) and 3.33 times ocellar span (2.875-3.71); preocellar area dull, finely and evenly microtomentose; parafrontal above tending to be thinly microtomentose, subshining; parafacial as in male, densely haired without changeable spot, width 0.47 times that of frons (0.43-0.52), 1.55 times ocellar span (1.25-1.71), and 0.81 times vibrissal interval (0.69-0.92/24); height of cheek 0.43 times maximum length of eye (0.40-0.47).

Apparent normal length of adults, 8-9.5 mm.

Third instar. One specimen examined, additional measurements derived from puparia. *Prothoracic fringe* 80 μ (50-120)/70. *Cephalopharyngeal skeleton* 950 μ (800-1050)/15 in length.

Puparium. Length 7.4 mm (7.2-7.5)/5; breadth 2.7 mm (2.4-3.3)/10; thick-walled, dull, brown. *Posterior region* (Fig. 57a,b): stigmatal plates 160 μ (150-175)/20 in diameter; distance between buttons 565 μ (500-625)/10, and across the stigmatal plates 1100 μ (1000-1200)/10; stigmatal ratio 0.52 (0.45-0.58)/10; stigmatal area outlined in folds and a few plaques; mesostigmatal folds moderately pronounced (Fig. 57a) to virtually absent (Fig. 57b), the former bisecting the stigmatal region; hyperstigmatal spines numerous, 33 μ (8-48)/100; hypostigmatal area with numerous, pronounced, irregular folds; meso- and hypostigmatal spines sparse, the latter 15 μ (8-30)/100; circumstigmatal folds absent (Fig. 57b) to pronounced concentric (Fig. 57a). *Dorsal cuticle* (Fig. 57c) with numerous, large spines, 35 μ (15-55)/100; cuticular ridges weak. *Ventral cuticle* (Fig. 57d): ventral band ratio 0.74 (0.58-0.87)/20; anterior band pronounced, anterior patch distinct, spines 15 μ (8-30)/100; medial band pronounced, spines 24 μ (15-50)/100; posterior band pronounced.

Type Series. Holotype male, allotype, and 16 paratypes (1 male, 15 females), **California**: Modoc Co., 7 mi N Adin, June 19, 1968, ex chickadee nest (D. L. Dahlsten). Holotype and allotype in the California Academy of Sciences, paratypes in the California Academy of Sciences, U.S. National Museum of Natural History, and University of California at Albany (Division of Biological Control). Additional paratypes: **California** [CAS, USNM, and U. Calif., Albany]: 5 males, 5 females, Contra Costa Co., Wildcat Canyon, June 1981, ex nest chestnut-backed chickadee (C. Gold); male, 3 females, El Dorado Co., U. California Blodgett Forest, 8 mi E Georgetown, nest of brown creeper, emerged July 6 and 11, 1978 (C. Gold); 6 males, 7 females, Blodgett Forest, emerged from nest of chestnut-backed chickadee May 29-July 11, 1978 (C. Gold); 2 males, 12 females, Blodgett Forest, ex nestboxes of mountain chickadee, July 1-22, 1978 (C. Gold); male, 20 females, Blodgett Forest, July 30-Aug. 6, 1977, ex nestboxes of chestnut-backed chickadee (C. Gold); 4 males, 18 females, Blodgett Forest, July 25-Aug. 6, 1977, ex nestboxes mountain chickadee (C. Gold); 3 females, Blodgett Forest, July 21, 1979, ex nestbox chestnut-backed chickadee (C. Gold); male, Lassen Co., Fredonyer Peak, 25 mi N Susanville, July 25, 1967, ex chickadee nest (F. M. Stephen); 8 males, 13 females, same locality as preceding, July 4, 1969, ex nest of red-breasted nuthatch (D. L. Dahlsten); female, Los Angeles Co., Tanbark Flat, June 22, 1950 (H. F. Robinson) [U. Calif., Davis]; 24 males, 49 females, Modoc Co., various localities (Roney Flat, Tom's Creek, Yellowjacket) and dates in 1978, chiefly July, ex nests of mountain chickadee (C. Gold); 22 males, 30 females, Modoc Co. near Adin, emerged July 25-Aug. 8, 1977, ex nestboxes mountain chickadee (C. Gold); 10 males, 9 females, Mono Co., Pilot Springs, 23 mi E Lee Vining, July 16, 1968, ex nest mountain chickadee (D. L. Dahlsten); 10 males, 12 females, Mono Co., 27 mi E Lee Vining, July 17, 1968, ex nest mountain chickadee (D. L. Dahlsten). **Utah**: 4 males, 1 female, Smithfield, July 11, 1970, nest of black-capped chickadee (T. L. Whitworth) [USNM, Whitworth Colln.].

Non-paratype Material: One female, Utah, and one male, Utah, Box Elder Co., July 21, 1969 (T. L. Whitworth, Malaise trap, and 21 males, 32 females, various localities in California, duplicated in the type series. Also 3 males, 6 females, California: Inyo National Forest, Sentinel Meadows, Aug. 6, 1963 (S. G. Herman) [U. Calif., Berkeley]. Also two isolated specimens, a male, California: Modoc Co., Adin Bridge, emerged July 27, 1978, ex barn swallow nest, and a female, probably *parorum*,

California: Lassen Co., Summit Camp, June 28, 1949 (Claude I. Smith) [U. Calif., Berkeley].

Specimens Examined. 126 males, 239 females.

Distribution. California and Utah.

Basis of description of immature stages: *Larva, 3rd instar* (4 specimens): Modoc Co., Calif., mountain chickadee, coll. Gold. *Puparium* (10 specimens): Modoc Co., Calif., mountain chickadee.

Hosts. Black-capped, chestnut-backed, and mountain chickadees (chiefly), brown creeper (few), red-breasted nuthatch (few), and barn swallow (one). Whitworth (1976) also recorded it from Oregon junco and house wren in Utah, and Gold and Dahlsten (1981, 1984) recorded it from Bewick's wren.

Ecology and Biology. The extensive available material was reared almost exclusively from nests of chestnut-backed and mountain chickadees, but finding specimens in Utah suggests a wider range when chickadee nests are investigated in other parts of the West. Gold and Dahlsten (1984) reported on the effects of *parorum* on chickadee nestlings. Cavity nesters in forest habitats appear to be the principal hosts.

Variation. Of the 360 specimens available, 347 have only one posterior bristle on the fore tibia. None have two on each fore tibia. Five males, four females show two bristles on one side or the other, but even most of these are not the usual well spaced bristles of species that characteristically have two. In some specimens the upper bristle is very short compared with the lower bristle, which is the standard one, and in two indivduals the two bristles are close together near the usual position of the lower bristle, not well spaced. Clearly, one posterior bristle is the regular character state for *parorum*.

Remarks. *P. parorum* is one of the complex of species with males characterized by a narrow curved surstylus. Single specimens will usually be difficult or impossible to identify with certainty. In the present case, the species is especially dark, a dark gray microtomentum in both sexes that gives it a different habitus from the other species, although this is not easily expressed in a key for those who do not have authoritatively identified examples before them. The three most difficult species to distinguish are *parorum*, *fallisi*, and *tundrae*, and the reader is referred to the general key. The first and last each have a characteristic habitus, once one is accustomed to the usual appearance of these flies. In addition, *P. tundrae* occurs only in the far north, in northern Canada and Greenland, as far as known at present, and *parorum* seems typically a parasite of chickadees, although some other small birds are known also as its hosts. Moreover the apparent frequency of chickadees as hosts of *parorum* could have been enhanced as a result of concentrated study on the chickadees by Dahlsten and colleagues.

Puparia of *parorum* can be confused with those of *bicolor* and *cuprina*. They can be separated from *bicolor* by the larger ventral band ratio and from *cuprina* by the larger prothoracic fringe. Great care should be taken with these species and a general evaluation of all available characters made before assigning a specific designation.

Storer's (1929) record of *Protocalliphora splendida* var. near *hirundo* from the nest of a mountain chickadee in the San Bernardino Mts., California, probably referred to *parorum*. The material is apparently lost (personal communication from Professor Storer). Storer recorded 76 larvae and 24 puparia in the nest, at 7400 ft altitude.

Etymology. The specific name is in the genitive plural, from the predominance of parasitism on species of the genus *Parus*, the chickadees.

Protocalliphora (*P.*) *sapphira* (Hall)
(Fig. 26)

Protocalliphora splendida var. *sialia*.--Shannon and Dobroscky, 1924: 252, in part [Naknek Lake, Alaska].

Apaulina sapphira Hall, 1948: 200 [Naknek Lake, Alaska; type series included in *sialia* in Shannon and Dobroscky 1924].

Protocalliphora sapphira.--Hall, 1965: 926 [Catalog].

Diagnosis: Sexes concolorous, monochromatic, shining metallic dark bluish; calypteres white; front tibia with 2 posterior bristles; frons of male relatively broad, 0.17 times head width; surstylus narrow and strongly curved.

Male (only holotype available). Body shining, dark metallic bluish; thorax and abdomen concolorous, former slightly darker, and duller with gray microtomentum between three broad shining stripes as viewed from behind, the gray stripes equal to or wider than the shining stripes; abdomen thinly microtomentose but highly shining, with linear black to purple median stripe that is especially evident on tergites 3 and 4 as viewed from behind at a low angle. Calypteres white, rims pale yellow.

Frons relatively broad for the genus, not narrowing anterior to median ocellus, at vertex 0.17 times head width and 1.9 times ocellar span; parafrontal and parafacial with light-brown-tinted microtomentum and sparse black hairs, width of parafacial 1.8 times breadth of 3rd antennal segment, 0.85 times width of frons, 1.6 times ocellar span, and 0.79 times vibrissal interval (not equal as stated in Hall's description; the ptilinal fissure is not completely closed in the holotype and the true lunule width is narrower); occipital hairs predominantly black above, in approximately three irregular rows; cheek relatively broad, its height 0.46 times maximum length of eye.

Thorax without accessory notopleural bristles; postalar wall not conspicuously haired, with one to three or four short, weak hairs visible under high magnification; tympanic membrane weakly setose (four on left, three on right). Fore tibia with two posterior bristles (a third bristle on right side is probably an aberration).

Terminalia: cerci distally acuminate, parallel (as in Fig. 5b); surstylus narrow, parallel sided, strongly curved, distally truncate (Fig. 26); aedeagus as usual for genus (cf. Fig. 6a).

Female (two paratypes). Color as in male; microtomentum of frons and parafacial more obviously brownish or brownish yellow; parafacial brownish above but silvery below. Frons at vertex 0.30-0.31 times head width and 1.73 times ocellar span, strongly broadening anteriorly, entirely microtomentose in preocellar area; reclinate orbital bristles present; parafacial broad, nearly twice breadth of 3rd antennal segment, over half (0.54) width of frons, 1.73 times ocellar span, and 1.08 times vibrissal interval; cheek height 0.49 and 0.50 times maximum length of eye. Fore tibia with two posterior bristles in one paratype, one in other.

Length of adults, 7.5 mm. (This does not necessarily indicate a small species; the few individuals might have been underfed).

Immature stages. Unknown.

Type Series. Holotype male and paratype female, Savonoski, Naknek Lake (now Lake Coville, near base of Alaska Peninsula), Alaska, Aug. 1, 1919 (J. S. Hine); paratype female, same locality and collector, June 1918. These specimens were part of the syntype series of *P. sialia* S. & D. (1924:252). Both Shannon and Dobroscky (1924) and Hall (1948) stated that all three specimens were collected Aug. 1, 1919; the different date on one of the paratypes was apparently overlooked. Type series in the U.S. National Museum of Natural History. There is also a female, same locality, Aug. 13, 1919 (A. J. Basinger) [CAS], which was not included in the type series.

Specimens Examined. 1 male, 3 females.

Distribution. Alaska, as far as known.

Hosts, and Ecology and Biology. Unknown.

Variation. The extreme variability in the short type series makes it impossible to properly assess the typical characteristics of the species. One paratype female has 2 posterior bristles on the fore tibia and a fairly good tympanic tuft of 6-8 hairs on each side, but only 2 or 3 very weak hairs on the postalar wall. The other paratype female has only one posterior bristle on the fore tibia, a fairly good tympanic tuft of 10-12 conspicuous black setae, but few (4) weak and inconspicuous hairs on the postalar wall. The holotype has two posterior bristles on the fore tibia, but both tympanic membrane and postalar wall only weakly haired (see description).

Remarks. In the holotype, the ptilinal fissure had not completely closed and the anterior part of the frons is not completely sclerotized. The specimen otherwise does not appear teneral and the upper half of the frons appears normal, hence the measure of the frontal width is accepted. This wide frons distinguishes *sapphira* from males of all other species having curved and narrow surstylus.

Females are not distinctive. The number of posterior bristles on the fore tibia probably varies, and further material is needed to show whether it is typically a two-bristle species, or whether the reverse is true. Our experience indicates that either is possible in these northern species, but that two bristles is the most common characteristic. The adequate series in *tundrae* demonstrates that even in a species where the two-bristle state is common, there will be a sizeable proportion of specimens that have only one bristle on the fore tibia.

The setose tympanic membrane and two posterior bristles on the fore tibia relate *sapphira* to Palearctic species, as might be expected from its far northern range, but it has male terminalia distinct from any Palearctic species. The chaetotactic characters just noted apply also to *chrysorrhoea*, in which the hair tufts are larger and stronger, but the wide frons and narrow and curved surstylus easily separate *sapphira*. The surstylus in profile is suggestive of *metallica*, but the latter is a much different species that would not be confused. In some respects, and especially in females, *sapphira* resembles *spatulata*, but the latter has, regularly, one posterior bristle on the fore tibia and no tympanic setae, and in the male conspicuously polished prevertical area.

Hall stated that *sapphira* differed from all other *Apaulina* in having apical bristles on the scutellum. In reality *sapphira* as well as most *Protocalliphora* lack well developed apical bristles although one pair of hairs is slightly stronger than surrounding hairs and sometimes, especially in large specimens, does appear bristlelike. In most specimens, this apical pair is actually subapical in varying degrees, less so in *sapphira* than in most species.

The only published record of *P. sapphira* other than those listed in the synonymy is that by Sailer and Lienk (1951), but later study showed these specimens to be *P. chrysorrhoea* (Meigen). The "cliff swallow" host was an error for bank swallow (R. I. Sailer, personal communication).

Protocalliphora (*P.*) *seminuda*, new species
(Figs. 25, 58, Map 6)

Protocalliphora chrysorrhaea [sic] (Meigen).--Walton, 1914: 175, in part [Koehler, N. Mex., mixed with *P. braueri*].
P. n. sp. A.--Horning and Barr, 1970: 73 [Idaho].
Protocalliphora sp. X.--Whitworth, 1976: 16ff. [Utah].

Diagnosis. Sexes concolorous, monochromatic, shining dark blue to bluish purple; calypteres whitish; male frons moderately narrow; parafacial and parafrontal of male conspicuously silvery microtomentose, with few hairs; surstylus in profile curved though not slender.

Male. Head with microtomentum of parafrontal and parafacial silvery. Body dark bluish to bluish purple, the thorax and abdomen concolorous, shining, the abdomen highly so, with thin gray microtomentum visible only from behind at an extremely low angle, the dorsum of thorax duller, more obviously microtomentose, especially on presutural area of scutum, the scutum viewed from behind with three broad shining stripes. Calypteres whitish, outer rims slightly pale-yellow tinted.

Frons moderately narrow, narrowing only slightly anterior to ocellar triangle, at narrowest 0.08 times (0.06-0.09/9) head width and barely wider than ocellar span (1.16x; 1.00-1.50/9); preocellar area shining, narrowly acuminate; frontal vitta at its narrowest less than twice width of a parafrontal; frontal bristles strong below, rapidly decreasing in strength above, the uppermost short, fine and hairlike; parafrontal silvery microtomentose dorsad nearly to median ocellus, and almost bare of hairs except for four or five in one row near lower frontal bristles (in one specimen a complete but inconspicuous row close to frontal row), prevertical area shining black; parafacial moderately wide, its width greater than breadth of 3rd antennal segment, 1.50 times

(1.33-1.83/9) width of frons, 1.74 times ocellar span (1.36-2.08/9), and 0.90 times (0.75-0.92/7) vibrissal interval, thinly haired along anterior half in about two to three irregular rows, posterior half bare; height of cheek 0.38 times (0.36-0.41/9) maximum length of eye.

Thorax without accessory notopleural bristles; tympanic membrane and sometimes postalar wall with a few scattered hairs. Fore tibia with one median posterior bristle.

Terminalia resembling those of *metallica*, the cerci straight and parallel, deeply notched, long, slender and acute at apex (cf. Fig. 5b); surstylus in profile curved, though not slender (Fig. 25); aedeagus as usual for the genus (cf. Fig. 6a).

Female. Microtomentum of parafacial and parafrontal light brownish yellow tinged. Body color as in male but mesonotum somewhat duller, with more abundant grayish microtomentum. Frons broad, about 1/3 width of head (0.33x; 0.32-0.36/6) and four times ocellar span (3.56-4.43/6); frontal vitta on each side often with several coarse setulae mesad of the frontal bristles (broken off in allotype and one paratype, but sockets evident); preocellar area broad, dull, subshining but evenly brownish gray microtomentose; parafrontal sparsely haired, in about two irregular rows, microtomentose throughout but prevertical area thinly so and subshining; parafacial fairly broad, 0.41 times (0.39-0.45/6) width of frons, 1.67 times ocellar span (1.50-1.875/6), and 0.92 times vibrissal interval (0.88-1.00/3), sparsely haired on mesal half, bare on outer half; height of cheek 0.44 times (0.42-0.48/6) maximum length of eye.

Apparent normal length of adults, 8-9.5 mm.

Third instar. No specimens examined, characters derived from puparium. *Prothoracic fringe* 50 μ (40-65)/26. *Cephalopharyngeal skeleton* 860 μ (800-950)/6 in length.

Puparium. Length 7.9 mm (7.4-8.3)/6; breadth 3.4 mm (3.1-3.7)/7; medium thick-walled, dull, brown. *Posterior region* (Fig. 58a): stigmatal plates 170 μ (150-175)/14 in diameter; distance between buttons 590 μ (575-625)/7, and across stigmatal plates 1175 μ (1000-1375)/7; stigmatal ratio 0.51 (0.40-0.63)/7; stigmatal area outlined in plaques and folds; mesostigmatal area with numerous pronounced folds and plaques, bisecting the stigmatal region; hyperstigmatal spines numerous, 22 μ (8-40)/70; hypostigmatal area rugose with numerous, pronounced, irregular folds; hypostigmatal spines scarce, 14 μ (18-25)/70; circumstigmatal folds pronounced; concentric. *Dorsal cuticle* (Fig. 58b) with numerous spines, 22 μ (15-40)/70; cuticular ridges absent or if present, very weak. *Ventral cuticle* (Fig. 58c):

ventral band ratio 0.63 (0.51-0.67)/9; anterior band pronounced, anterior patch distinct, with numerous small spines, 12 μ (8-25)/70; medial band pronounced, spines 20 μ (15-30)/70; posterior band pronounced on all segments.

Type Series. Holotype male, and allotype, Koehler, N. Mex. (W. R. Walton; Webster No. 7707; reared from horned lark). Type in the National Museum of Natural History [USNM]. Paratypes: 3 males, 2 females, same data as holotype [USNM]; 2 males, California: Goodale Creek, N of Lone Pine, Inyo Co., Apr. 3, 1953 (H. B. Leech) [CAS]; 2 males, Craters of the Moon National Monument, Idaho, June 29, July 23, 1965 (D. S. Horning, Jr.), both collected in a Malaise trap [CAS]; 3 males, 6 females, Pocatello Valley, Oneida Co., Idaho, July 5, 1970 (T. L. Whitworth, nest of Brewer's blackbird [Whitworth Colln., USNM]; 3 males, Benson, Cache Co., Utah, July 9, 1970 (T. L. Whitworth), nest of barn swallow, in mixed infestation with *P. halli* [USU]; male, Green Lake, Uinta Mts., Utah, Aug. 17, 1940 (D. G. Hall) [USNM]; male, female, Stratton Experimental Watershed near Saratoga, Wyo., May 26-June 2, 1976 (J. M. Schmid), from pit and Malaise traps in sagebrush [USNM].

Non-paratype Material: Male, Box Elder Co., Utah, July 9, 1969 (K. J. Capelle) [Whitworth Colln.]; male, Summit Co., Bear River, Utah, Aug. 5-12, 1971 (W. Hanson, G. F. Knowlton, Malaise trap) [Whitworth Colln.].

The holotype and the Craters of the Moon paratypes are of the usual size for most *Protocalliphora*, and we assume that this is the normal size for the species. The other specimens are in general smaller, and some are more or less teneral, especially the New Mexico specimens, most of which have the abdomen more or less collapsed.

Specimens Examined. 18 males, 10 females.

Distribution. California, Idaho, New Mexico, Utah, Wyoming.

Basis of description of immature stages: *Puparium* (7 specimens): Koehler, N. Mex., horned lark (3); Pocatello Valley, Oneida Co., Idaho, Brewer's blackbird (4).

Hosts. Brewer's blackbird, horned lark, and barn swallow.

Ecology and Biology. Two few hosts are known to give any reliable picture of the host associations. Walton (1914), as already discussed under *braueri*, stated that the larvae of *P. "chrysorrhaea"* [sic], as it was then identified, "were contained in purulent sores on the sides of the body near the legs and on the neck." The embedding of the larvae in the skin is well known for *braueri* and perhaps it is also true for *seminuda*. Walton's brief published notes say nothing of rearing any larvae from the nest, but rather they suggest that all maggots were taken directly from a weak and emaciated fledgling. Possibly some larvae or puparia may have been found in the nest material and reared as if all were the one series. Certainly the infestation was mixed; *braueri* and *seminuda* are decidedly different species and cannot be confused in either sex, nor in the mature larvae and puparia.

Variation. The allotype has the chaetotactic aberration of four orbital bristles on the left side and three on the right.

Remarks. The head of the male is similar in proportions and silvery microtomentum to that of *sialia*--indeed the New Mexico series stood in the collection under *sialia* and was so identified by Shannon and Dobroscky (1924)--but the parafacials are narrower and have distinctly fewer hairs, and the parafrontals are almost bare. The male terminalia resemble those of *metallica* and *fallisi*. For males, the thinly haired and partly bare parafacials are distinctive, also the narrow frons with very narrow frontal vitta. The females are not well separated from those of *hirundo* and other western species. The coarse setulae towards the outer margins of the frontal vitta might offer a means of recognition, but the sample is too small to be sure of the consistency of the character.

Puparia of *seminuda* can be confused with those of *hesperia*, *hesperioides*, and *spenceri*, which they broadly resemble. They can be distinguished, however, from *hesperioides* on the basis of the shorter prothoracic fringe, and from *spenceri* on the basis of the concentric circumstigmatal folds and absence of dorsal cuticular ridges.

Etymology. The specific name, an adjective, is compounded from the Latin prefix *semi*, half or partly, plus *nudus*, bare.

Protocalliphora (*P.*) *shannoni*, new species
(Figs. 13, 59, 60, Map 12)

Protocalliphora sialia, in part.--Shannon and Dobroscky, 1924: 252 [White Mts., N.H., Farmingdale and Ithaca, N.Y.].

P. splendida form *sialis* [sic].--Johnson, 1925a: 216 [Peterboro, N. H. specimen].

P. splendida splendida.--Johnson, 1929a: 29 [Ashland, N.H., from robin nest].

P. splendida.--Mason, 1944: 232-247, in part [Mass.].

P. sp. (or n. sp.) near *sialia*.--Judd, 1954: 123 and 1957: 32 [London, Ont.]; Boyes, 1961: 557 [Cytology]; Boyes and van Brink, 1965: 543 [Cytology]; Boyes and Shewell, 1975: 442, 446 [Cytology].
P. sp. VII.--Whitworth, 1976: 16 ff. [Utah].

Diagnosis. Sexes concolorous, monochromatic, shining dark metallic blue to bluish black; calypteres white; frons of male very narrow; surstylus digitate, of intermediate width.

Male. Head black in ground color; parafrontal and parafacial gray microtomentose, bright but not silvery, rather a slightly pale brown tint. Thorax shining dark metallic blue to bluish purple or bluish black, dorsum thinly microtomentose, especially anteriorly, the microtomentum not demarcating stripes and scutum appearing entirely shining to the unaided eye. Abdomen concolorous with thorax, but generally a brighter blue, sometimes tinged with green, and very shining, thinly microtomentose when viewed from behind at a low angle. Calypteres opaque whitish, at most the outer rims tinged with yellow.

Frons very narrow, 0.06 times width of head (0.05-0.07), and equal to (or nearly so) ocellar span (0.985x; 0.83-1.18)); parafrontal with only one row of hairs, on lower fourth, merging with parafacial hairs of which there are two or sometimes three irregular rows; parafacial comparatively narrow, equal to or barely wider than 3rd antennal segment, nearly twice width of frons (1.78x; 1.46-2.20), 1.76 times ocellar span (1.50-2.20), and 0.70 times vibrissal interval (0.58-0.815); cheek relatively narrow, about 1/3 maximum length of eye (0.37x; 0.34-0.40).

Thorax without accessory notopleural bristles and without hairs on tympanic membrane, and with only an occasional hair or two on postalar wall. Fore tibia with one posterior bristle.

Terminalia: cerci distally acuminate, narrow and parallel; surstylus digitate, of intermediate width (Fig. 13), somewhat as in *sialia*, broadly rounded at distal end; aedeagus as usual in the genus (cf. Fig. 6a).

Female. Concolorous with male; microtomentum of parafrontal and parafacial slightly more brownish yellow than in male; dorsum of thorax slightly more microtomentose, weakly demarcating three shining stripes, but to the unaided eye scutum only slightly less shining than in male.

Frons 1/4 width of head (0.25x; 0.23-0.27) and 3.275 times ocellar span (3.00-3.54), widening slightly anteriorly; frontal vitta with a few short and weak hairs, never with rows of bristlelike hairs; triangular preocellar area often rugose, usually polished black but sometimes brownish microtomentose and subshining; parafacial narrow as in male, its width equal to or barely wider than 3rd antennal segment, about half width of frons (0.495x; 0.44-0.54), 1.62 times ocellar span (1.43-1.78), and 0.81 times vibrissal interval (0.72-0.88); height of cheek 0.39 times maximum length of eye (0.37-0.41).

Apparent normal length of adults, 8.5-9.5 mm.

First instar. Length up to 3 mm. *Prothoracic fringe* 14 μ (6-23)/50. *Cephalopharyngeal skeleton* 240 μ (255-260)/7 in length. *Posterior region:* stigmatal plates 26 μ (21-29)/12 in diameter; peritreme absent; mesostigmatal spines absent; hyperstigmatal area with numerous spines, 6 μ (2-10)/50; hypostigmatal spines scarce, as tubercles, 2 μ (1-4)/50; anal lobes small. *Dorsal cuticle* with numerous spines, 8 μ (4-14)/50. *Ventral cuticle:* spines of anterior patch typically three- to four-pronged, 4 μ (2-6)/50; medial band present, spines 6 μ (2-12)/50; posterior band present.

Second instar. *Prothoracic fringe* 41 μ (24-62)/19. *Cephalopharyngeal skeleton* 500 μ (405-530)/16 in length. *Posterior region:* stigmatal plates 77 μ (69-92)/19 in diameter; button indistinct; mesostigmatal spines scarce; hyperstigmatal spines 10 μ (6-16)/50; hypostigmatal spines numerous (Cf. 3rd instar), as tubercles, 3 μ (2-5)/50. *Dorsal cuticle* with numerous spines, 12 μ (8-19)/50. *Ventral cuticle:* spines of anterior patch 7 μ (4-10)/50; medial band pronounced, spines 9 μ (6-14)/50; posterior band pronounced.

Third instar. Peritreme usually closed; spines small and scarce between stigmatal plates and on hypostigmatal area. *Prothoracic fringe* 67 μ (41-82)/75. *Cephalopharyngeal skeleton* 780 μ (580-940)/123 in length.

Puparium. Length 7.4 mm (5.5-9.4)/156; breadth 3.5 mm (2.6-4.2)/171; thick-walled, shiny, black-brown. *Posterior region* (Figs. 59a,b,; 60a,b): stigmatal plates 160 μ (130-195)/364 in diameter; distance between buttons 550 μ (375-730)/180, and across stigmatal plates 1200 μ (850-1420)/180; stigmatal ratio 0.46 (0.43-0.52)/180; stigmatal area outlined in plaques, frequently bisected; mesostigmatal spines small and scarce; mesostigmatal plaques present, mesostigmatal folds infrequent; hyperstigmatal spines usually as spines, 11 μ (4-24)/130, sometimes as tubercles; hypostigmatal area variable, folds normally present but sometimes absent; hypostigmatal spines as tubercles, appearing as fine sandpaper, scarce, 6 μ (2-10)/130; circumstigmatal folds highly variable, normally rectangular with pronounced folds (Figs. 59a, 60a), frequently with folds surrounding stigmatal area (Figs. 59b, 60b), or, rarely, absent. *Dorsal cuticle* (Figs. 59c, 60c) with numerous spines, 18 μ (6-41)/130; cuticular ridges pronounced (eastern forms, Fig. 59c) or weak (western forms, Fig. 60c). *Ventral cuticle* (Figs. 59d, 60d); ventral band ratio 0.54 (0.51-0.67)/100; spines of anterior patch 11 μ (6-27)/130; medial band pronounced, spines 14 μ (6-27)/130; posterior band weak but present on all segments. The ventral cuticle of the eastern forms (Fig. 59d) has pronounced cuticular ridges while that of the western forms (Fig. 60d) has weak cuticular ridges.

Type Series. Holotype male, allotype, and 12 male, 12 female paratypes, Algonquin Park, Ont., 1955 (G. F. Bennett), nest of American robin. In the U.S. National Museum of Natural History.

Additional paratypes: **Canada** [chiefly CNC, USNM]: 95 males, 96 females; **Nova Scotia**: Cariboo. **Ontario**: Algonquin Park (nests of American robin, common grackle, gray catbird, red-winged blackbird), London (nests of American robin, red-winged blackbird) [U. Western Ontario]. **Quebec**: 4 mi N Eardley, Laurentide Park, La Verendrye Provincial Park. **Yukon Territory**: Dawson. Major series, 57 males, 53 females, Algonquin Park, Ont. (Bennett), chiefly from nests of American robin.

Contiguous United States: 150 males, 138 females: **Connecticut**: Colebrook [Boston U.]. **Delaware**: Newark [U. Del., USNM]. **Maine**: Amherst and Aurora in Hancock Co., Dryden, Enfield in Penobscot Co. [U. Maine, Field Mus. Nat. Hist.]. **Massachusetts**: Blue Hills Reservation and Groton [MCZ], Concord and Martha's Vineyard [Boston U.]. **Montana**: Hamilton [Wash. State U., USNM]. **New Hampshire**: Ashland and Peterboro [Boston U.], White Mts. [USNM]. **New York**: Babylon [CAS], Granby Center in Oswego Co., Ithaca [Cornell U.]. **Ohio**: Burton Township in Geauga Co. [Ohio State U.]. **Pennsylvania**: Butler [Preston Colln.], Pittsburg [USNM, Mehner Colln.]. **Rhode Island:** Buttonwoods in Kent Co. [Boston U.]. **Utah**: Franklin Basin and Malibu Campground in Cache Co. [USNM, Whitworth Colln.].

Non-Paratype Material: 140 males, 154 females. Many specimens are in poor condition and have not been included in the type series, although they can be recognized for additional locality and host records. **British Columbia:** Aleza Lake and Trinity Valley just E of Vernon, latter in nest of a junco. **Ontario:** Algonquin Park in nests of white-throated sparrow and hermit thrush, Dunrobin, Petawawa. **Prince Edward Island:**

Souris, in nest of common grackle. **Maine:** Levant, in American tree swallow nest; Stetson in Penobscot Co. **Maryland:** Bethesda, in nest of American robin. **Michigan:** Channing in Dickinson Co. **New York:** Ballston Spa, Farmingdale, Yonkers.

Specimens Examined. 398 males, 401 females.

Distribution. Chiefly southeastern Canada (Ontario to Nova Scotia) and northeastern United States (Michigan to Maine, south to Ohio and Maryland), plus scattered records from British Columbia and Yukon Territory, Montana, and Utah.

Basis of description of immature stages (all Algonquin Park, Ont., some from type series): *Larvae, 1st instar* (7 specimens): red-winged blackbird, July 1954 (3); common grackle, June, 1955 (4). *Second instar* (16 specimens): red-winged blackbird, June, 1954 (3); American robin, August, 1954 (8); common grackle, June, 1955 (5). *Third instar* (10 specimens): red-winged blackbird, June, 1954 (4); American robin, June, 1955 (6). *Puparium* (180 specimens): common grackle, June, 1952 (17) and June, 1954 (27); red-winged blackbird, June 1954 (11); American robin, August, 1954 (23), June, 1955 (67) and August, 1955 (35).

Hosts. Red-winged blackbird, eastern bluebird, gray catbird, black-capped chickadee, common grackle, a junco, American robin, white-throated sparrow, American tree swallow, brown thrasher, hermit and wood thrushes, and house wren, probably brown-headed cowbird (one of three fledglings from infested wood thrush nest). Also, from puparia only, American goldfinch, blue jay, gray [Canada] jay, eastern kingbird, song sparrow, myrtle warbler, and cedar waxwing.

Ecology and Biology. Primarily in nests at intermediate or canopy levels in forests. It has been found especially frequently in robin nests, although this may in part reflect the facts that robin nests are relatively common and easily observed and sampled.

Variation. A long series from Hamilton, Mont., June 1951 (R. B. Spurrier) from robins' nests showed variation in the number of posterior bristles on the fore tibia. Of those which could be counted (i.e., both fore legs present), 9 males and 10 females had one posterior bristle on both sides, one male and two females had two on each side, and six males had two bristles on one side only, left or right.

Remarks. *P. shannoni* is easily confused with *sialia*, especially with males of the latter having relatively narrow frons. Undoubtedly some past records of *sialia* are misidentifications, especially of females. *P. shannoni* falls in a group of intermediate species with few or no outstanding features;

indeed its most distinctive character can be said to be the absence of distinctive features in both sexes. However, the combination of its narrow frons with the digitate surstylus will distinguish it from others in the group. *P. sialia* has broader parafacials and usually a broader frons. Males of *aenea* could easily be confused, but the yellowish calypteres of male *aenea* will be a clue, and any associated females of *aenea* will assure one beyond a doubt. Brownish calypteres will separate some western species with equally narrow frons and parafacials.

Females of *shannoni* are even less distinctive than males, but the combination of polished or shining preocellar area, narrow parafacials, and white calypteres will separate them from other females of the *sialia* complex. If *shannoni* and *sialia* are mixed, as they sometimes were in the study at Algonquin Park, the obvious difference in parafacial width in both sexes will separate the two reasonably well, aided by the more densely haired parafacials of *sialia*. Associated puparia will also be definitive; the densely long-spined and ridged puparium of *sialia*, with its unusually long prothoracic fringe, is outstanding among Nearctic *Protocalliphora*.

Puparia of *shannoni* are similar to those of *deceptor* and *metallica* but can be separated by the presence of spines, rather than tubercles, in the hyperstigmatal area. *Protocalliphora shannoni* occurs across the continent and some differences between the eastern (Fig. 59) and western (Fig. 60) forms can be recognized. Eastern forms have rather pronounced cuticular ridges while the western forms have, at best, weak cuticular ridges.

Etymology. The specific name is in the genitive case, honoring R. C. Shannon, calliphorid specialist whose seminal work with Irene Dobroscky on the North American species of *Protocalliphora* marked the first real attention to the rich North American fauna in the genus.

Protocalliphora (*P.*) *sialia* Shannon and Dobroscky (Figs. 15, 37, 61, Map 13)

Protocalliphora chrysorrhoea (Meigen).--Henshaw, 1908: 87-8 [Mass.; in type series of *sialia*], Miller, 1909: 3-7 [Penn.]; Banks, 1912: 21, pl. 3 [Mass., Penn.; description of larva].

Phormia chrysorrhea [sic].--Lloyd, 1922: 116 [Que.].

Protocalliphora splendida var. *sialia* Shannon and Dobroscky, 1924: 249, 251-2 [Widespread United States, especially northeastern, and Ontario].

P. splendida form *sialis* [sic].--Johnson, 1925a: 216 [New England; mixture of *sialia*, *shannoni*, *metallica*]; Johannsen, 1928: 827 [N.Y.; part *metallica*, part not verified].

P. splendida form *sialia*.--Johnson, 1927a: 1-3 [Mass.]; Johnson, 1927b: 33 [Mass.]; Johnson, 1927c: 77-9 [Mass.]; Johnson, 1927d: 204 [Maine, not verified].

P. splendida var. *sialia*.--Johnson, 1929a: 29-30 [Mass.].

P. splendida.--Lloyd, 1927: 111 [Correction of identification in Lloyd, 1922]; McAtee, 1929: 105-111, in part [Md.].

P. sialia.--Johnson, 1920: 169-170 [Mass.]; Johnson, 1931:21-4 [Review of previous records]; Johnson, 1932: 26-8 [Mass., Vt., Ohio]; Wetherbee, 1932: 73-4 [Mass., Conn.].

P. splendida.--Mason, 1944: 232-247 [Mass.; mixture of *sialia*, *shannoni*, *metallica*].

Apaulina sialia.--Hall, 1948: 201 [Northeastern U.S., s. Canada].

Protocalliphora spp.--Dodge and Seago, 1954: 53-4, in part [Ga.].

P. sialia.--Kenaga, 1961: 91-3 [Mich]; Boyes, 1961: 557 [Cytology]; Boyes and van Brink, 1965: 543 [Cytology]; Hall, 1965: 926 [Catalog, widespread]; Zeleny, 1970: 12 [Damage, control]; Barbara Ann, 1970: 85-6 [Md.]; Boyes and Shewell, 1975: 442, 446 [Cytology]; Whitworth, 1976: 16ff. [Utah]; Zeleny, 1976: 123 [Biology, effect on birds]; Whitworth, 1977: 4933-B.

Diagnosis. Sexes concolorous, monochromatic, shining dark blue to bluish purple; male frons of medium width; female frons typically with polished rugose preocellar area.

Male. Head black in ground color; parafrontal and parafacial silvery microtomentose. Thorax shining metallic dark blue to bluish purple, dorsum thinly microtomentose, viewed from behind the microtomentum demarcating three broad shining stripes, microtomentum especially conspicuous anterior to transverse suture. Abdomen concolorous with thorax, but not as dark, appearing more shining, tergites 2-4 with microtomentum most evident when viewed at a low angle from behind. Calypteres whitish.

Frons (Fig. 37) 0.09 times head width (0.08-0.11) and only 1.37 times ocellar span (1.13-1.67); frontal vitta narrow posteriorly, widening anteriorly, at narrowest easily twice width of an adjacent parafrontal; each parafrontal distinct throughout but greatly narrowed above, the row of hairs on upper portion almost merged with the row of frontal bristles, below becoming two irregular rows and merging into the numerous hairs on the parafacial; parafacial width approximately 1.5 times width of 3rd antennal segment and 1.43 times width of frons (1.20-1.62), nearly twice ocellar span (1.96x; 1.69-2.18), and about equal to vibrissal interval (1.04x; 0.93-1.13); cheek height 0.42 times maximum length of eye (0.38-0.45).

Thorax typically without accessory notopleural bristles and without hairs on tympanic membrane and postalar wall. Fore tibia with one posterior bristle.

Terminalia: cerci distally acuminate, parallel (as in Fig. 5b); surstylus digitate, slightly wider basally, beyond the sides subparallel, apex rounded (Fig. 15); aedeagus as usual in the genus (cf. Fig. 6a).

Female. Concolorous with male, but thorax more heavily microtomentose, viewed from behind the microtomentum and three shining stripes evident to unaided eye; parafrontal and parafacial grayish to light brown microtomentose, commonly with a spot of changeable reflections on upper surface of parafacial, opposite lunule, viewed from above the spot appearing dark brown against the gray or light brown of parafacial and parafrontal, from below bright yellowish gray against a black background.

Frons at vertex 0.265 times width of head (0.24-0.29), widening anteriorly and 3.31 times ocellar span (3.06-3.73); frontal vitta parallel sided, usually with two rows of bristlelike setae, each just mesad of a row of frontal bristles, rarely only a few; preocellar area polished black, rugose, rarely dull and microtomentose or the polished area greatly reduced in size; parafacial moderately broad, width 0.555 times that of frons (0.52-0.62), 1.84 times ocellar span (1.65-2.00), and 1.12 times vibrissal interval (1.00-1.32); cheek height 0.44 times maximum length of eye (0.42-0.46).

Apparent normal length of adults, 8.5-9.5 mm.

Egg. Length 1130 μ (1020-1205)/25; breadth 380 μ (275-455)/25; cuticle white in color and ornamented with minute spines.

First instar. Length up to 4 mm. Prothoracic fringe 13 μ (8-21)/50. *Cephalopharyngeal skeleton* 220 μ (205-240)/6 in length. *Posterior region:* stigmatal plates 19 μ (16-25)/6 in diameter; peritreme absent, mesostigmatal spines absent; hyperstigmatal spines numerous, 6 μ (2-8)/50; hypostigmatal spines scarce, 4 μ (2-6)/50; anal lobes large. *Dorsal cuticle* with numerous spines, 8 μ (4-12)/50. *Ventral cuticle:* spines of anterior patch typically three- to four-pronged, 3 (1-7)/50; medial band present, spines 6 (4-12)/50; posterior band present.

Second instar. *Prothoracic fringe* 52 (36-76)/50. *Cephalopharyngeal skeleton* 525 (475-550)/13 in length. *Posterior region:* button distinct; mesostigmatal spines scarce; hyperstigmatal spines numerous, 14 (8-26)/50; hypostigmatal spines numerous, 6 (2-12)/50. *Dorsal cuticle* with numerous spines, 16 (8-31)/50. *Ventral cuticle:* spines of anterior patch 7 (4-10)/50; medial band pronounced, spines 11 (6-16)/50; posterior band pronounced.

Third instar. Peritreme highly variable, open to closed; meso- and hypostigmatal spines present. *Prothoracic fringe* long, 150 (80-200)/150. *Cephalopharyngeal skeleton* 975 (675-1100)/105 in length.

Puparium. Length 8.5 mm (5.8-10.8)/100; breadth 3.8 mm (2.2-5.2)/100; thick-walled, dull, black brown. *Posterior region* (Fig. 61a,b): stigmatal plates 190 (140-235)/250 in diameter; distance between buttons 450 (235-690)/128, and across stigmatal plate 1000 (655-1390)/128; stigmatal ratio 0.46 (0.43-0.56)128 (in parastigmate forms, i.e., poorly fed, the stigmatal ratio ranges from 0.36 to 0.44); stigmatal area not usually outlined in plaques; mesostigmatal spines scarce and small; mesostigmatal folds and/or plaques present; hyperstigmatal spines numerous, 35 (20-56)/150; hypostigmatal area with numerous, irregular cuticular folds; hypostigmatal spines numerous, 12 (4-22)/150; circumstigmatal folds present, variable, folds normally pronounced and irregular (Fig. 61a,b) but sometimes few and confined to hypostigmatal area and absent in parastigmate forms. *Dorsal* cuticle (Fig. 61c) with numerous spines, 34 (20-56)/150; cuticular ridges numerous and prominent, frequently with two rows of spines per ridge, one directed posteriorly and one anteriorly. *Ventral cuticle* (Fig. 61d): ventral band ratio 0.89 (0.80-1.0)/100; spines of anterior patch 18 (6-33)/150; medial band entire and pronounced, spines 27 (16-45)/150; posterior band pronounced; ventral cuticle with pronounced cuticular ridges.

Puparia formed by larvae which have fed insufficiently are frequently found in infestations of *sialia.* These puparia are shorter, light brown in color and thin-walled; the stigmatal plates are usually close together, giving a stigmatal ratio of 0.37 (0.36-0.44) - the parastigmate forms. The spines on the dorsal and ventral cuticle are crowded, giving the impression of more than the usual number of spines; the three ventral spine bands are frequently elided.

Type Series. For some reason, no holotype was designated by the authors, although they did so for all other new taxa (including new varieties) described in their 1924 revision. Hall (1948) recorded the failure to establish a holotype, but did not designate a lectotype. Shannon and Dobroscky specified that *sialia* was based only on females, and detailed records were given for a number of specimens. These females, being the material on which the variety was based, constitute the syntype series of *sialia*, in the absence of any contrary or positive statement by the authors. Many of these have been located in collections, a lectotype has been selected, and lectoparatypes labeled. A few of these original females must be referred to other species.

Lectotype and two paralectotypes, female, Shawnee, on Delaware River, Penn., July 31, 1908, "on young bluebird," and labeled "*Protocalliphora splendida* > *sialia* S. & D." by Shannon himself. Published by Miller (1909) as *P. chrysorrhoea.*

Paralectotypes, all females [USNM except as noted]: Two, Wellesley, Mass., one labeled Aug. 10, 1907 (Mrs. E. F. Everett); one, Blue Hill, Mass. (N. Banks) [MCZ]; one, "Ottawa, Ont.", July 23, 1922 (H. Lloyd), "parasitic on nestling bluebird" (apparently one of the specimens published by Lloyd, 1922, as *Phormia chrysorrhoea* from 5 mi W of Hull, Quebec).

Syntypes of *sialia* referred elsewhere [all USNM]: two, Alaska: Savonoski, Naknek Lake (J. S. Hine), paratypes of *Apaulina sapphira* Hall (1948); one, New Hampshire: White Mts. [H. K. Morrison] to *shannoni* n. sp.; one, New Mexico: top of Las Vegas Range, 11000 ft, end of June (T. D. A. Cockerell) to *spatulata* n. sp. (probable); many, Virginia: East Falls Church (E. A. Chapin), and one, Rosslyn, May 29, 1913 (R. C. Shannon), brown thrasher, to *metallica*; two, Washington: Almota (A. L. Melander) to *hirundo.*

Specimens Examined. 798 males, 746 females, many unassociated puparia.

Distribution. Widespread, Alaska and Northwest Territories to New Brunswick, south to California, New Mexico, Iowa, and Virginia, also northern Georgia, Mexico (Durango, Puebla).

Alaska: Eagle, Kasilov and Sterling on Kenai Peninsula, Palmer.

Canada: British Columbia: Haney, Kamloops, Penticton, Quesnel, Rike Creek in Becher's Prairie. **New Brunswick**: Acadia Forestry Station, Chamcook, Komhibouguac National Park. **Northwest Territories**: Norman Wells in District of Mackenzie. **Ontario**: Algonquin Park (143 males, 178 females, many puparia), Elgin, Guelph, Mallorytown Landing, Marmora, Ogoki, Osgoode,

Ottawa, Toronto. **Quebec**: Barrière St. Anne in Laurentides Park, La Trappe, La Verendrye Provincial Park, Masham Township in Gatineau Co., Montreal, Parc du Mont Tremblant, St. Pierre de Wakefield. **Yukon Territory**: Kluane Lake.

Contiguous United States, East: Connecticut: Pamfret, Stamford, Waterbury. **Delaware**: Yorklyn. **Georgia**: Mt. Enotah in Towns and Union Counties, Rabun Bald in Rabun Co. **Iowa**: Cedar Falls. **Maine**: Aurora, Enfield, Levant, Mariaville. **Maryland**: Bell, Beltsville, Bowie, Catonsville, Laurel. **Massachusetts**: Babson Park, Belmont, Blue Hill, Dover, Groton, Needham, Rock, Waban, Wellesley, West Bridgewater, Worcester. **Michigan**: Ann Arbor, Cheboygan Co., East Lansing, Midland, Vicksburg, "Upper Peninsula" (no locality). **New York**: Ballston Spa, Mt. Hurricane in Essex Co., Voorheesville in Albany Co. **Ohio**: Columbus, Leetonia in Columbiana Co., Oak Openings Park in W Lucas Co., Toledo. **Pennsylvania**: Butler, Clarksville, Kennett Square near Philadelphia, Shawnee-on-Delaware. **South Dakota**: No locality. **Virginia**: mouth of Aquia Creek in Stafford Co., East Falls Church, Falls Church, Front Royal, Hopewell. **West Virginia**: Ona.

Contiguous United States, West: Arizona: Chiricahua Mts., Portal, Ramsey Canyon in Huachuca Mts. **California**: Fredonyer Peak in Lassen Co. (25 mi N Susanville), Carmel Valley in Monterey Co., French Camp in San Joaquin Co., O'Neals in Madera Co., Roney Flat in Modoc Co., Sagehen near Hobart Mills in Lassen Co. **Colorado**: Colorado Springs, Ft. Collins, Weld County. **Idaho**: Moscow Mt. **Montana**: Hamilton, top of Mt. Sentinel in Missoula Co., Ravalli Co. **New Mexico**: Tomay in Valencia Co. **Oregon**: Klamath Forest National Wildlife Refuge in Klamath Co. **Utah**: Benson, Mendon, Newton, Smithfield and Wellsville in Cache Co., Moroni, near Roosevelt in Duchesne Co., Snowville in Box Elder Co. **Washington**: Dartford, Kamiak Butte near Pullman, Pullman, Seattle. **Wyoming**: La Barge in Lincoln Co.

Mexico: Male, 11 m W Durango, 7000 ft, June 20, 1964 (J. F. McAlpine) [CNC]. Puebla: 3 males, San Martin, Texmelucán, May 26, 1922 (E. G. Smyth) [USNM].

Basis of description of immature stages (all material from Algonquin Park, Ont.): *Egg*: from caged female, July 1955. *Larvae, 1st instar* (6 specimens): common grackle, June, 1953. *Second instar* (13 specimens): common grackle, June, 1955. *Third instar* (10 specimens): common grackle, June, 1955 (6); European starling, June, 1952 (2) and June, 1955 (2). *Puparium* (150 specimens): common grackle, June, 1955 (59); European starling, June, 1955 (10); rough-winged swallow, August, 1950 (6); downy woodpecker, July, 1955 (10); barn swallow, July, 1955 (65).

Hosts (Eastern): Eastern bluebird, yellow-shafted flicker (puparia only), great crested flycatcher, common grackle, eastern kingbird, purple martin, white-breasted nuthatch, eastern screech-owl (puparia only), eastern phoebe (puparia only), common raven, American robin, house and song sparrows, European starling, swallows (bank, barn, cliff, rough-winged, American tree), downy woodpecker, and house wren. Shannon and Dobroscky (1924) recorded *sialia* from the common crow at Ithaca, N.Y., but we have not seen such material; there might have been confusion with females of *avium*.

Hosts (Western). Brewer's blackbird, mountain and western bluebirds, mountain chickadee, American crow, ash-throated flycatcher, American kestrel [sparrow hawk], western kingbird, white-breasted nuthatch, flammulated and great horned owls, western screech-owl, Say's phoebe, common raven, American robin, house sparrow, European starling, various swallows (bank, barn, cliff, American tree, and violet-green), and house wren. *P. sialia* was also reared from a ringed turtle-dove nest in Colorado by R. M. Stabler, but the hosts were caged birds in experimental studies of *Trichomonas* and not a feral established host population.

Part of the type series (syntypes) of *sialia*, from Ithaca, N.Y., was recorded as reared from "crows and robins." Many of the specimens may actually have been part of the series from crows herein recorded as *avium*. *P. sialia* occurs typically in cavity nests, or in low-nesting birds, and the common crow would be a most unusual host.

Ecology and Biology. This species occurs frequently in both cavity and non-cavity nests in man-made structures, and in cavity nests anywhere. The biology has been studied extensively by Bennett in Algonquin Park, Ont., where it occurred commonly in nests of common grackle and barn swallow.

Variation. The ratios given in the description are those of *sialia* from the East, the typical range of the species. Specimens from the western United States and southwestern Canada varied slightly from the typical form and were observed closely to see whether they might represent a western subspecies, or even a closely related species, similar to the *asiovora-avium* pairing. No consistent or significant differences could be found and all are here referred to *sialia*. Ratios of eastern and western forms are given here for comparison, based on 25 of each sex from each region (Table IX).

Occasional specimens show some slight aberration, which we have come to expect in these flies. One male from Mt. Enotah, Ga., has accessory notopleural bristles like those of *braueri*, and females from Kamloops, B.C., and Hamilton, Mont., have two posterior bristles on one fore tibia but only one on the other, the former having the two on the right tibia, the latter two on the left.

Remarks. *P. sialia* is one of the commonest species of the genus. It is frequently found in nests of swallows and the common grackle, but it has also been recorded from a number of other birds. The broad silvery parafacial of the male is distinctive, as is the polished black preocellar area of the female, at least those of the East. A polished black preocellar area is found only in *hesperioides* and some *shannoni*, in addition to *sialia*, so this is an especially useful characteristic for females. In the West females often have the preocellar area microtomentose, but no other differences could be found. In view of the distinctive puparium like that of eastern specimens, with strong spines and especially long prothoracic fringe, the western specimens are here recorded as *sialia* but with the reservation that further attention should be given to the western population.

Shannon and Dobroscky's original series of *sialia* was greatly mixed, since the name was applied to blue females, which brought together under the one name several distinct species.

P. sialia and *P. hirundo* are similar, but the latter has distinctly gray microtomentose preocellar area in both sexes, whereas that area is shining black in both sexes in *sialia*. Separation of females of western *sialia* and *hirundo* will be difficult or impossible, however, unless there are associated males or puparia. Likewise, unassociated females of *avium* and *sialia* will be difficult to distinguish, although *avium* has slightly broader parafacials that are usually slightly convex and appearing bulging. *P. avium* is also usually larger and more robust.

P. sialia can also be confused with the new species *shannoni*, especially since the two occupy similar ranges in the East. The wider frons of the male and wider parafacial in both sexes are characteristic of *sialia*, but teneral males of *sialia* may have a narrow frons that is suggestive of *shannoni*.

Table IX. Mean ratios and ranges of individual ratios of eastern and western *Protocalliphora sialia*

	frons/head	frons/ocellar span	parafacial/ frons	parafacial/ ocellar span	parafacial/ vibrissal interval	Cheek height/ eye length
			MALES			
Eastern	.09(.08-.11)	1.37(1.13-1.67)	1.43(1.20-1.62)	1.96(1.69-2.18)	1.04(.93-1.13)	.42(.38-.45)
Western	.08(.07-.09)	1.10(1.00-1.25)	1.71(1.50-1.87)	1.88(1.67-2.07)	1.00(.84-1.18)	.40(.38-.44)
			FEMALES			
Eastern	.265(.24-.29)	3.31(3.06-3.73)	.555(.52-.62)	1.84(1.65-2.00)	1.12(1.00-1.32)	.44(.42-.46)
Western	.27(.25-.30)	3.58(3.06-4.00)	.53(.48-.60)	1.91(1.73-2.29)	1.03(.90-1.20)	.43(.41-.47)

Puparia of *sialia* are similar to those of *avium*, *chrysorrhoea*, and *hirundo*. They can be separated from these and all other *Protocalliphora* by the unusually long prothoracic fringe.

One doubtful record should be mentioned because of the myiasis involved. Lincoln (1931) recorded "*spendida*" larvae from a barn swallow nestling at Fairdale, N. Dak. The maggots were clinging to its eyelids and throat, and some in the nostrils. Unfortunately these maggots are very small 3rd-stage larvae and one can only say that they are probably *sialia*, but possibly *hirundo*.

Protocalliphora (*P.*) *spatulata*, new species
(Fig. 27, 62, Map 12)

Protocalliphora "*metallica*."--James, 1955: 25 [Calif.].
"Blowfly larvae," nest of horned lark.--Verbeek, 1967: 216 [Wyo.].
"Blowfly larvae," nest of water pipit.--Verbeek, 1970: 440 [Wyo.; larvae only, apparently *P. spatulata*].

Diagnosis. Sexes concolorous, monochromatic, shining metallic dark blue to bluish purple; calypteres opaque white; frons of male relatively broad; parafacials broad in both sexes; male with surstyli in profile narrow, parallel sided, curved, becoming spatulate distally (Fig. 27).

Male. Head dark, entirely black or nearly so, parafrontal and parafacial obscurely brownish microtomentose, microtomentum on parafrontals thinning posteriorly to the chiefly polished black prevertical area. Thorax shining, dark metallic blue to bluish purple, scutum thinly dark gray microtomentose, especially anteriorly, but shining stripes not clearly delineated and scutum appears entirely dark and shining, especially to unaided eye, and weakly microtomentose only when viewed from behind. Abdomen approximately concolorous with thorax, but usually a brighter dark metallic blue to bluish purple, appearing shining in direct view but thinly dark gray microtomentose when viewed from behind at a low angle, 5th tergite shining. Calypteres opaque white, outer rims tinged with yellow.

Frons relatively broad, 0.12 times head width (0.10-0.13) and obviously wider than ocellar span (1.60x; 1.33-1.79); parafrontal with hairs outside the frontal bristles on its full length at least to level of ocellar bristles, parafrontal not narrowed above but continued of same width to vertex; prevertical area smooth and polished, bare of microtomentum, rarely thinly microtomentose about bases of bristles and hairs or across the middle; preocellar area narrow, elongate and acuminate, rugose and subshining; parafacial broad, obviously much wider than 3rd antennal segment, its width slightly greater than the broad frons (1.19x; 1.08-1.38), nearly twice ocellar span (1.90x; 1.71-2.29), and 0.90 times vibrissal interval (0.79-1.09), with numerous coarse black hairs in four irregular rows; cheek broad, its height nearly half (0.47) maximum length of eye (0.42-0.49).

Thorax without accessory notopleural bristles and without hairs on tympanic membrane, a few inconspicuous hairs sometimes present on postalar wall. Fore tibia with one posterior bristle.

Terminalia: cerci distally acuminate, parallel (as in Fig. 5b); surstylus (Fig. 27) moderately narrow, parallel-sided for most of length, curved, usually widening rather abruptly at distal end to appear spatulate; aedeagus as usual in the genus (cf. Fig. 6a).

Female. Concolorous with male; parafrontal and parafacial decidedly brownish microtomentose; dorsum of thorax slightly more microtomentose than in male, but microtomentum still not delineating the three shining stripes usually evident in females of the genus, the thorax appearing to unaided eye as entirely shining and unstriped.

Frons broad, nearly twice width of eye and nearly 1/3 width of head (0.31x; 0.29-0.33) and 3.56 times ocellar span (3.10-4.00), parallel sided but gradually widening anteriorly; frontal vitta with a number of coarse black bristlelike and mesoclinate setae; preocellar area sometimes subshining but always evenly brownish microtomentose; parafacial broad, as in male, its width nearly half as broad as head (0.485x; 0.42-0.54), 1.73 times ocellar span (1.40-2.00), and 0.98 times vibrissal interval (0.85-1.10); height of cheek about half maximum length of eye (0.50x; 0.45-0.54).

Apparent normal length of adults, 9-10 mm.

Third instar. Peritreme closed; spines sparse between stigmatal plates and in hypostigmatal area. *Prothoracic fringe* 58 μ (34-90)/100 (measurements from both larvae and puparia). *Cephalopharyngeal skeleton* 715 μ (670-840)/16 in length.

Puparium. Length 8.1 mm (7.2-8.8)/16; breadth 3.6 mm (3.1-3.9)/16; medium thick-walled, dull, dark brown. *Posterior region* (Fig. 62a): stigmatal plates 178 μ (150-200)/32 in diameter; distance between buttons 434 μ (320-500)/16, and across stigmatal plates 1010 μ (910-1230)/16; stigmatal ratio 0.43 (0.34-0.51)/16; stigmatal area outlined in plaques and weak folds; mesostigmatal spines virtually absent; mesostigmatal folds and plaques pronounced, bisecting the stigmatal region; hyperstigmatal spines rather slender, 30 μ (17-47)/160 in length; hypostigmatal area with numerous, irregular, pronounced folds; hypostigmatal spines rather sparse, 20 μ (8-43)/160 in length, most spines short but a few long ones on each puparium; circumstigmatal folds concentric, rather weak. *Dorsal cuticle* (Fig. 62b) with numerous spines, shortest towards the center line, 28 μ (14-46)/160; cuticular ridges usually weak. *Ventral cuticle* (Fig. 62c): ventral band ratio 0.83 (0.72-0.92)/16; spines of anterior patch 17 μ (8-29)/160, rather slender; medial band pronounced, frequently almost contiguous with anterior band, spines slender, 24 μ (11-38)/160; posterior band pronounced.

Type series. Holotype male, allotype, and 26 male, 25 female paratypes, Two Ocean Mt., near Togwotee Pass, Teton Co., Wyo. (N. R. French), parasite on rosy finch, larvae collected July 29, 1951, pupated about middle of August, and adults emerged Sept. 9, 1951. In the U. S. National Museum of Natural History paratypes in French Collection, U. Colo., CNC, CAS, and Amer. Mus. Nat. Hist., by courtesy of the collector. Additional paratypes: **Alaska**: male, female, Sheenjek River, NE Alaska, 68° 26'-32' N, 143° 45'-55' W, June 2, 1956 (G. B. Schaller) [USNM]; male, Crowbill Mountain, Cape Thompson [N of Kotzebue], June 7, 1960 (W. C. Hanson) [Wash. State U.]. **Canada: Northwest Territories**: male, Tununuk [Mackenzie District], Aug. 15, 1930 (O. Bryant) [USNM]; male, Wrigley [On Mackenzie River north of Ft. Simpson], June 7, 1969 (G. E. Shewell) [CNC]. **Ontario**: male, female, Ogoki [At juncture of Ogoki and Albany Rivers, NE of Lake Nipigmy, N of Lake Superior] [CNC]. **Yukon Territory**: male, Mile 87, Dempster Highway, Aug. 4-8, 1973 (G. and D. M. Wood) [CNC]. **Contiguous United States: California** [chiefly U. Calif., Berkeley]: male, Lassen Peak, 7500 ft, Shasta Co., July 18, 1949 (E. L. Atkinson); male, Sentinel Dome, Yosemite, 8117 ft, June 22, 1931 ["*metallica*" in James 1955]; male, Yosemite, 3880-4000 ft, June 22, 1931; male, Sonora Pass, 9-10000 ft, Mono Co., July 11, 1957 (J. W. MacSwain): male, Sonora Pass, Tuolumne Co., June 26, 1960 (J. F. Lawrence) [CAS]. **Colorado**: female, Doolittle Ranch, 9800 ft, Mt. Evans, Aug. 3, 1961 (J. G. Chillcott) [CNC]. **Montana**: 2 males, top of Squaw Peak, 7996 ft, near Missoula, July 14, 1951 and July 28, 1952 (J. A. Chapman) [USNM]; male, West Fork, Ravalli Co., June 10, 1934 (W. L. Jellison) [USNM]. **New Mexico**: female, top of Las Vegas Range [Lower end of Sangre de Cristo Mts., NW of Las Vegas], 11000 ft, end of June (T. D. A. Cockerell) [USNM] [paratype of *sialia*].

Wyoming: male, Riverside, Yellowstone Park, Aug. 4, 1918 (A. L. Melander) [USNM]; 5 males, Park Co., 10500 ft, above timberline on Beartooth Plateau, about 1 km N of Beartooth Pass, larvae collected Aug. 20, 1964, nest of horned lark (N. A. M. Verbeek), and 4 males, 5 females, same data except larvae collected July 20, 1964 [USNM].

Non-paratype Material: male, Summit Mt. Moscow, Idaho, July 1, 1932 (J. M. Aldrich) [USNM]; 2 females, Park Co., Wyo., larvae collected Aug. 20, 1964 (Verbeek) [USNM]; male, Yukon Territory, Mile 87, Dempster Highway, Aug. 4-8, 1973 (G. and D. M. Wood) [CNC]. Also, 12 larvae from the same Park Co., Wyo., locality, collected July 29, 1963 (Verbeek) from nest of a water pipit [USNM], appear to be *spatulata*.

Specimens Examined: 53 males, 37 females.

Distribution: Western, apparently chiefly at high elevations, typically in rocky alpine tundra, Alaska to western Ontario, south to California and New Mexico,

Basis of description of immature stages: *Larva, 3rd instar* (4 specimens): Park Co., Wyo., horned lark (Verbeek, No. XV). *Puparium* (16 specimens): Park Co., Wyo., horned lark (Verbeek No. VII) (9) and (Verbeek No. XII) (7).

Hosts: Rosy finch, horned lark, water pipit (probable, larvae only).

Ecology and Biology: The known hosts, two and probably three, are ground-nesting birds, characteristically in rocky alpine tundra. Most of the known specimens have been taken at high elevations in the West.

The type series from Two Ocean Mt. was reared from a rosy finch nest. Dr. French (personal communication) has supplied the following information: "The parasites were first noticed on July 29, 1951, when four young birds were recovered from the nest hole for banding. At this time six larvae were seen clinging to the bare areas of the lower breast and abdomen of two of the young birds. On August 2, 1951, when the empty nest was collected it was found to contain a large number of larvae." They pupated about the middle of August and the first adult flies appeared about September 9th.

Variation. Normally there is only one posterior bristle on the fore tibia, but specimens have been seen in which either a right or a left tibia had none or two, the other tibia one. Such examples would confuse the user of a key, if he had only one specimen, and there would be even more confusion if two or none occurred on both fore tibiae. Luckily in males the relatively wide frons and parafacial and the characteristic surstyli will overcome any difficulty with the variation.

The non-paratype male from Yukon Territory is a "runt", obviously an underfed individual, only 6.5 mm in length compared with the normal of 9-10 mm.

Remarks. *P. spatulata* is one of the group of species with narrow curved surstylus. Its relatively broad frons, broad parafacial, and broad and very setose parafrontal make it one of the most distinctive species in that group. It is closest to *parorum*, and the differences are not great.

The puparia of this species can be confused with *chrysorrhoea* and to a lesser extent with *avium* and *sialia*. *P. spatulata* can be best separated from the latter two by the lack of irregular circumstigmatal folds and from the first by the longer prothoracic fringe; it differs from all three in lacking cuticular ridges.

Etymology. The specific name, an adjective referring to the shape of the surstylus, is derived from the Latin *spatula*, diminutive of *spatha*, a flat instrument for stirring.

Protocalliphora (*P.*) *spenceri*, new species
(Figs. 32, 63, Map 14)

Protocalliphora sp. XV.--Whitworth, 1976: 20, 33, 38 [Utah].

Diagnosis. Sexes concolorous, monochromatic, shining metallic blue to bluish purple, female somewhat more microtomentose and slightly duller; calypteres white; male with surstylus curved and strongly narrowed.

Male. Parafrontal and parafacial with fine shining microtomentum, yellowish to light brownish tinted, not forming a distinct changeable spot on upper part of parafacial. Body shining, metallic blue to dark bluish purple or bluish black, scutum very thinly microtomentose, delineating three shining stripes anterior to transverse suture, the abdomen highly shining but thinly microtomentose when viewed at a low angle from behind, 5th tergite more shining. Calypteres opaque whitish, outer rims faintly yellowish.

Frons narrow, 0.07 times head width (0.06-0.8/17) and usually slightly wider than ocellar span (1.14x; 1.00-1.30/71); preocellar area a polished line or narrow ridge, sometimes rugose; lower parafrontal and parafacial sparsely to considerably haired, in two or three irregular rows; width of parafacial about equal to breadth of 3rd antennal segment, slightly greater than width of frons (1.48x; 1.29-1.67/17), 1.68 times ocellar span (1.50-1.90/17), and 0.66 times vibrissal interval (0.57-0.79/17); height of cheek 0.38 times maximum length of eye (0.36-0.41/17); occiput with two irregular rows of black hairs behind row of postocular setae.

Thorax without accessory notopleural bristles, tympanic hairs, and hairs on postalar wall. Fore tibia with a single posterior bristle.

Terminalia: cerci distally acuminate, parallel (as in Fig. 5b); surstylus curved, strongly narrowed on distal portion (Fig. 32); aedeagus as usual in the genus (cf. Fig. 6a).

Female. Concolorous with male, but gray microtomentum denser on scutum and abdomen, the shining scutal stripes more distinct.

Frons not as wide as in many females of *Protocalliphora*, 1/4 width of head (0.25x; 0.24-0.275/24) and 3.22 times ocellar span (2.86-3.62/24); frontal vitta with small irregular polished spot adjacent to median ocellus, sometimes microtomentose but subshining; prevertical area dull, microtomentose like rest of parafrontal; parafacial about equal to breadth of 3rd antennal segment, 0.485 times width of frons (0.45-0.52/24), 1.56 times ocellar span (1.31-1.69/24), and 0.75 times vibrissal interval (0.68-0.85/24); height of cheek 0.40 times maximum length of eye (0.37-0.435/24).

Apparent normal length of adults, 7.5-8 mm.

Third instar. No specimens examined, characters derived from puparium. *Prothoracic fringe* 47 μ (23-80)/80. *Cephalopharyngeal skeleton* 860 μ (770-935)/11 in length.

Puparium. Length 7.5 mm (6.7-8.3)/6; breadth 3.4 mm (3.0-4.0)/15; medium thick-walled, dull, brown. *Posterior region* (Fig. 63a): stigmatal plates 160 μ (135-190)/30 in diameter; distance between buttons 475 μ (380-540)/15, and across stigmatal plates 1030 μ (870-1300)/15; stigmatal ratio 0.49 (0.46-0.53)/15; stigmatal area outlined in large cuticular plaques; mesostigmatal area with pronounced plaques; mesostigmatal spines present, but scarce and small; hyperstigmatal spines numerous, stout, 21 μ (10-50)/150; hypostigmatal area with reduced folds, sometimes pronounced; hypostigmatal spines generally scarce, 9 μ (4-25)/150; circumstigmatal folds generally absent or at best weak, concentric. *Dorsal cuticle* (Fig. 63b) with numerous spines, 20 μ (10-50)/150; cuticular ridges weak. *Ventral cuticle* (Fig. 63c): ventral band ratio 0.70 (0.40-0.80)/15; anterior band pronounced with anterior patch distinct, spines 9 μ (4-17)/150; medial band pronounced, spines 14 μ (6-30)/150; posterior band pronounced, present on all segments, with numerous spines.

Type Series. Holotype male, allotype, and 11 male, 8 female paratypes, Okanagan Landing, B.C., July 19-28, 1941 (G. J. Spencer), nest of yellow warbler collected July 12. Holotype and allotype in the U.S. National Museum of Natural History, by courtesy of the late Professor Spencer; paratypes in the Canadian National Collection and the University of British Columbia. Additional paratypes: 5 males, 17 females, Ft. St. James, B.C., American redstart nest collected July 16, 1947 (J. Grant), flies reared in laboratory Aug. 10, 1947 (G. J. Spencer) [UBC and USNM]: male, Waha, Idaho, Aug. 12, 1923 (A. L. Melander) [USNM, labeled, but not published, as a paratype of *cuprina*]; 3 males, 3 females, Franklin Basin, Cache Co., Utah, July 27, 1971 (T. L. Whitworth), nest of warbling vireo [USNM and USU]; male, female, Ninette, Manitoba, May 5, 1958 (J. F. McAlpine), "at bleeding stump *Acer negundo*" [CNC].

Non-paratype Material: 3 males, 1 female, Okanagan Landing, B.C., from series from nest of yellow warbler; male, female, Ft. St. James, B.C., from series from nest of American redstart; Trinity Valley, B.C., puparia only, from nest of slate-colored junco [UBC].

Specimens Examined. 26 males, 32 females.

Distribution. British Columbia, Manitoba, Idaho, Utah.

Basis of description of immature stages: *Puparium* (15 specimens): Trinity Valley, B.C., slate-colored junco, August, 1946 (3); Fort St. James, B.C., American redstart, July, 1947 (5); Okanagan Landing, B.C., yellow warbler, July, 1941 (2); Franklin Basin, Cache Co., Utah, July 1971, warbling vireo (5).

Hosts. American redstart, warbling vireo, yellow warbler; also, from puparia only, slate-colored junco.

Ecology and Biology. Unknown.

Variation. An extremely aberrant variation was found among the paratypes from Okanagan Landing, B.C. Two females lacked reclinate fronto-orbital bristles, a characteristic considered to be typical of and confined to the subgenus *Trypocalliphora* and its single included species *braueri*. In fact, these two specimens would cause difficulty in the very first couplet of the key to females. One male and one female of the Cache Co., Utah series have two posterior bristles on the fore tibia.

Remarks. The curved surstylus associates *spenceri* with a certain group of species, at least a grouping of convenience. Males of these species, except for one or two, are difficult, and unassociated females are virtually impossible. For best identification, the sexes must be associated and preserved with puparia. The extremely narrow frons and parafacial will separate *spenceri* males from the other species. *P. bicolor* and *P. metallica* also have such a narrow frons but wider parafacials, and the former is further distinguished by having the 5th tergite cupreous in the male.

The main series from Okanagan Landing, B.C., was a mixed infestation of *spenceri* and *cuprina*, but luckily both sexes of these two species are easily distinguished. Males of the two have distinct surstyli, and in females of *cuprina* the 5th tergite is cupreous.

Puparia of *spenceri* are separated with difficulty from those of *halli* and *seminuda*. Separation is based on a consideration of all the characters.

Etymology. We take pleasure in dedicating this species to the late Professor G. J. Spencer of the University of British Columbia, whose long interest in *Protocalliphora* led to the accumulation of fine series of adults and immature stages, which he generously loaned to us for study.

Protocalliphora (*P.*) *tundrae*, new species
(Figs. 23, 36, 64, Map 12)

Calliphora azurea (Fallén).--Lundbeck, 1895: 114 [Hekla-Havn, East Greenland; same record repeated in several subsequent compilations on the insects of Greenland].
C. azurea.--Nielsen, 1907: 395 [Same record, with addition of date of collection, June 11, 1892].
Protocalliphora sp.--Oliver, 1963: 178 [Ellesmere I., N.W.T.].

Diagnosis. Sexes concolorous, monochromatic, shining metallic dark bluish; calypteres white; frons of male of medium width, preocellar area broadly triangular, bright gray microtomentose; surstylus narrow, curved.

Male. Head chiefly black in ground color, parafrontal and parafacial silvery gray microtomentose. Body shining, metallic dark bluish to bluish purple, thorax darker than abdomen, scutum viewed from behind with two stripes of gray microtomentum that demarcate a broad median shining stripe and the broad lateral shining areas; abdomen brighter than scutum, shining in direct view but gray microtomentose on tergites 2-4 when viewed from behind at a low angle. Calypteres white.

Frons of moderate width (Fig. 36), at narrowest 0.135 times head width (0.12-0.16) and 1.43 times ocellar span (1.22-1.73); ocellar tubercle typically with 6-8 small bristles behind the strong ocellar bristles; preocellar area broadly triangular, entirely and evenly bright gray microtomentose; upper end of row of frontal bristles usually opposite anterior end of preocellar triangle; parafrontal narrowed just before vertex, silvery gray microtomentose, parafrontal hairs in two irregular rows below, narrowing to one row above and ending well before end of row of frontal bristles; parafacial silvery gray microtomentose, with numerous fine black hairs in three irregular rows, its width obviously greater than breadth of 3rd antennal segment, equal to width of frons (1.00x; 0.91-1.13), 1.43 times ocellar span (1.22-1.62), and 0.86 times vibrissal interval (0.73-1.00); height of cheek 0.46 times maximum length of eye (0.43-0.48); occiput with 3-4 irregular rows of black hairs behind row of postocular cilia.

Thorax lacking accessory notopleural bristles; tympanic membrane and postalar wall with few short weak and inconspicuous hairs. Fore tibia typically with two posterior bristles.

Terminalia: cerci long and narrow to acute apices (cf. Fig. 5b); surstylus long and narrow, strongly curved (Fig. 23); aedeagus as usual in the genus (cf. Fig. 6a).

Female. Concolorous with male, but slightly more gray microtomentose. Frons nearly 1/3 width of head (0.31x; 0.30-0.32) and over three times ocellar span (3.16x; 2.89-3.62); preocellar area entirely and evenly dull, heavily yellowish gray to brownish gray microtomentose, as is microtomentum of parafrontal and parafacial; parafacial with numerous fine black hairs in 3-4 irregular rows, definite changeable spot along upper margin, width nearly twice breadth of 3rd antennal segment, 1/2 breadth of frons (0.50x; 0.44-0.54), 1.59 times ocellar span (1.29-1.75), and nearly equal to vibrissal interval (0.99x; 0.76-1.04); height of cheek nearly half maximum length of eye (0.49x; 0.47-0.51).

Apparent normal length of adults, 7.5-8.5 mm. (The two smallest specimens are only 5.5 mm).

Third instar. No specimens examined, characters derived from puparium. *Prothoracic fringe* 54 μ (26-84)/75. *Cephalopharyngeal skeleton* 860 μ (770-1000)/10 in length.

Puparium. Length 7.5 mm (6.8-7.8)/12; breadth 3.1 mm (2.6-3.4)/12; thin-walled, dull, light brown. *Posterior region* (Figs. 64a,b): stigmatal plates 180 μ (160-195)/23 in diameter;

distance between buttons 565 μ (470-630)/12; and across stigmatal plates 1080 μ (760-1190)/12; stigmatal ratio 0.51 (0.48-0.52)/12; stigmatal area somewhat depressed, faintly outlined in small plaques; numerous mesostigmatal spines present; mesostigmatal folds and plaques weak; hyperstigmatal spines numerous, 31 μ (16-46)/160; hypostigmatal area with few cuticular ridges or folds; hypostigmatal spines numerous, 23 μ (8-42)/160; circumstigmatal folds weak and concentric. *Dorsal cuticle* (Fig. 64c) with numerous, weak, cuticular ridges; spines numerous, 29 μ (10-46)/160. *Ventral cuticle* (Fig. 64d): ventral band ratio 0.57 (0.51-0.62)/10; spines of anterior patch slender, 17 μ (8-28)/150; medial band pronounced, spines 20 μ (8-38)/160; posterior band reduced on anterior 5-6 segments and vestigial or absent on segments six to twelve.

Type Series (all but a few reared and accompanied by puparia). Holotype male, allotype, and 20 paratypes (8 males, 12 females), *Northwest Territories:* Ellesmere Island, Hazen Camp, 81° 49' N, 71° 18' W, various dates of emergence July 20-Aug. 3, 1963 (D. R. Oliver), ex snow bunting nest. In the Canadian National Collection, Ottawa; some paratypes in USNM. Additional paratypes [CNC except for representatives in USNM and one in Copenhagen Museum]: **Canada, Northwest Territories: Ellesmere I.**, same locality as holotype, 12 males, 20 females, various dates, Aug. 2-19, 1963 (R. E. Leech), ex snow bunting nest; 5 males, 5 females, same locality, various dates July 27-Aug. 1, 1962 (R. B. Madge), reared but host not given; one female, same locality, June 19, 1962 (J. F. McAlpine); 2 males, 1 female, same locality, Aug. 7, 1961 (D. R. Oliver), ex snow bunting nest; 2 males, Hazen Lake, John's I., June 20, 1963 (B. Hocking) and June 12, 1964 (R. E. Leech); 2 males, same locality, June 20, 1963 (C. D. Sharplin). **Melville I.**: male, Bailey Point, July 24, 1965 (J. E. H. Martin). **Victoria I.** 11 males, 8 females, 71° 17' N, 114° W, June 25-30, 1975 (G. and D. M. Wood). **Keewatin District**: male, Repulse Bay, June 25, 1950 (J. E. H. Martin); female, Chesterfield, July 18, 1950 (J. G. Chillcott). **Mackenzie District**: 4 males, 2 females, 21 mi E Tuktoyaktuk, July 17, 21, 1971 (D. M. Wood). **Labrador**: male, female, Hebron, Aug. 5, 1974 (J. F. McAlpine), with puparia, nest of savannah sparrow on rocky ledge in tundra. **Quebec**: male, Ft. Chimo, Aug. 13, 1948 (H. N. Smith). **Yukon Territory**: 6 males, 7 females, Dempster Highway, mi 40, 51, 87, km 140.5, 155, various dates June 17-Aug. 8, 1973, 1980, and 1982 (G. and D. M. Wood); 2 females, Dempster Highway, km 155, June 29-July 3, 1980 (D. M. Wood, D. Lafontaine); male, 4 females, Dempster

Highway, km 465, June 23-25, 1980 (D. Lafontaine). **Greenland**: male, "Hekla-Havn," East Greenland, June 11, 1892 (H. Deichmann) [Copenhagen Mus.]; male, Sondrestrom Air Base, June 11, 1952 (W. J. Brown)[CNC]; 37 males, 31 females, Pearyland, Nedre Midsommer Sö [Lake, N of head of Independence Fjord], 82°14' N, 33°20' W, ex two snow bunting nests, puparia July 5, 1966, adults July 10-20, 1966 (J. E. H. Martin) [CNC].

Non-paratype Material: Two males, three females from the Hazen Camp locality, a male from Pearyland, and a female from Tuktoyaktuk, small or in poor condition.

Specimens Examined: 100 males, 100 females.

Distribution: Arctic Canada, Yukon Territory to southern Greenland, Labrador and northern Quebec, north to Melville and Ellesmere Islands and Pearyland in northern Greenland.

Basis of description of immature stages: *Puparium* (12 specimens): Hebron, Labrador, savannah sparrow, (2); Hazen Camp, Ellesmere Island, N. W. T., snow bunting, (10).

Hosts: Snow bunting, savannah sparrow.

Ecology and Biology. Although records are few and only two host species are known, this is clearly a species of the far north, a tundra species, living in the nests of ground-nesting birds. The records from snow bunting in Ellesmere Island and Pearyland in northern Greenland are especially interesting, as this bird nests "farther north than any other land bird" (AOU Checklist, 1957), and our records are from the northernmost extent of its breeding range.

Variation. Apparently the usual number of posterior bristles on the fore tibia is two, but enough specimens have only one that the variation will give some trouble to users of the keys. Of the 185 specimens that could be counted, 69 males and 50 females have two posterior bristles on each fore tibia, and another 8 males and 30 females have two on one side or the other, but 6 males and 22 females have only one posterior bristle on each fore tibia. One female (Dempster Highway) has one bristle on one tibia but none at all on the other.

Remarks. This species is very close to *fallisi*, but the characteristic appearance of the frons and especially the preocellar area is so consistent that we consider the two to be distinct species. When puparia are compared, the hyperstigmatal area in *fallisi* has stout, well spaced spines (Fig. 50), whereas in *tundrae* these spines are relatively slender and appear to be closer together (Fig. 64). Puparia of *tundrae* are most easily confused with those of *halli*. They can be separated by the fact that the posterior band, although reduced, is present on the anterior 5-6 segments of *tundrae* but absent on all segments of *halli*. In addition, *tundrae* appears to be limited, distributionally, to the arctic and tundra regions while the other species has only been recorded in western North America.

The series of *tundrae* was compared with the type series (3 males, 7 females) of *P. nuortevai* Grunin from Finland, loaned from the Zoological Museum of the University of Helsinki by courtesy of B. Lindeberg. Unfortunately the sample of *nuortevai* is small and its reared specimens are not in good condition. The Nearctic and Palearctic specimens are very similar. The most striking difference is the breadth of the parafacial in both sexes of *nuortevai*, approximately twice the breadth of the 3rd antennal segment, much greater than the ocellar span (2.00-2.14x in the four measured females, 1.86 and 2.00x in the two measured males), and in the four females slightly greater than breadth of frons (0.54 and 0.555x). The surstyli are the same curved type. The puparia are similar in gross appearance, although the few specimens of *nuortevai* show stronger hypostigmatal spines. Obviously *tundrae* and *nuortevai* are closely related, if indeed they are not the same species with subspecific relationships. Until a more adequate sample of *nuortevai* is available and puparia are also studied, it seems best to recognize *P. tundrae* as distinct. The far northern range of this species is remarkable. The specimens from Ellesmere Island and Pearyland in northern Greenland were reared from puparia found in snow bunting nests at latitudes 81° 49' N and 82° 14' N, respectively.

Etymology. The specific name is in the genitive case, based on the tundra habitat.

Calliphora splendida Macquart, species dubia

Calliphora splendida Macquart, 1846: 324 (1846: 196) [Galveston, Tex.].

Beginning with Harbeck (1907), this species was referred to *Protocalliphora*, and was considered the senior synonym of *P. metallica* (Townsend). The type of *splendida*, a female, has apparently been lost. Townsend (1919), recognizing some discrepancies of description, and Hall (1948), having regard to the then unknown occurrence of *Protocalliphora* so far south, both regarded *splendida* as unrecognizable. Today the distribution is not quite as strong as argument, although the genus is still unknown from Texas (Cf. Map 1), and the likelihood of its occurrence in coastal plain Texas is opposed by all we know of the group. Furthermore, even if *splendida* were a *Protocalliphora*, the association with *metallica* is by no means certain. The principal color character, the cupreous apex of the abdomen, is now known to be present in females of several species of *Protocalliphora*, besides Neotropical flies of other genera and families that might be involved. We prefer, with Townsend and Hall, to leave *Calliphora splendida* as a species dubia.

Appendix
Natural Hosts of *Protocalliphora* in North America

Common and scientific names are those adopted in the "Check-List of North American Birds", 6th edition (1983), published by the American Ornithologists' Union. In a few cases, well known former family names or common names for species are cited in brackets immediately after the presently adopted names. In addition to the natural hosts, the common chicken and the budgerigar have been reported as hosts of *Protocalliphora* (see under *P. braueri*), and *P. sialia* was reared from a nest of caged ringed turtle-doves used in experimental studies in Colorado, far from any feral established population of the doves.

Birds are recorded by species only, not by subspecies, except for the flickers, juncos, and yellow-rumped warbler, for which relatively recent taxonomic changes have altered the long-used classification and the common names. New host records are starred (*), with the name of the collector(s) credited in brackets at the end of the entry. The new records include those found in the studies by Bennett and Whitworth, plus the hosts of reared material identified by Sabrosky from various sources. References are given for records overlooked in or published since the last host list for North America (Hall 1948). Many of the published hosts have also been recorded by Bennett or by Whitworth, who reared *Protocalliphora* from at least 43 and 48 species, respectively.

Blackbirds: Emberizidae, Icterinae [Icteridae]
- Brewer's, *Euphagus cyanocephalus* (Wagler) [Wood 1953, Bent 1958]
- Red-winged, *Agelaius phoeniceus* (Linnaeus) [Miller 1909, Judd 1954]
- *Yellow-headed, *Xanthocephalus xanthocephalus* (Bonaparte) [Whitworth, C. S. Gold]

Bluebirds: Muscicapidae, Turdinae [Turdidae]
- Eastern, *Sialia sialis* (Linnaeus)
- *Mountain, *Sialia currucoides* (Bechstein) [G. J. Spencer, Whitworth]
- *Western, *Sialia mexicana* Swainson [Vasiliki Demas, C. S. Gold]

Buntings: Emberizidae [Fringillidae, part], Cardinalinae (Indigo, Lazuli) and Emberizinae (Snow)
Indigo, *Passerina cyanea* (Linnaeus)
*Lazuli, *Passerina amoena* (Say) [Whitworth]
*Snow, *Plectrophenax nivalis* (Linnaeus) [J. F. McAlpine, R. B. Madge, J. E. H. Martin, D. R. Oliver, D. M. Wood]

Bushtits: *Aegithalidae*
Bushtit, *Psaltriparus minimus* (Townsend)

Cardinals: Emberizidae [Fringillidae, part], Cardinalinae
Northern, *Cardinalis cardinalis* (Linnaeus)

Catbirds: Mimidae
Gray, *Dumetella carolinensis* (Linnaeus)

Chats: Emberizidae, Parulinae [Parulidae]
Yellow-breasted, *Icteria virens* (Linnaeus)

Chickadees: Paridae
Black-capped, *Parus atricapillus* Linnaeus
*Carolina, *Parus carolinensis* Audubon [E. T. McKnight, W. A. Tarpley]
Chestnut-backed, *Parus rufescens* Townsend [Gold & Dahlsten 1984]
Mountain, *Parus gambeli* Ridgway [Storer 1929, Gold & Dahlsten 1984]

Cowbirds: Emberizidae, Icterinae [Icteridae]
*Brown-headed, *Molothrus ater* (Boddaert) [V. E. Davis]

Creepers: Certhiidae
Brown, *Certhia americana* Bonaparte [Gold & Dahlsten 1984]

Crows: Corvidae
American [Common], *Corvus brachyrhynchos* Brehm

Cuckoos: Cuculidae
*Black-billed, *Coccyzus erythropthalmus* (Wilson) [Bennett]

Dippers: Cinclidae
*American [Water Ouzel], *Cinclus mexicanus* Swainson [C. L. Remington, Whitworth, J. A. Halstead]

Doves: Columbidae
Mourning, *Zenaida macroura* (Linnaeus)

Eagles: Acciptridae
Bald, *Haliaeetus leucocephalus* (Linnaeus) [Bortolotti 1985, A. H. Grewe, Jr., B. E. Cooper]
Golden, *Aquila chrysaetos* (Linnaeus) [Philip 1938, Hill and Work 1947]

Falcons and Kestrels: Falconidae
American Kestrel [Sparrow Hawk], *Falco sparverius* Linnaeus [Hill and Work 1947]
*Gyrfalcon, *Falco rusticolus* Linnaeus [K. G. Poole]
*Peregrine Falcon, *Falco peregrinus* Tunstall [J. A. Hagar]
Prairie Falcon, *Falco mexicanus* Schlegel [possible, White 1963]

Finches: Fringillidae
*Cassin's, *Carpodacus cassinii* Baird [Whitworth]
House, *Carpodacus mexicanus* (Muller)
Purple, *Carpodacus purpureus* (Gmelin)
Rosy, *Leucosticte arctoa* (Pallas) [Black Rosy, *L. atrata* Ridgway] [French 1959]

Flickers: Picidae
Northern [Common], *Colaptes auratus* (Linnaeus)
*Red-shafted, *C. auratus cafer* (Gmelin) [Whitworth]
Yellow-shafted, *C. auratus auratus* (Linnaeus)[Kenaga 1961]

Flycatchers: Tyrannidae
*Alder [Traill's, in part], *Empidonax alnorum* Brewster [Bennett]
*Ash-throated, *Myiarchus cinerascens* (Lawrence) [Vasiliki Demas]
*Dusky, *Empidonax oberholseri* (Phillips) [C. S. Gold]
Great Crested, *Myiarchus crinitus* (Linnaeus)
*Western, *Empidonax difficilis* Baird [N. A. M. Verbeek, C. S. Gold]
*Willow [Traill's, in part], *Empidonax traillii* (Audubon) [C. S. Gold, J. A. Halstead]

Goldfinches: Fringillidae
American, *Carduelis tristis* (Linnaeus)
Lesser, *Carduelis psaltria* (Say)

Goshawks. See under Hawks

Grackles: Emberizidae, Icterinae [Icteridae]
*Common, *Quiscalus quiscula* (Linnaeus) [Bennett, D. M. Davies, F. M. Stiner, Jr.]

Grosbeaks: Emberizidae, Cardinalinae, and Fringillidae (Pine Grosbeak)
*Black-headed, *Pheucticus melanocephalus* (Swainson) [D. C. Rentz]
Blue, *Guiraca caerulea* (Linnaeus) [Stabler 1959]
*Pine, *Pinicola enucleator* (Linnaeus) [Whitworth]

Harriers: See Hawks

Hawks: Acciptridae
Broad-winged, *Buteo platypterus* (Vieillot) [Meng 1954]
Cooper's, *Accipiter cooperii* (Bonaparte)
*Ferruginous, *Buteo regalis* (Gray) [Whitworth]
Northern Goshawk, *Accipiter gentilis* (Linnaeus) [Meng 1954]
Northern Harrier [Marsh Hawk], *Circus cyaneus* (Linnaeus) [F. & F. Hamerstrom 1954]
Red-shouldered, *Buteo lineatus* (Gmelin)
Red-tailed, *Buteo jamaicensis* (Gmelin)
Sparrow. See Falcons and Kestrels
Swainson's, *Buteo swainsoni* Bonaparte

Jays: Corvidae
*Blue, *Cyanocitta cristata* (Linnaeus) [Bennett]
*Gray [Canada], *Perisoreus canadensis* (Linnaeus) [Bennett]
*Pinyon, *Gymnorhinus cyanocephalus* Wied [K. A. Walker]

Juncos: Emberizidae [Fringillidae, part], Emberizinae
Dark-eyed, *Junco hyemalis* (Linnaeus)
*Oregon, *J. hyemalis oreganus* (Townsend) [Whitworth]
Slate-colored, *J. hyemalis hyemalis* (Linnaeus)

Kestrels: See Falcons and Kestrels

Kingbirds: Tyrannidae
Cassin's, *Tyrannus vociferans* Swainson
*Eastern, *Tyrannus tyrannus* (Linnaeus) [Bennett, Whitworth]
*Western [Arkansas], *Tyrannus verticalis* Say [Bennett, Whitworth, G. J. Spencer]

Larks: Alaudidae
Horned, *Eremophila alpestris* (Linnaeus)

Magpies: Corvidae
Black-billed, *Pica pica* (Linnaeus)

Martins: Hirundinidae
Purple, *Progne subis* (Linnaeus)

Meadowlarks: Emberizidae, Icterinae [Icteridae]
Eastern, *Sturnella magna* (Linnaeus) [Miller 1909]
Western, *Sturnella neglecta* Audubon [Arnold 1919]

Mockingbirds: Mimidae
Northern, *Mimus polyglottos* (Linnaeus)

Nuthatches: Sittidae
Brown-headed, *Sitta pusilla* Latham [Norris 1958]
Pygmy, *Sitta pygmaea* Vigors [probable, Law 1929]
Red-breasted, *Sitta canadensis* Linnaeus [Gold & Dahlsten 1984]
*White-breasted, *Sitta carolinensis* Latham [D. L. Dahlsten; Museum of Comparative Zoology, collector unknown]

Ospreys: Acciptridae, Pandioninae [Pandionidae]
*Osprey, *Pandion haliaetus* (Linnaeus) [F. and F. Hamerstrom]

Ovenbirds: Emberizidae, Parulinae [Parulidae]
*Ovenbird, *Seiurus aurocapillus* (Linnaeus) [Bennett]

Owls: Strigidae
*Eastern Screech-, *Otus asio* (Linnaeus) [Bennett]
*Flammulated, *Otus flammeolus* (Kaup) [Whitworth]
Great Horned, *Bubo virginianus* (Gmelin) [Lee and Ryckman 1954, Tirrell 1978]
Long-eared, *Asio otus* (Linnaeus)
*Western Screech-, *Otus kennicottii* (Elliot) [Sabrosky]

Pewees: Tyrannidae
*Western Wood-, *Contopus sordidulus* Sclater [Whitworth]

Phoebes: Tyrannidae
*Black, *Sayornis nigricans* (Swainson) [H. B. Leech]
Eastern, *Sayornis phoebe* (Latham) [Brimley 1938]
Say's, *Sayornis saya* (Bonaparte)

Pipits: Motacillidae
*Water, *Anthus spinoletta* (Linnaeus) [N. A. M. Verbeek]

Ravens: Corvidae
*Common, *Corvus corax* Linnaeus [Whitworth]

Redstarts: Emberizidae, Parulinae [Parulidae]
American, *Setophaga ruticilla* (Linnaeus) [Johnson 1930, George and Mitchell 1948]

Robins: Muscicapidae, Turdinae [Turdidae]
American, *Turdus migratorius* Linnaeus

Shrikes: Laniidae
Loggerhead, *Lanius ludovicianus* Linnaeus

Siskins: Fringillidae
*Pine, *Carduelis pinus* (Wilson) [J. D. Tiner]

Solitaires: Muscicapidae, Turdinae [Turdidae]
Townsend's, *Myadestes townsendi* (Audubon) [Munro 1949]

Sparrows: Emberizidae [Fringillidae, part], Emberizinae, and Passeridae (House Sparrow)
American Tree, *Spizella arborea* (Wilson)
Chipping, *Spizella passerina* (Bechstein)
*Field, *Spizella pusilla* (Wilson) [V. Nolan, Jr.]
*Fox, *Passerella iliaca* (Merrem) [Whitworth]
*Grasshopper, *Ammodramus savannarum* (Gmelin) [D. F. Owen]
House [English], *Passer domesticus* (Linnaeus)
Savannah, *Passerculus sandwichensis* (Gmelin) [Munro 1949, Bedard and McNeil 1979]
Song, *Melospiza melodia* (Wilson)
Swamp, *Melospiza georgiana* (Latham)
*Vesper, *Pooecetes gramineus* (Gmelin) [Bennett]

Sparrows: (continued)
White-crowned, *Zonotrichia leucophrys* (Forster)
White-throated, *Zonotrichia albicollis* (Gmelin)

Starlings: Sturnidae
European, *Sturnus vulgaris* Linnaeus

Swallows: Hirundinidae
Bank, *Riparia riparia* (Linnaeus)
Barn, *Hirundo rustica* Linnaeus
Cliff, *Hirundo pyrrhonota* (Vieillot)
*Northern Rough-winged, *Stelgidopteryx serripennis* (Audubon) [*S. ruficollis* (Vieillot)] [Bennett]
Tree, *Tachycineta bicolor* (Vieillot)
*Violet-green, *Tachycineta thalassina* (Swainson) [Whitworth, R. E. Leech, S. Marino and W. L. Jellison]

Tanagers: Emberizidae, Thraupinae [Thraupidae]
Scarlet, *Piranga olivacea* (Gmelin)

Thrashers: Mimidae
Brown, *Toxostoma rufum* (Linnaeus)

Thrushes: Muscicapidae, Turdinae [Turdidae]
Hermit, *Catharus guttatus* (Pallas) [George and Mitchell 1948]
Swainson's [Olive-backed, Russet-backed], *Catharus ustulatus* (Nuttall)
*Wood, *Hylocichla mustelina* (Gmelin) [Bennett, V. E. Davis, C. Strehl, F. M. Stiner, Jr.]

Towhees: Emberizidae [Fringillidae, part], Emberizinae
Brown, *Pipilo fuscus* Swainson
Rufous-sided, *Pipilo erythrophthalmus* (Linnaeus)

Vireos: Vireonidae
*Red-eyed, *Vireo olivaceus* (Linnaeus) [Bennett]
Solitary [Blue-headed], *Vireo solitarius* (Wilson) [Johnson 1930]
*Warbling, *Vireo gilvus* (Vieillot) [Whitworth]

Warblers: Emberizidae, Parulinae [Parulidae]
Black-throated Blue, *Dendroica caerulescens* (Gmelin)
*Canada, *Wilsonia canadensis* (Linnaeus) [C. E. Hope]

Warblers: (continued)
Chestnut-sided, *Dendroica pensylvanica* (Linnaeus) [Johnson 1929a]
*Nashville, *Vermivora ruficapilla* (Wilson) [Bennett]
Prairie, *Dendroica discolor* (Vieillot) [Nolan 1955, 1959]
Wilson's [Pileolated], *Wilsonia pusilla* (Wilson) [Rausch 1972]
Worm-eating, *Helmitheros vermivorus* (Gmelin)
Yellow, *Dendroica petechia* (Linnaeus)
*Yellow-rumped, *Dendroica coronata* (Linnaeus)
*Audubon's, *D. coronata auduboni* (Townsend) [Whitworth]
*Myrtle, *D. coronata coronata* (Linnaeus) [Bennett, C. E. Hope]

Waterthrushes: Emberizidae, Parulinae [Parulidae]
*Louisiana, *Seiurus motacilla* (Vieillot) [A. Wetmore]
*Northern, *Seiurus noveboracensis* (Gmelin) [Bennett]

Waxwings: Bombycillidae
Cedar, *Bombycilla cedrorum* Vieillot

Woodpeckers: Picidae
*Downy, *Picoides pubescens* (Linnaeus) [Bennett]

Wrens: Troglodytidae
Bewick's, *Thryomanes bewickii* (Audubon) [Gold & Dahlsten 1981, 1984]
Carolina, *Thryothorus ludovicianus* (Latham)
House, *Troglodytes aedon* Vieillot
*Marsh [Long-billed Marsh], *Cistothorus palustris* (Wilson) [Whitworth]

Yellowthroats: Emberizidae, Parulinae [Parulidae]
*Common, *Geothlypis trichas* (Linnaeus) [Bennett]

Annotated References

Annotations (by Sabrosky) in brackets give (1) the species as recorded in the publication; (2) locality and host, unless these are indicated in the title; and (3) the actual identity of the parasite species as determined by rechecking original material, or the probable identity if that can be deduced with reasonable assurance from the published biology and distribution. The species of blow flies referred to are all in *Protocalliphora* (abbreviated as *P.*), unless otherwise stated. Only publications that contain original records on Nearctic species are so annotated; those that merely repeat earlier published records or information, such as text books and some state lists, have not been included unless they are cited for other reasons.

A few publications not specifically referred to in the text are included because their host or locality records were used for lists, tables, or maps.

Aldrich, John M.

1901. Synonymische Notiz. Wiener Entomologische Zeitung, 20: 68. [*Avihospita* Hendel 1901 = *Protocalliphora* Hough 1899].

1905. A catalogue of North American Diptera. Smithsonian Miscellaneous Collections, 46: 680 pp.

Allen, Robert W., and Margaret M. Nice

1952. A study of the breeding biology of the purple martin (*Progne subis*). American Midland Naturalist, 47: 606-665. [Larvae of *P. splendida*, George Reserve near Ann Arbor, Mich.; unverified].

American Ornithologists' Union, Committee on Classification and Nomenclature

1957. Check-List of North American birds (5th edition). America Ornithologists' Union, pp. xiii + 691.

1983. Check-list of North American birds (6th edition). American Ornithologists' Union, Washington, D.C., pp. xxix + 877.

Arnold, W. W.

1919. Maggot infested birds. Auk, 36: 147-148. [*P. azurea*, Colo., various hosts; = *P. braueri*, paratypes of *hirudo*].

Banks, Nathan
1912. The structure of certain dipterous larvae with particular reference to those in human foods. United States Department of Agriculture, Bureau of Entomology, Technical Series, 22: 44 pp. [*P. chrysorrhoea*, Penn. and Mass., on bluebirds; = *P. sialia*.

Barbara Ann, Sister
1970. Five years' experience with a bluebird population. Maryland Birdlife, 26: 81-92. [Md.; *P. sialia* reared; det. Sabrosky].

Bedard, Jean, and Jeremy N. McNeil
1979. *Protocalliphora hirudo* (Diptera: Calliphoridae) infesting Savannah sparrow, *Passerculus sandwichensis* (Aves: Fringillidae), in eastern Quebec. Canadian Entomologist 111: 111-112. [det. G. F. Bennett and B. Cooper].

Bennett, Gordon, F.
1957. Studies on the genus *Protocalliphora* (Diptera: Calliphoridae). University of Toronto, Canada, Unpublished Ph.D. Dissertation: 194 pp., 33 pls.

Bent, Arthur C.
1937. Life histories of North American birds of prey. Order Falconiformes (Part 1). Bulletin of the United States National Museum, 167: viii + 409, 102 pls. [*Protocalliphora* larvae in ears of young red-shouldered hawks, probably N.H.; undoubtedly *P. avium*].

1942. Life histories of North American flycatchers, larks, swallows, and their allies. Bulletin of the United States National Museum, 179: xi + 555, 70 pls. [p. 395, *P. splendida* var. *sialia*, nest of tree swallow; p. 418, *P. splendida*, Maine, bank swallow; p. 455, nest of barn swallow].

1948. Life histories of North American nuthatches, wrens, thrashers and their allies. Bulletin of the United States National Museum, 195: xi + 475, 90 pls. [*Protocalliphora splendida sialia*, citing Mason (1936)].

1958. Life histories of North American blackbirds, orioles, tanagers, and allies. Bulletin of the United States National Museum, 211: ix + 549, 37 pls. [*Protocalliphora* larvae on nestlings of Brewer's blackbird].

Bergtold, W. H.
1927. A house finch infected by fly larvae. Auk, 44: 106-107. [P. *splendida*, Colo., larvae in abscess; undoubtedly *P. braueri*).

Bezzi, Mario
1922. On the dipterous genera *Passeromyia* and *Ornithomusca*, with notes and bibliography on the non-pupiparous Myiodaria parasitic on birds. Parasitology, 14: 29-46.

Bezzi, Mario and Paul Stein
1907. Band III. Cyclorrapha (part), 828 pp., *in* Becker, Bezzi, Kertész, and Stein, Katalog der paläarktischen Dipteren. 4 vols. Budapest.

Bohm, Robert T.
1978. *Protocalliphora* infestation in great horned owls. Wilson Bulletin, 90: 297. [*P. avium*, det. Sabrosky, near Foley, Minn., in ear cavities].

Bortolotti, Gary R.
1985. Frequency of *Protocalliphora avium* (Diptera: Calliphoridae) infestations on bald eagles (*Haliaeetus leucocephalus*). Canadian Journal of Zoology, 63: 165-168. [*P. avium*, det. Sabrosky, Sask., common in ear cavities].

Boyd, A. W.
1935. A fly parasite on the swallow. British Birds, 28: 225.
1936. Report on the swallow enquiry, 1935. Ibid., 30: 98-116.

Boyes, J. W.
1961. Somatic chromosomes of higher Diptera. V. Interspecific and intra-specific variation in the Calliphoridae. Canadian Journal of Zoology, 39: 549-570, 1 plate. [6 spp. of *Protocalliphora*, Algonquin Park, Ontario; "n. sp. near *sialia*" = *shannoni*].

Boyes, J. W., and Janny M. van Brink
1965. Chromosomes of Calyptrate Diptera. Canadian Journal of Genetics and Cytology, 7: 537-550. [6 spp. of *Protocalliphora;* "n. sp. near *sialia*" = *shannoni*].

Boyes, J. W., and G. E. Shewell
1975. Cytotaxonomy of Calliphoridae (Diptera). Genetica, 45: 435-488. [6 spp. of *Protocalliphora;* "n. sp. nr. *sialia*" = *shannoni*].

Brimley, C. S.
1938. The insects of North Carolina. North Carolina Department of Agriculture, Division of Entomology, Raleigh, N.C., 560 pp. [*P. avium*, Raleigh, N.C., nest of phoebe; = *P. deceptor*, paratypes].

Britton, W. E.
1920. Check-list of the insects of Connecticut. Bulletin of the Connecticut State Geological and Natural History Survey, 31: 397 pp. [*P. azurea*, Conn., apparently on authority of C. W. Johnson].
Brower, A. E.
1966. Management of bird houses to reduce nesting loss to *Protocalliphora*, a parasitic fly. Maine Field Naturalist, 22: 73-74. [No spp. mentioned; general on use of pyrethrum powder in bird houses].
Burks, B. D.
1979. Family Pteromalidae, pp. 768-835, *in* Krombein et al. (eds.), Catalog of Hymenoptera in America North of Mexico, Vol. 1: 1198 pp. Washington, D.C.
Burtch, Verdi
1920. Maggots in the ears of nestling Cooper's hawks (*Accipiter cooperi*). Auk, 37: 293. [Larvae "possibly" *Compsomyia macellaria* (Fabricius), Branchport, N.Y.; presumably = *P. avium*].
Cais, Leszek
1965. [*Protocalliphora hirudo* Shannon and Dobroscky, 1924 (Diptera, Calliphoridae) in Poland] (In Polish). Fragmenta Faunistica (Polska Akademia Nauk, Instytut Zoologiczny), 12: 183-191.
Chapman, B. R.
1973. The effects of nest ectoparasites on cliff swallow populations. Texas Technological University, Unpublished Ph.D. Dissertation: 70 pp. [Dissertation Abstracts, B: 73-23753]. [No mention of *Protocalliphora*].
Chapman, L. B.
1935. Studies of a tree swallow colony. Bird-Banding, 6:45-57. [*Protocalliphora* larvae, Princeton, Mass.; not checked].
Cole, Frank R., and A. L. Lovett
1921. An annotated list of the Diptera (flies) of Oregon. Proceedings of the California Academy of Sciences, Series 4, 11: 197-344. [*P. azurea*, Corvallis, Oreg.; = *P. braueri*].
Coutant, Albert F.
1915. The habits, life history, and structure of a blood-sucking muscid larva (*Protocalliphora azurea*). Journal of Parasitology, 1: 135-150. [Ithaca, N.Y., crow; = *P. avium*].

Crocoll, Scott, and James W. Parker
1981. *Protocalliphora* infestation in broad-winged hawks. Wilson Bulletin, 93: 110. [Chautauqua Co., N.Y.; *P. avium*].
Crosskey, R. W.
1981. Simuliid taxonomy--the contemporary scene. pp. 3-18, *in* Laird, M. (ed.). Blackflies. Academic Press, New York, pp. xiv + 399.
Dobroscky, Irene D.
1925. External parasites of birds and the fauna of birds' nests. Biological Bulletin, 48: 274-281 [*P. avium*, from crows, and *P. splendida*, N.Y.; *avium* probably correct, latter probably *metallica*].
Dodge, Harold R., and John M. Seago
1954. Sarcophagidae and other Diptera taken by trap and net on Georgia mountain summits in 1952. Ecology, 35: 50-59. [*Protocalliphora*, 2 spp., not named; = *P. sialia* and *P. braueri*].
Dufour, L.
1845. Histoire des métamorphoses de la *Lucilia dispar*. Annales de la Société Entomologique de France, Séries 2, 3: 205-214. [Earliest biological study of bird blow flies, now considered to be *P. azurea* (Fallén)].
Emden, F. I. van
1954. Diptera Cyclorrhapha, Calyptrata (I), Section (a) Tachinidae and Calliphoridae. pp. 133, *In* Handbooks for the Identification of British Insects, vol. X, Part 4a.
Engel, Erich O.
1920. Dipteren, die nicht Pupiparen sind, als Vogelparasiten. Zeitschrift für Wissenschaftliche Insektenbiologie, 15: 249-258, 9 figs.
Eshuis-van der Voet, C. W.
1972. Parasitism by *Protocalliphora azurea* (Fall.). Verhandelingen der Koninklijke Nederlandsche Akademie van Wetenschappen, Afdeeling Natuurkunde, Tweede Reeks (Series 2), 61(3): 73-74.
1975. Parasitism by *Protocalliphora* spp. Ibid. 66: 81-84.
Eshuis-van der Voet, C. W., and E. Houwink
1976. Parasitism by *Protocalliphora* spp. Ibid. 67: 116.
Eshuis-van der Voet, C. W., and H. N. Kluyver
1971. Parasitism by *Protocalliphora azurea* Fall. Ibid. 60(2): 50-51.

Eshuis-van der Voet, C. W., and R. H. de Reede
1974. The effect of parasitism by *Protocalliphora*. Ibid. 63: 77-79.
Fallén, C. F.
1817. Beskrifning öfver de i Sverige funna Fluge Arter, som kunna föras till slägtet *Musca*. Första Afdelningen. Kongliga Vetenskaps- Academiens Handlingar (Stockholm), [ser. 3], 1816: 226-254. [*Musca azurea*, n. sp., p. 245].
1821. Monographia Muscidum Sveciae (part). pp. 41-48. Lund, Sweden.
Fan, C. T.
1965. [Key to the common synanthropic flies of China] (In Chinese). Academy of Science, Peking, pp. xv + 330, 40 pls.
French, Norman R.
1959. Life history of the Black Rosy Finch. Auk, 76: 161-180. [NW Wyoming; *Protocalliphora* new species = *P. spatulata* n. sp.].
Gahan, A. B.
1933. Description of a chalcidoid parasite of *Protocalliphora* (Hymenop.). Canadian Entomologist, 65: 31-33. [*Morodora armata* new genus and species, from pupae of *P. avium*, Ravalli County, Mont.; = *P. asiovora*].
George, J. L., and R. T. Mitchell
1948. Notes on two species of Calliphoridae (Diptera) parasitizing nestling birds. Auk, 65: 549-552. [*Apaulina metallica* and *A. hirudo*, Lake Clear Junction, N.Y.; former is chiefly *P. metallica*, partly *P. bicolor; P. hirudo* (now *braueri*) is correct].
Gibson, Arthur
1917. The entomological record, 1916. Annual Report of the Entomological Society of Ontario, (1916) 47: 137-171. [*P. azurea*, Ottawa, Ont.; = *P. metallica*].
1919. The entomological record, 1918. Annual Report of the Entomological Society of Ontario, (1918) 49: 97-123. [*Phormia azurea*, Outremont, Que.; not checked].
Gill, Gordon D.
1956. Filth flies of central Alaska. Journal of Economic Entomology, (1955) 48: 648-653 [*P. avium*, det. M. T. James; not checked].

Goddard, Jerome, and Paul K. Lago
1983. An annotated list of the Calliphoridae (Diptera) of Mississippi. Journal of the Georgia Entomological Society, 18(4): 481-484 [*P.* n. sp. = *P. deceptor*].
Gold, C. S., and D. L. Dahlsten
1981. A new host record for *Tachinaephagus zealandicus* [Hym.: Encyrtidae]. Entomophaga, 26: 459-460 [Host: *P.* n. sp., puparia only].
1984. Effects of parasitic flies (*Protocalliphora* spp.) on nestlings of mountain and chestnut-backed chickadees. (1983) Wilson Bulletin, 95(4): 560-572, 4 tables [*P.* n. sp. no. 1 = *P. parorum*].
Gregor, František, and Dalibor Povolný
1959. Kritischer Beitrag zur Kenntnis der Tribus Phormiini (Diptera, Calliphoridae). Casopis Ceskoslovenské Spolecnosti Entomologické, 56: 26-51, 5 plates.
Grimshaw, Percy H.
1901. Diptera. *In* Fauna Hawaiiensis (David Sharp, editor), 3 (part i):1-77, 3 pls. [*Calliphora azurea*, Kona, Hawaii; = *Phormia regina* (Meigen)].
Grunin, K. J.
1966. [New and little-known Calliphoridae (Diptera), mainly blood-sucking or subcutaneous parasites of birds] (In Russian). Entomologicheskoe Obozrenie, 45: 897-903 [English translation, Entomological Review 45: 503-506].
1970a. 108. Fam. Calliphoridae. pp. 607-624, *in* [Flies, Fleas] (Stackelberg and Nartshuk, editors), volume 5, part 2, of Bei-Bienko, G. J. (editor), [Identification of the insects of the European part of the USSR] (In Russian). Zoological Institute of the Academy of Sciences, Leningrad, 943 pp.
1970b. [New species of Calliphoridae (Diptera) for the fauna of the USSR] (In Russian). Entomologischeskoe Obozrenie, 49: 471-483.
1971a. Ergebnisse der zoologischen Forschungen von Dr. Z. Kaszab in der Mongolei. 277. Calliphoridae (Diptera) II. Acta Zoologica Academiae Scientiarum Hungaricae, 17: 255-259.
1971b. [*Phormiata* Grunin, gen.n.--The sixth genus of the tribe Phormiini (Diptera, Calliphoridae)] (In Russian). Entomologicheskoe Obozrenie, 50: 444-445.

1972. Beschreibung einer ornithoparasitischen Fliege, *Protocalliphora nuortevai* Grunin, sp. n. (Dipt., Calliphoridae) aus Nord-Finnland. Annales Entomologici Fennici, 38: 156-158.

1975. [Gasterophilidae, Calliphoridae, Oestridae, Hypodermatidae (Diptera) of the Soviet-Mongolian expedition, 1969-1971] (In Russian). [Insects of Mongolia], 3: 620-627. Zoological Institute, Academy of Sciences of the USSR.

Grunin, K. J., P. Nuorteva, and P. Rajala

1969. *Trypocalliphora lindneri* Peus (Dipt., Calliphoridae) as a subcutaneous parasite of the wheatear in northern Finland. Annales Entomologici Fennici, 35: 56-57.

Guberlet, J. E., and H. H. Hotson

1940. A fly maggot attacking young birds, with observations of its life history. Murrelet, 21: 65-68. [*P. hesperia*, Seattle, Wash., subcutaneous infestation of house sparrow; = *P. braueri*].

Hagar, Joseph A.

1969. History of the Massachusetts peregrine falcon population, 1935-57. pp. 123-131, *in* Hickey, J. J. (editor), Peregrine falcon populations: their biology and decline. University of Wisconsin Press, Madison, Wis., pp. xxii + 596. [*Protocalliphora* larvae in ears; undoubtedly *P. avium*].

Hakanen, R., K. J. Grunin, and P. Nuorteva

1974. Larvae of *Trypocalliphora lindneri* Peus (Dipt., Calliphoridae) as subcutaneous pathogens on nestlings of the meadow pipit and common redpoll in the subarctic. Annales Entomologici Fennici, 40: 15-18.

Hall, David G.

1948. The blowflies of North America. Thomas Say Foundation (Entomological Society of America), Lafayette, Ind., 477 pp., 51 plates. [10 spp. of *Apaulina*, i.e., *Protocalliphora;* see records under individual species; exact date of publication noted in Sabrosky, 1950].

1965. Family Calliphoridae [except Melanomyini]. pp. 922-933, *in* Stone, Alan, et al., A catalog of the Diptera of America north of Mexico. United States Department of Agriculture, Agriculture Handbook 276: pp. iv + 1696.

Hamerstrom, Frances, and Frederick Hamerstrom

1954. Myiasis of the ears of hawks. Falconry News and Notes 1 (3): 4-8 [*Protocalliphora*, Iowa, Wis., Mich.; undoubtedly *P. avium*].

Harbeck, H. S.
1907. [Seven species of Diptera new to the New Jersey state list of insects]. Entomological News, 18: 216. [*P. splendida*, Ashland, N.J.; not checked].

Hardy, D. Elmo
1981. Diptera Cyclorrhapha IV. pp. vii + 491. *In* Insects of Hawaii, vol. 14. University Press of Hawaii, Honolulu.

Hearle, Eric
1938. Insects and allied parasites injurious to livestock and poultry in Canada. Canada Department of Agriculture, Farmers' Bulletin, 53: 108 pp. [*P. hirudo cuprea*, eastern Canada, in tumorlike swellings on swallows, *P.* sp. on baby chicks, and *P. avium*, eastern Canada, on swallows; none checked: first two undoubtedly *braueri*, but *avium* doubtful on swallows].

Heinz, H. J.
1954. Weitere Vogelblutfliegen (Gattung *Protocalliphora, Dipt.*) in Deutschland und Österreich. Die Vogelwarte, 17: 39-42, 3 figs. [*P. hirudo* (now *braueri*) in Austria, Steiermark].

Hendel, Friedrich
1901. Beitrag zur Kenntnis der Calliphorinen (Dipt.). Wiener Entomologische Zeitung, 20: 28-33.

Hennig, Willi
1939. Über Namen und Artenzahl der deutschen "Vogelblutfliegen." Arbeiten über Physiologische und Angewandte Entomologie aus Berlin-Dahlem, 6: 359-364.
1952. Die Larvenformen der Dipteren, Part 3. Akademie-Verlag: Berlin, pp. vii + 628.

Henshaw, H. W.
1908. A parasitic fly injurious to our native birds. Auk, 25: 87-88. [*P. chrysorrhoea*, Wellesley Hills, Mass., on bluebirds; = *P. sialia*, part of type series].

Hickey, Joseph J.
1942. Eastern population of the duck hawk. Auk, 59: 176-204. [*Protocalliphora* larvae, southern N.Y., on peregrine falcon; probably *P. avium*].
1969(editor). Peregrine falcon populations: Their biology and decline. University of Wisconsin Press, Madison, Wis., pp. xii + 596, 60 plates. [Plate 43: *Protocalliphora* larvae, Vt., in auditory meatuses of young peregrine falcons; undoubtedly *P. avium*].

Hicks, Ellis A.
[1959]. Check-list and bibliography on the occurrence of insects in birds' nests. Iowa State College Press, Ames, Iowa, 681 pp.
1962. Ibid. Supplement I. Iowa State Journal of Science, 36(3): 233-348.
1971. Ibid. Supplement II. Iowa State Journal of Science, 46(2): 123-338.

Highhouse, W. L.
1963. Operation bluebird (1962) in Warren County, Penn. The Kingbird, 13(2): 24. [*Apaulina* larvae; use of flea powder in nests].

Hill, H. M.
1948. Raptorial hosts of *Protocalliphora*. Condor, 50: 131. [Corrections to Hill and Work (1947)].

Hill, H. M., and T. H. Work
1947. *Protocalliphora* larvae infesting nestling birds of prey. Condor, 49: 74-75. [*Protocalliphora* larvae, Alameda Co., Calif., sparrow hawk and golden eagle; probably *P. asiovora*].

Horning, Donald S., Jr., and William F. Barr
1970. Insects of Craters of the Moon National Monument Idaho. Miscellaneous Series of the University of Idaho College of Agriculture, no. 8: 118 pp. [*Protocalliphora* n. spp. A., B, and sp. = *P. seminuda*, *P. beameri*, and *P. brunneisquama*, respectively].

Hough, Garry de N.
1899a. Some North American genera of the dipterous group, Calliphorinae Girschner. Entomological News, 10: 62-66 (March).
1899b. Synopsis of the Calliphorinae of the United States. Zoological Bulletin, 2: 283-290 (Sept.).

ICZN (International Code of Zoological Nomenclature)
1961 (1st edition), 1985 (third edition). International Trust for Zoological Nomenclature, London, England.

Ishijima, Hideo
1967. Revision of the third stage larvae of synanthropic flies of Japan (Diptera: Anthomyiidae, Muscidae, Calliphoridae and Sarcophagidae). Japanese Journal of Sanitary Zoology, 18: 47-100, 12 plates. [*P. azurea:* brief description of larva, figures of cephalopharyngeal sclerites and posterior spiracles].

Itämies, Juhani, and Mikko Ojanen
1975. *Trypocalliphora lindneri* Peus. (Dipt. Calliphoridae) infesting a great tit nestling (*Parus major* L.). Annales Entomologici Fennici, 41: 42.
James, Maurice T.
1955. The blowflies of California. Bulletin of the California Insect Survey, 4 (1): 34 pp. [7 species of *Protocalliphora*, with Calif. records for 5 (see under individual species); "*hesperia*" is a mixture of three species].
1969. A study in the origin of parasitism. Bulletin of the Entomological Society of America, 15: 251-253.
Jellison, William L.
1949. Biological studies on the faunae of nests of birds and rodents in relation to disease of animals and man. University of Minnesota, Summaries of Ph.D. Theses, 3: 53-57. [Thesis of 1940, *Protocalliphora* on crow and magpie fledglings; = *P. asiovora*].
Jellison, William L., and Cornelius B. Philip
1933. Faunae of nests of the magpie and crow in western Montana. Canadian Entomologist, 65: 26-31 [*P. avium;* = *P. asiovora*].
Johannsen, O. A.
1928. Diptera, pp. 687-868, *in* Leonard, M. D. (editor), A list of the insects of New York. Memoir of the Cornell University Agricultural Experiment Station, 101: 1121 pp. [*Protocalliphora*, p. 827, N.Y. records for *P. avium*, *P. splendida* form *splendida*, and *splendida* form *sialis* (sic); *P. avium* verified, form *splendida* not checked, of form *sialia* the males of the McLean, Caroline, and Fall Creek records are *P. metallica*, the Farmingdale specimen not seen].
Johnson, Charles W.
1910. Order Diptera, pp. 703-814, *in* Smith, John B., The insects of New Jersey. Annual Report of the New Jersey State Museum 1909: 15-888 [*P. splendida*, Ashland, N.J., July 16; not checked].
1925a. Fauna of New England. 15. List of the Diptera or two-winged flies. Occasional Papers of the Boston Society of Natural History, 7: 326 pp. [*P. splendida* = part *P. metallica*, part *P. bicolor; P. hirudo cuprea* = *P. braueri; P. splendida sialis* [sic] = part *P. sialia*, *P. shannoni*, *P. metallica*].

1925b. Insects that infest birds. Bulletin of the Northeastern Bird-Banding Association, 1: 51-53. [*P. hirudo cuprea*, Middleboro, Mass., in neck of sparrow; = *P. braueri*].

1927a. The infestation of bluebirds' nests by *Protocalliphora*. Bulletin of the Northeastern Bird-Banding Association, 3: 1-3 [*P. splendida* form *sialia*, Rock and Needham, Mass.; = *P. sialia*].

1927b. Dipterological notes. Psyche, 34: 33-35. [Same records as in 1927a].

1927c. Further studies of *Protocalliphora* infesting nestling birds. Bulletin of the Northeastern Bird-Banding Association, 3: 77-82. [*P. splendida splendida*, Ashland, N. H., black-throated blue warbler, and *P. splendida* form *sialia*, various localities and hosts; former = *P. bicolor*, latter = *P. sialia*].

1927d. The insect fauna. Biological survey of the Mount Desert region, Part 1: 247 pp. [*P. splendida* form *sialia*, Southwest Harbor, Maine; not checked; same record repeated in enlarged editions of 1938 and 1946].

Johnson, Charles W.

1929a. Some additional notes on *Protocalliphora*. Bulletin of the Northeastern Bird-Banding Association, 5: 29-30. [*P. splendida splendida*, N.H., chestnut-sided warbler and robin = *P. bicolor* and *P. shannoni*, respectively; *P. splendida* var. *sialia*, Mass., bluebird and barn swallow, verified as *sialia*; *P. hirudo* var. *cuprea*, Mass. = *braueri*, repetition of Johnson, 1925b].

1929b. The injury to nestling birds by the larvae of *Protocalliphora*. Annals of the Entomological Society of America, 22: 131-135. [Review of the 1925-1928 seasons].

1930. Notes on *Protocalliphora* during the summer of 1930. Bird-Banding, 1: 169-173. [*P. sialia*, Mass., nests of bluebird and tree swallow = *P. sialia; P. splendida splendida*, N.H., nests of redstart and blue-headed vireo = *P. bicolor*].

1931. Nestling birds destroyed by the larvae of *Protocalliphora*. Bulletin of the Boston Society of Natural History, 59: 21-24. [Brief review of previous records].

1932. Notes on *Protocalliphora* during the summer of 1931. Bird-Banding, 3: 26-29 [*P. sialia*, Mass., Vt., Ohio = *P. sialia; P. splendida*, Mass. = *P. metallica*].

Judd, W. W.

1944. Dipterous larvae parasitic on nestlings of the song sparrow. Canadian Field Naturalist, (1943) 57: 146. [*P. splendida*, Que.; = *P. metallica*].

1954. Insects collected from birds' nests at London, Ontario. Canadian Field Naturalist, 68: 122-123. [*P.* sp. near *sialia*, red-winged blackbird; = *shannoni*].

1957. Diptera (Calliphoridae, Heleidae, and Chloropidae) collected from birds' nests at London, Ontario. Entomological News, 68: 32-34. [*P.* sp. (*sialia*?), robin and brown thrasher; = *P. shannoni*].

Kamal, Adel S.

1958. Comparative study of thirteen species of *sarcosaprophagous* Calliphoridae and Sarcophagidae (Diptera). I. Bionomics. Annals of the Entomological Society of America, 51: 261-271. [*P. avium*, Moscow, Idaho, on magpies; undoubtedly *P. asiovora*].

Kano, Rokuro, and Satoshi Shinonaga

1966. Notes on flies of medical importance in Japan, Part XXV. Descriptions of two species belonging to the genus *Protocalliphora* (Diptera: Calliphoridae). Japanese Journal of Sanitary Zoology, 17: 164-168.

Kenaga, Eugene E.

1961. Some insect parasites associated with the eastern bluebird in Michigan. Bird-Banding, 32: 91-94. [*P. sialia*, det. Sabrosky; also *Protocalliphora* larvae recorded on other birds].

Kluyver, H. N., and C. W. Eshuis-van der Voet

1970. Parasitism by *Protocalliphora* sp. Verhandelingen der Koninklijke Nederlandsche Akademie van Wetenschappen, Afdeeling Natuurkunde, Tweede Reeks (Series 2), 59(3): 48.

Knowlton, George F.

1936. Some Utah flesh flies. Proceedings of the Utah Academy of Sciences, Arts and Letters, 13: 237-239. [*P. splendida*, Hooper and Salt Lake City, Utah; Hooper specimen not located but possibly *cuprina*, Salt Lake City specimen = *Phormia regina* (Meigen)].

Krug, Howard H.

1941. Bluebird banding at Chesley, Ontario. Bird-Banding, 12: 23-26. [Maggots, probably *Protocalliphora*, in nests of bluebirds and tree swallows].

Laskey, Amelia R.
1948. Some nesting data on the Carolina wren at Nashville, Tennessee. Bird-Banding, 19: 101-121. [Pupa cases of *Protocalliphora*].
Law, J. Eugene
1929. Down-tree progress of *Sitta pygmaea*. Condor, 31: 45-51. [Flies, presumably *Protocalliphora*, entering nests, Calif.].
Lee, Robert D., and Raymond E. Ryckman
1954. Coleoptera and Diptera reared from owl nests. Bulletin of the Brooklyn Entomological Society, 49: 23-24. [*Protocalliphora* sp., Flagstaff, Ariz., nest of great horned owl; = *P. asiovora*].
Lehrer, Andy Z.
1970. Considérations phylogénétiques et taxonomiques sur la famille Calliphoridae (Diptera). Annotationes Zoologicae et Botanicae, no. 61: 1-51, 44 figs.
Lincoln, F. C.
1931. *Protocalliphora* infestation of barn and bank swallows. Bird-Banding, 2: 33-34. [*P. splendida* larvae, Fairdale, N. Dak., barn swallows; very small 3rd-instar larvae, probably *sialia* but possibly *hirundo*].
Lindner, Erwin
1957. Vogelblutfliegen (*Protocalliphora*, Calliphorinae, Diptera): Ein Beitrag zur Kenntnis ihrer Systematik und Biologie. Die Vogelwarte, 19: 84-90.
Linsdale, J. M.
1937. The natural history of magpies. Cooper Ornithological Club, Pacific Coast Avifauna, 25: 234 pp.
Lloyd, Gary D., and Cornelius B. Philip
1966. The "wingless" fly, *Carnus hemapterus* Nitzsch (Milichiidae), on hawk fledglings in northern Utah. Journal of Parasitology, 52: 414. [*Protocalliphora* larvae, heavy infestations in ears of nestling red-tailed hawks, Summit Co., Utah; undoubtedly *P. asiovora*].
Lloyd, Hoyes
1922. Larvae of *Phormia chrysorrhea* [sic] Macq. found upon nestling bluebirds. Canadian Field Naturalist, 36: 116. [Near Hull, Que.; = *P. sialia*].
1927. Identification of flies found infesting bluebirds. Canadian Field Naturalist, 41: 111. [Correction of 1922 identification to *Protocalliphora splendida; = sialia*].

Lundbeck, William

1895. Østgrønlandske Insekter. Fortegnelse over de indsamlede Insekter. Meddelelser om Grønland (Copenhagen), 19: 105-120. [*Calliphora azurea* (Fallén), Hekla Havn, East Greenland; = *P. tundrae*. Same record repeated by Lundbeck, 1900, on the Diptera of Greenland, and by Henriksen and Lundbeck, 1917, on the insects of Greenland].

Macquart, Jean

1846. Diptères exotiques nouveaux ou peu connus. [Ier] Supplément. Mémoires de la Société Royale des Sciences, de l'Agriculture et des Arts de Lille, 1844: 133-364, 20 plates (Also separate publication, 1846, Paris, pp. 5-238, 20 plates). [*Calliphora splendida*, n. sp. Texas; = species dubia].

Marshall, A. G.

1981. The ecology of ectoparasitic insects. New York, Academic Press. Pp. xvi + 459.

Mason, Edwin A.

1936. Parasitism of birds' nests by *Protocalliphora* at Groton, Massachusetts. Bird-Banding, 7: 112-121. [Larvae of *Protocalliphora* spp., various hosts].

1944. Parasitism by *Protocalliphora* and management of cavity-nesting birds. Journal of Wildlife Management, 8: 232-247. [*P. splendida*, Groton, Mass., various hosts; = mixture of *sialia*, *shannoni*, and *metallica*].

McAlpine, J. F., et al.

1981. Manual of Nearctic Diptera. Volume 1. Research Branch, Agriculture Canada, Monograph 27: 674 pp.

1987. Ibid. Volume 2. Research Branch, Agriculture Canada, Monograph 28: 1332 pp. (Calliphoridae by G. E. Shewell, pp. 1133-1145).

McAtee, W. L.

1927. Notes on the insect inhabitants of bird houses. Proceedings of the Entomological Society of Washington, 29: 87-90. ["Sarcophagid puparia" found in many nests, Bell, Md.; = chiefly *Protocalliphora*, according to McAtee 1929].

1929. Further notes on insect inhabitants of bird houses. Proceedings of the Entomological Society of Washington, 31: 105-111. [*P. splendida*, Bell, Md., 5 hosts; = *P. sialia* in nests of starling, house sparrow, and purple martin, other hosts not checked; *P.* sp., probably new, from nest of house wren, = *P. deceptor*, n. sp.].

1940. An experiment in song bird management. Auk, 57: 333-348. [*Protocalliphora* larvae, Glenn Dale, Md., in purple martin house].

Meigen, Johann W.

1824. Systematische Beschreibung der bekannten europäischen zweiflügeligen Insekten. Vol. 4: xii + 428 pp, pls. 34-41. Hamm.

1826. Systematische Beschreibung der bekannten europäischen zweiflügeligen Insekten. Vol. 5: xii + 412 pp., pls. 42-54. Hamm.

Meng, Heinz

1954. The bird nest screw-worm fly (*Apaulina avium*). Falconry News and Notes, 1(5): 21-22. [Myiasis by *A. avium* in ears of hawks, N.Y.; = undoubtedly *P. avium*].

Miller, Charles W.

1909. The occurrence of the larvae of a parasitic fly, *Protocalliphora chrysorrhoea*, Meigen, in bird nests in the vicinity of Shawnee, Pa. Bulletin of the Worthington Society for the Study of Bird Life, 2: 8 pp. [Attacking various birds; reared adults from bluebird nests = *P. sialia*].

Miller, Robert C.

1929. Subcutaneous parasitism of a young warbler by fly larvae. Murrelet, 10(1): 13-14. [*P. splendida* var. *hesperia* infesting yellow warbler, Seattle, Wash., = *P. braueri*].

Munro, J. A.

1949. The birds and mammals of the Vanderhoof region, British Columbia. American Midland Naturalist, 41: 1-138. [*P. hirudo* (now *braueri*) abundant; adults reared from song sparrow = *braueri*, but larvae from some dead fledglings proved to be *Phaenicia sericata* (Meigen) and *Phormia regina* (Meigen)].

Neff, Johnson A.

1945. Maggot infestation of nestling mourning doves. Condor, 47: 73-76. [*P. metallica* and a n. sp., Calif.; = *cuprina* and *asiovora* (synonym, *basingeri* Hall), respectively].

Nielsen, J. C.

1907. The insects of East-Greenland. Meddelelser om Grønland, 29: 365-409. [*Calliphora azurea* (Fallén), record of Lundbeck (1895) repeated, with addition of exact date, 11 June 92; = *P. tundrae*].

Nielsen, Peder, Oscar Ringdahl, and S. L. Tuxen
1954. Diptera 1 (exclusive of Ceratopogonidae and Chironomidae). 189 pp., *in* The Zoology of Iceland, Vol. III, Part 48a. Copenhagen and Reykjavik.

Nolan, Val, Jr.
1955. Invertebrate nest associates of the prairie warbler. Auk, 72: 55-61. [*P. metallica*, Bloomington, Ind.; det. Sabrosky].
1959. Additional invertebrate nest associates of the prairie warbler. Auk, 76: 352-357. [*P. metallica* and a n. sp., Bloomington, Ind.; = *P. metallica* and *P. deceptor*].
1978. The ecology and behavior of the prairie warbler, *Dendroica discolor*. Ornithological Monographs (American Ornithologists' Union), 26: xxii + 595, frontispiece (color), 42 figs. [Updating of records of species recorded in Nolan (1955, 1959)].

Norris, Robert A.
1958. Comparative biosystematics and life history of the nuthatches *Sitta pygmaea* and *Sitta pusilla*. University of California Publications in Zoology, 56: 119-300. [*P.* "near *sialia*" (det. W. W. Wirth), one puparium, Irwinville, Ga., brown-headed nuthatch nest; not checked].

Nuorteva, Pekka
1960. *Protocalliphora chrysorrhoea* (Meig.) kärpäsen toukat törmäpääskyn loisina (*Protocalliphora chrysorrhoea* (Meig.) as a parasite of *Riparia riparia* (L.) in Finland). Ornis Fennica, 37: 122-124.

Nuorteva, Pekka, and Unto Järvinen
1961. The insect fauna of the nests of the sand martin (*Riparia riparia* L.) in Finland. Annales Entomologici Fennicae, 27: 197-204. [*Protocalliphora chrysorrhoea* (Meigen)].

Oliver, Don R.
1963. Entomological studies in the Lake Hazen area, Ellesmere Island, including lists of species of Arachnida, Collembola, and Insecta. Arctic (Journal of the Arctic Institute of North America), 16: 175-180. [*P.* sp.; = *P. tundrae*].

Owen, D. F.
1954. *Protocalliphora* in birds' nests. British Birds, 47: 236-243.

Owen, D. F., and J. S. Ash
1955. Additional records of *Protocalliphora* (Diptera) in birds' nests. British Birds, 48: 225-229.

Peck, O.
1963. A catalogue of the Nearctic Chalcidoidea (Insecta: Hymenoptera). Canadian Entomologist-Supplement, 1963: 1092 pp.
Petch, C. E., and J. B. Maltais
1932. [Revision of] A preliminary list of the insects of the province of Quebec, Part II, Diptera, by A. F. Winn and G. Beaulieu. Annual Report of the Quebec Society for the Protection of Plants, 24, Supplement: 100 pp. [*P. splendida*, Aylmer, Que. = *P. metallica; P. azurea*, Montreal, not checked].
Peterson, Roger Tory
1947. A field guide to the birds (ed. 2). Houghton Mifflin Co., Boston, pp. xxvi + 290.
Peus, Fritz
1960. Zur Kenntnis der ornithoparasitischen Phormiinen (Diptera, Calliphoridae). Deutsche Entomologische Zeitschrift, Neue Folge, 7: 193-235.
Philip, Cornelius B.
1938. A parasitological reconnaissance in Alaska with particular reference to varying hares. II. Parasitological data. Journal of Parasitology, 24: 483-488. [*Protocalliphora* larvae, Rapids, Alaska, golden eagle; = *P. braueri*].
Plath, O. E.
1919a. Parasitism of nestling birds by fly larvae. Condor, 21: 30-38 (January). [*P. azurea*, San Francisco Bay area, various hosts; = *hesperia* complex].
1919b. A muscid larva of the San Francisco Bay region which sucks the blood of nestling birds. University of California Publications in Zoology, 19: 191-200 (February). [*P. azurea*, same material as in Plath, 1919a].
1919c. The prevalence of *Phormia azurea* Fallen (larva parasitic on nestling birds) in the Puget Sound region and data on two undescribed flies of similar habit. Annals of the Entomological Society of America, 12: 373-378 (December)(*Phormia metallica* Townsend described on pp. 379-380. See Townsend 1919). [*Phormia azurea* and *Phormia metallica;* part of *azurea* = paratypes of *Protocalliphora hesperia* and *hesperioides*, n. sp.; *metallica* from robin, Puget Sound, formed type series of *Protocalliphora hirudo* var. *cuprea*, which equals *braueri*, plus several females of *cuprina*].

Pletsch, Don D.
1948. Parasitic dipterous larvae from the nasal cavity of a nestling magpie. Auk, 65: 296-297. [*P. avium*, Bozeman, Mont.; = *P. asiovora*].

Pont, Adrian C.
1976. Calliphoridae, pp. 107-108, *in* Kloet and Hincks, A Check List of British Insects, 2nd edition, Volume XI, Part 5: Diptera and Siphonaptera.

Poole, K.G., and R.G. Bromley
1988. Natural history of the gyrfalcon in the Central Canadian Arctic. Arctic, 41: 31-38. [*P. avium*, Kilgavik, Mackenzie District, Northwest Territories, larvae].

Rausch, Robert L.
1972. Cutaneous myiasis in a bird by the larval *Protocalliphora h. hirudo* Shannon and Dobroscky, 1924 (Diptera: Calliphoridae). Aquilo, Series Zoologica, 13: 1-4. [*P. hirudo*, Alaska, on Wilson's warbler; = *braueri*].

Ringdahl, Oscar
1937. Bidrag till kännedomen om de svenska tachinidernas utbredning. Entomologisk Tidskrift (Stockholm), 58: 31-38.
1945. Förteckning över de av Zetterstedt i Insecta Lapponica och Diptera Scandinaviae beskrivna tachiniderna med synonymer jämte anteckningar över en del arter. Opuscula Entomologica (Lund, Sweden), 10: 26-35.

Robert, Adrien
1962. Les parasites et commensaux des nids de l'hirondelle à front blanc, *Petrochelidon pyrrhonota*, au Lac Monroe, parc du Mont Tremblant. Annals of the Entomological Society of Quebec, 7: 87-92 [*P. sialia*, Quebec, cliff swallow; not verified].

Rodhain, J., and J. Bequaert
1916. Histoire de "*Passeromyia heterochaeta*" Villen. et de "*Stasisia* (*Cordylobia*) *rodhaini*" Ged. Bulletin Scientifique de la France et de la Belgique, séries 7, 49: 236-289.

Rognes, Knut
1985. Revision of the bird-parasitic blowfly genus *Trypocalliphora* Peus, 1960 (Diptera: Calliphoridae). Entomologica Scandinavica, 15: 371-382, 23 text figs.

Rohdendorf, E. B.
1957. [On the parasitic flies, injurious to nestling singing birds](In Russian, English summary). Entomologicheskoe Obozrenie, 36: 116-124.

Rossi, F. W.
1848. Systematisches Verzeichniss der zweiflügelichten Insecten (Diptera) des Erzherzogthumes Österreich. iii-x + 86 pp. Wien. [rearing of *Musca azurea*].
Sabrosky, Curtis W.
1950. Date of publication of Hall's "Blowflies of North America". Proceedings of the Entomological Society of Washington, 52: 204.
1955. The taxonomic status of the armyworm parasite known as *Archytas piliventris* (Van der Wulp)(Diptera: Larvaevoridae). Florida Entomologist, 38: 77-83.
1956. The nomenclature of *Protocalliphora* (Diptera: Calliphoridae). Proceedings of the Royal Entomological Society of London, Series B, 25: 175-179.
1967. Corrections to A Catalog of the Diptera of America North of Mexico. Bulletin of the Entomological Society of America, 13: 115-125. [*Trypocalliphora* Peus 1960, which included *Protocalliphora hirudo* (now *braueri*), accepted as subgenus of *Protocalliphora*].
1983. A synopsis of the world species of *Desmometopa* Loew (Diptera, Milichiidae). Contributions of the American Entomological Institute, 19(8): 1-69.
1984. A proposed amendment to Article 70b of the International Code of Zoological Nomenclature on misidentified type species. Bulletin of Zoological Nomenclature, 41(3): 152-155.
1989. *Protocalliphora* Hough, 1899 (Insecta, Diptera) and its type species Musca azurea Fallén, 1817: Proposed conservation of usage by designation of a replacement lectotype. Bulletin of Zoological Nomenclature 46(2): 126-128.
Sabrosky, Curtis W., and Gordon F. Bennett
1958. The utilization of morphological, ecological, and life history evidence in the classification of *Protocalliphora* (Diptera, Calliphoridae). Proceedings of the Tenth International Congress of Entomology (Montreal, 1956), 1: 163-164.
1971. The bird blow flies (*Protocalliphora*): Taxonomy, biology and distribution. (Abstract). Proc. XIII International Congress of Entomology (Moscow, 1968), 1: 194.

Sailer, Reece I., and Siegfried E. Lienk
1951. Blow flies (Calliphoridae) in Alaska. Canadian Entomologist, 83: 208-211. [*Apaulina sapphira*, Tok, Alaska, "cliff swallow" (error for bank swallow: Sailer in litt.); = *P. chrysorrhoea*].

Sargent, William Dunlap
1938. Nest parasitism of hawks. Auk, 55: 82-84. [*P. splendida sialis* (sic), Ithaca, N.Y., maggots in ears; undoubtedly *P. avium*].

Schumann, H.
1986. Family Calliphoridae, pp. 11-58, *in* Soó(ed.), Catalogue of Palaearctic Diptera, Vol. 12: 265 pp. Budapest.

Séguy, Eugene
1928. Études sur les mouches parasites. Tome I. Conopides, OEstrides et Calliphorines de l'Europe occidentale. Encyclopédie Entomologique, Sér. A, 9: 251 pp., 6 plates.
1929. Étude sur les diptères à larves commensales ou parasites des oiseaux de l'Europe occidentale. Encyclopédie Entomologique, Sér. B, II. Diptera, 5: 63-82.
1955. Introduction à l'étude biologique et morphologique des insectes diptères. Publicações Avulsas Museu Nacional Rio de Janeiro, 17: 1-260. [*Protocalliphora;* general notes].

Seidensticker, John C., IV, and Harry V. Reynolds III
1971. The nesting, reproductive performance, and chlorinated hydrocarbon residues in the red-tailed hawk and great horned owl in south-central Montana. Wilson Bulletin, 83: 408-418. [*Protocalliphora* sp., on hawks; probably *asiovora*].

Shannon, Raymond C.
1923. Genera of Nearctic Calliphoridae, blowflies, with revision of the Calliphorini. Insecutor Inscitiae Menstruus, 11: 101-118.
1926. Synopsis of the American Calliphoridae (Diptera). Proceedings of the Entomological Society of Washington, 28: 115-139.

Shannon, Raymond C., and Irene D. Dobroscky
1924. The North American bird parasites of the genus *Protocalliphora* (Calliphoridae, Diptera). Journal of the Washington Academy of Sciences, 14: 247-253. [Numerous records, referred to under the various species].

Shewell, G. E.
1987. Calliphoridae. pp. 1133-1145, *in* J. F. McAlpine et al., Manual of Nearctic Diptera, Vol. 2.
Sømme, L.
1961. On the overwintering of house flies (*Musca domestica* L.) and stable flies (*Stomoxys calcitrans* (L.)) in Norway. Norsk Entomologisk Tidsskrift, 11: 192-223. [*Protocalliphora azurea* overwintering in crevices in mountain huts].
Stabler, Robert M.
1959. Nesting of the blue grosbeak in Colorado. Condor, 61: 46-48. [*P. metallica;* det. Sabrosky, M. T. James].
Stein, Paul
1924. Die verbreitetsten Tachiniden Mitteleuropas nach ihren Gattungen und Arten. Archiv für Naturgeschichte, Abteilung A, 90(6): 1-271.
Stiner, Frederic M., Jr.
1969. Overwintering by *Protocalliphora metallica* (Diptera: Calliphoridae). Annals of the Entomological Society of America, 62: 1205-1206. [Newark, Del.; det. Sabrosky].
Stoner, Dayton
1926. Observations and banding notes on the bank swallow--II. Auk, 43: 198-213. [*Protocalliphora* sp., Lake Okoboji, Iowa; one example checked = *hirundo*].
1936. Studies on the bank swallow, *Riparia riparia riparia* (Linnaeus) in the Oneida Lake region. Annals of the Roosevelt Wild Life Forest Experiment Station of the New York State College of Forestry at Syracuse University, 4: 126-233. [*P. splendida*; not checked].
Storer, Tracy I.
1929. *Protocalliphora* in the nest of a mountain chickadee. Condor, 31: 227. [*P. splendida* var. near *hirundo*, San Bernardino Mts., Calif.; material apparently lost, possibly = *P. parorum* n. sp.]
Taverner, P. A.
1934. Birds of Canada. Bulletin of the Canadian National Museum, 72: 445 pp.
Theowald, Br.
1955. Calliphorinae (Dipt. Brach.) uit Vogelnesten. Natuurhistorisch Maandblad, 44: 59-62.

Thomas, Edward S.
1936. Additional records of *Protocalliphora*. Bird-Banding, 7: 46-47. [*Protocalliphora* spp., Ohio, 5 hosts; = *P. deceptor* from Carolina wren and prairie and worm-eating warblers, *metallica* from yellow-breasted chat and Mississippi song sparrow].

Thompson, Patrick H.
1966. Arthropods from nests of the house sparrow. Proceedings of the Entomological Society of Washington, 68: 44-48. [*Protocalliphora* larvae, Madison, Wis.].

Tirrell, Peter B.
1978. *Protocalliphora avium* (Diptera) myiasis in great horned owls, red-tailed hawks, and Swainson's hawks in North Dakota. Raptor Research, 12 (1/2): 21-27, 2 figs. [det. Sabrosky].

Townsend, Charles H. T.
1919. Description of the new species of *Phormia*. Annals of the Entomological Society of America, 12: 379-380 (Appended to paper by Plath 1919c). [*Phormia metallica* n. sp., holotype N.H.; paratypes are a mixture of *Protocalliphora metallica*, *P. aenea*, and *P. braueri*].
1931. Notes on Old-World OEstromuscoid types, Part I. Annals and Magazine of Natural History, series 10, 8: 369-391. [Notes on "holotype" of *Musca azurea* Fallén in Stockholm].
1935. Manual of Myiology. Part II, Muscoid classification and habits. Charles Townsend & Filhos, Itaquaquecetuba, São Paulo, Brasil, 296 pp., 9 plates.

Van Tyne, J., and A. J. Berger
1976. Fundamentals of ornithology (ed. 2). John Wiley & Sons, New York, xviii + 808 pp.

Verbeek, Nicolaas A. M.
1967. Breeding biology and ecology of the horned lark in alpine tundra. Wilson Bulletin, 79: 208-218, 2 figs. ["blowfly larvae," Park Co., Wyo. = *P. spatulata*].
1970. Breeding ecology of the water pipit. Auk, 87: 425-451, 6 figs. ["blowfly larvae," Park Co., Wyo., apparently *P. spatulata*].

Villeneuve, Joseph
1911. Notes diptérologiques. Wiener Entomologische Zeitung, 30: 84-87.
1918. Sur *Phormia sordida* Zett. [Dipt.]. Bulletin de la Société entomologique de France 1918 (10): 158-159.

1928. Quelques mots sur les Calliphorinae paléarctiques. Bulletin & Annales de la Société Entomologique de Belgique, 68: 147-151. [*Protocalliphora* a subgenus of *Phormia;* notes on confusion between European species *Phormia azurea* Fallén and *Phormia sordida* Zetterstedt].

1931. Aperçus critiques sur le mémoire de P. Stein: "Die verbreitetsten Tachiniden Mitteleuropas." Konowia, 10: 47-74. [Reiterates opinion that *Protocalliphora* can only be a subgenus of *Phormia*].

Walsh, B. D.

1866a. Answers to correspondents: Rev. Jas. B. Fisher, N.Y. Practical Entomologist, 1(10): 102. July 30. [Larvae found attached to head and body of young swallow, identified as "*Oestrus* family" but undoubtedly larvae of *Protocalliphora*, cf. Walsh 1866b].

Walsh, B. D.

1866b. Answers to correspondents: Rev. Jas. B. Fisher, N.Y. Practical Entomologist, 2(1): 8. October. [Larvae from Walsh 1866a identified by Osten Sacken as of "*Musca* family in the vicinity of *Musca* or *Sarcophaga*." Undoubtedly *Protocalliphora*].

Walton, William R.

1914. Report on some parasitic and predaceous Diptera from northeastern New Mexico. Proceedings of the United States National Museum, 48: 171-186. [*P. chrysorrhaea* (sic), horned lark; part = *P. braueri*, part in type series of *P. seminuda*].

Wetherbee, Mrs. Kenneth B.

1932a. Two pairs of tree swallows mated during two successive seasons. Bird-Banding, 3: 72-73. [*P. splendida* var. *sialia*, Pomfret, Conn.; *P. sialia*, Worcester, Mass., pupae only].

1932b. Nesting bluebirds successful in spite of *Protocalliphora*. Bird-Banding, 3: 73-74. [*P. sialia*, Worcester, Mass., and Conn.; verified].

White, Clayton M.

1963. Botulism and myiasis as mortality factors in falcons. Condor, 65: 442-3. [*Calliphora* sp., in heads of prairie falcons, NW. Utah; possibly *P. asiovora*].

Whitehead, W. E.

1933. Parasites from a bird's nest. Journal of Economic Entomology, 26: 292-293. [*P. splendida*, Macdonald College, Que., purple martin; not checked].

Whiting, A. R.
1967. The biology of the parasitic wasp *Mormoniella vitripennis* [= *Nasonia brevicornis*] (Walker). Quarterly Review of Biology, 42: 333-406, 45 figs.
Whitworth, Terry L.
1971. A study of the biology of the species of *Protocalliphora* in the northern Wasatch Range. Utah State University, Logan, Utah, unpub. M.S. thesis, 58 pp.
1976. Host and habitat preferences, life history, pathogenicity and population regulation in species of *Protocalliphora* Hough (Diptera: Calliphoridae). Utah State University, Logan, Utah, unpub. PhD. dissertation pp. xii + 144.
1977. [same title as Whitworth 1976]. Dissertation Abstracts International, B, 37 (10): 4933-B.
Wing, Leonard W.
1956. Natural history of birds. A guide to ornithology. Ronald Press Co., New York, xi + 539 pp.
Wood, E. D.
1953. Observations and report of bird-banding in the vicinity of Victoria, B.C. Victoria Naturalist, 9: 76-78. [*Protocalliphora* sp. or spp.; = probably *P. braueri*, at least in part].
Woods, H. E.
1926. A new host for the blood-sucking larva fly, *Protocalliphora splendida*. Bulletin of the Northeastern Bird-Banding Association, 2: 12-13. [Huntington, Mass., barn swallow; = *P. hirundo*].
Zeleny, Lawrence
1968. Bluebirds for posterity. National Association for the Protection and Propagation of the Purple Martins and Bluebirds of America, Crawfordsville, Ind., 16 pp., 2 figs. [Popular brochure; *Apaulina sialia*].
1970. Most serious bluebird parasite is the blowfly. Purple Martin Capital News, July 29, 1970: 12. [Popular article in a commercial publication; *P. sialia*, det. Sabrosky].
1976. The bluebird. Indiana University Press, Bloomington, Ind., pp. xix + 170. [*P. sialia*, etc., biology and effect on birds].
Zetterstedt, Johann W.
1838. Diptera. pp. 477-868, in Zetterstedt, 1838-1840, Insecta Lapponica, Lipsiae [= Leipzig], 1140 pp.

Zumpt, F.
1956. Calliphorinae. Fam. 64i, 140 pp., 10 plates, *in* Lindner, E. (editor), Die Fliegen der palaearktischen Region. E. Schweizerbart'sche Verlagsbuchhandlung, Stuttgart. Lieferungen 190, 191, 193.
1965. Myiasis in man and animals in the Old World. Butterworth & Co., London, England, pp. xv + 267, 1 plate. [*Protocalliphora*, 9 spp. summarized].

Index to Bird Hosts

Index to *Protocalliphora* and Miscellaneous Names

(Family-group names are in capitals, valid genus-group names in boldface when cited alone, valid species-group names in roman, and all synonyms in italics. Specific names standing alone are in *Protocalliphora*, or have been at some time.)

Figures

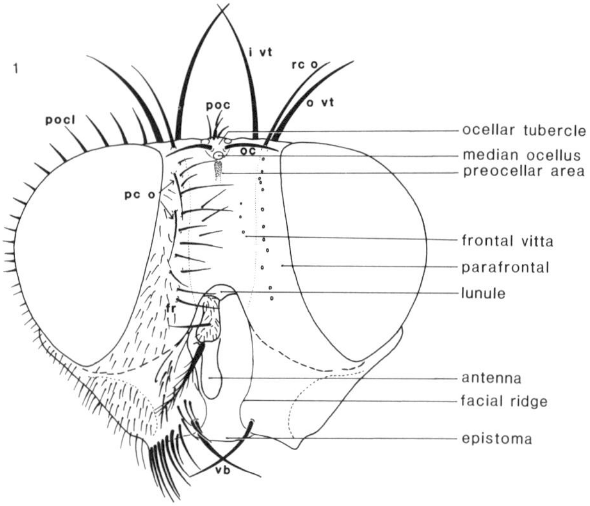

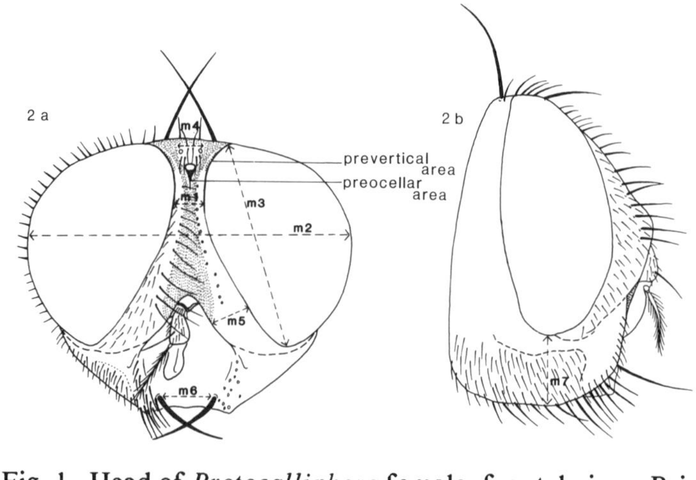

Fig. 1. Head of *Protocalliphora* female, frontal view. Bristles: fr, frontal; i vt, inner vertical; oc, ocellar; o vt, outer vertical; pc o, proclinate orbital; poc, postocellar; pocl, postocular; rc o, reclinate orbital; vb, vibrissa.

Fig. 2. Head of *Protocalliphora* male(a, frontal view, b, side view), showing the measures used. ml, width of frons at narrowest part; m2, maximum width of head (measured from dorsal view); m3, maximum length of eye; m4, ocellar span; m5, width of parafacial (opposite parafrontal-parafacial angle); m6, vibrissal interval; m7, height of cheek.

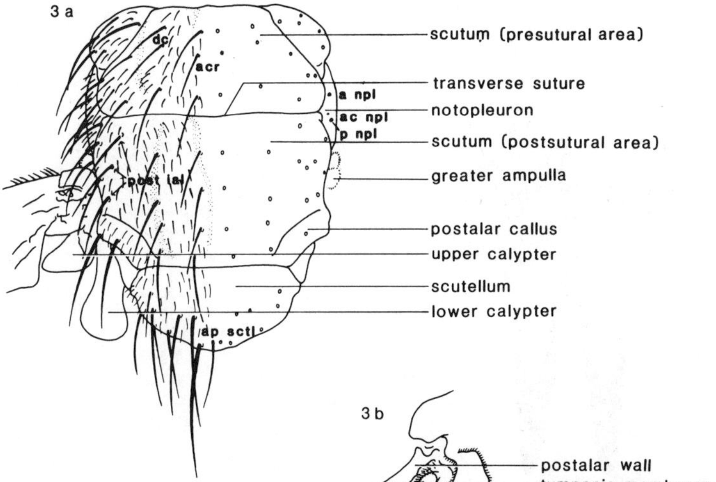

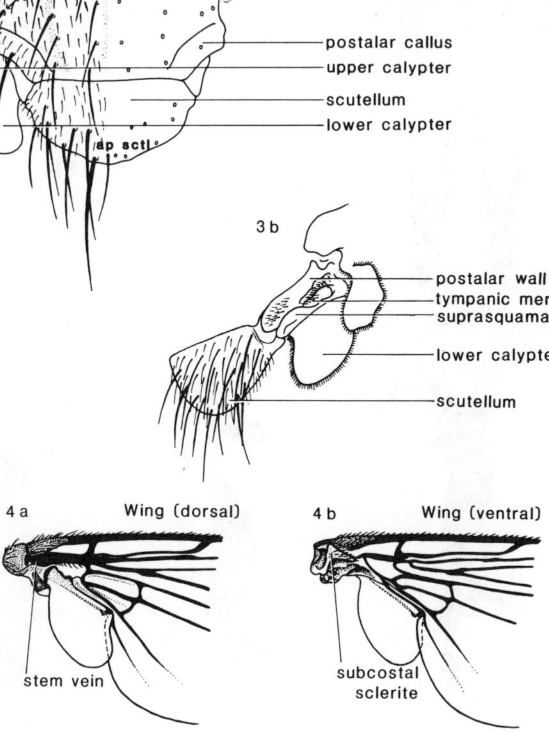

Fig. 3. Thorax of *Protocalliphora* (a, dorsal view, b, detail of postalar and suprasquamal areas). Bristles: acr, acrostical; ac npl, accessory notopleural; a npl, anterior notopleural; ap sctl, apical scutellar; dc, dorsocentral; p npl, posterior notopleural; post ia, postsutural intraalar.

Fig. 4. Base of wing of *Protocalliphora*. a, dorsal; b, ventral.

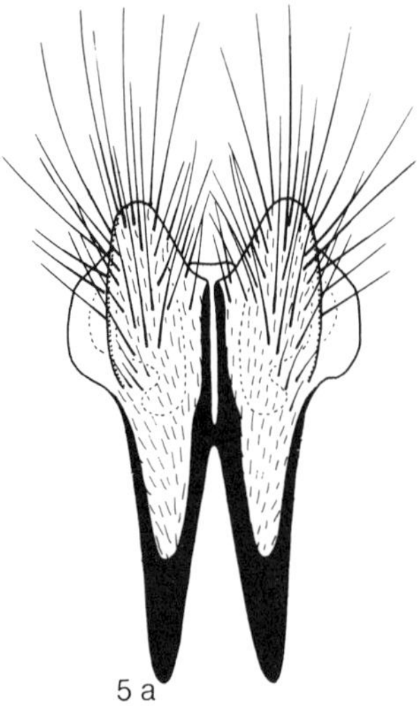

5 a
avium

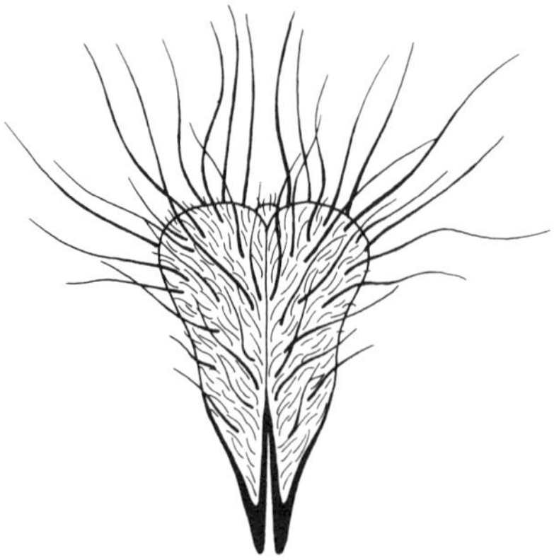

5 b
deceptor

6 a
avium

6 b
chrysorrhoea

6 c
braueri

Fig. 5. Cerci of *Protocalliphora*, posterior view.

Fig. 6. Aedeagus, of *Protocalliphora*, side view.

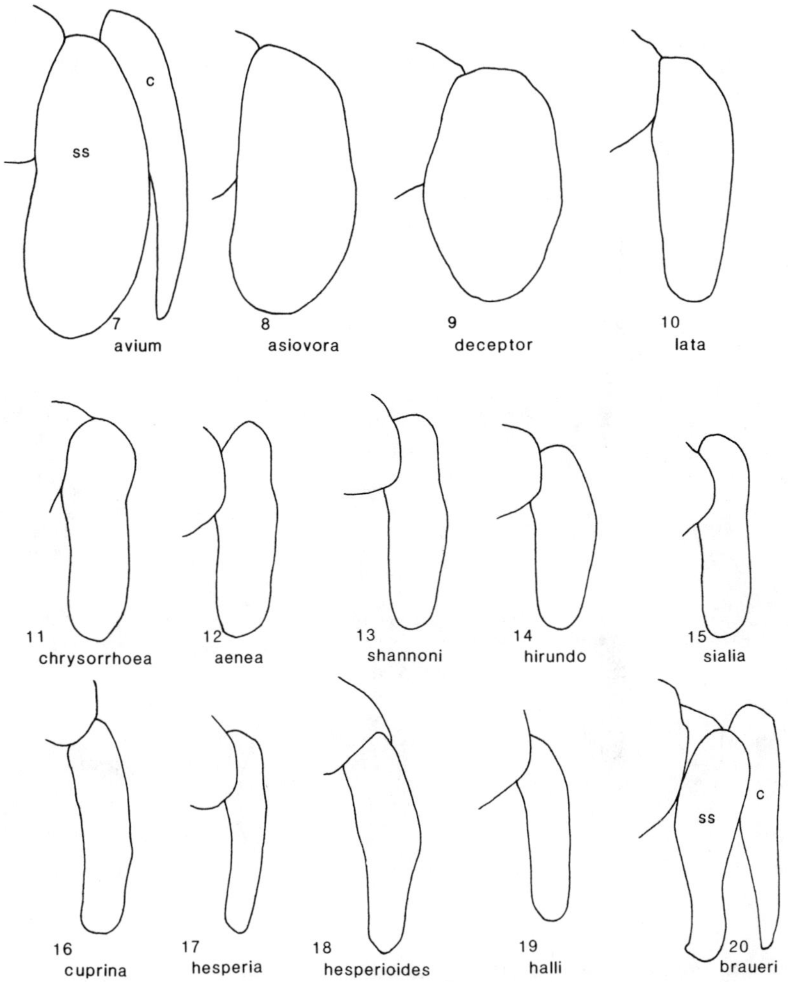

Figs. 7-20. Left surstylus, in profile, of *Protocalliphora* spp. (s, surstylus, and c, cercus, on figs. 7 and 20).

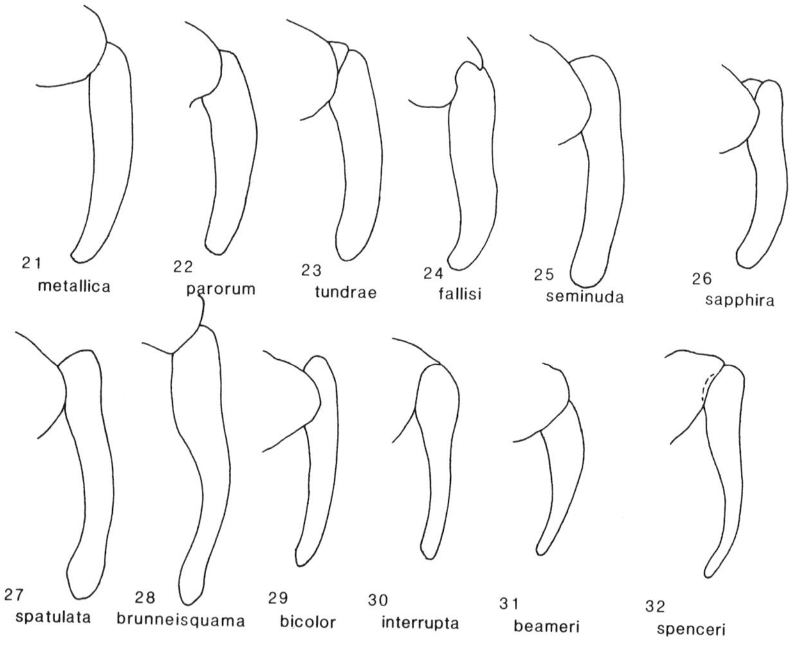

Figs. 21-32. Left surstylus, in profile, of *Protocalliphora* spp.

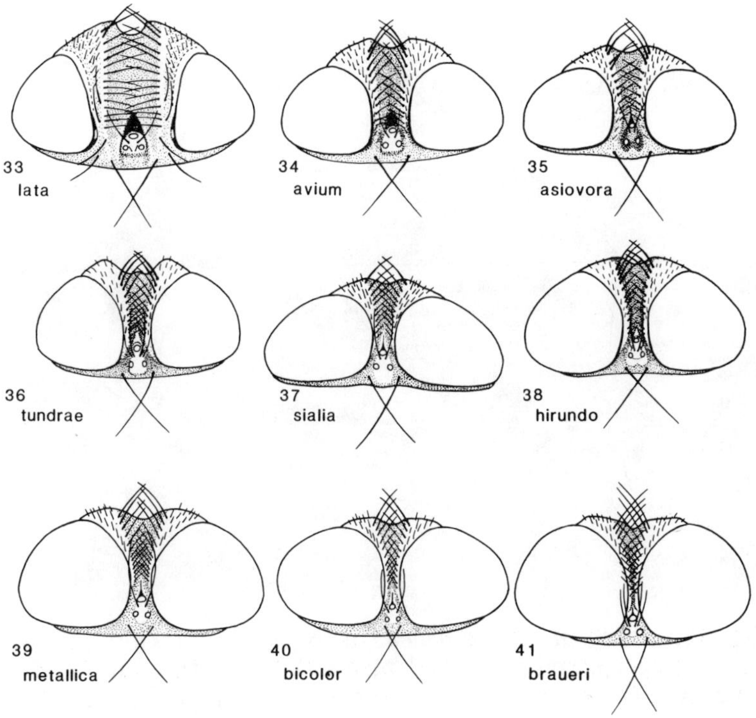

Figs. 33-41. Dorsal view of heads of *Protocalliphora* males.

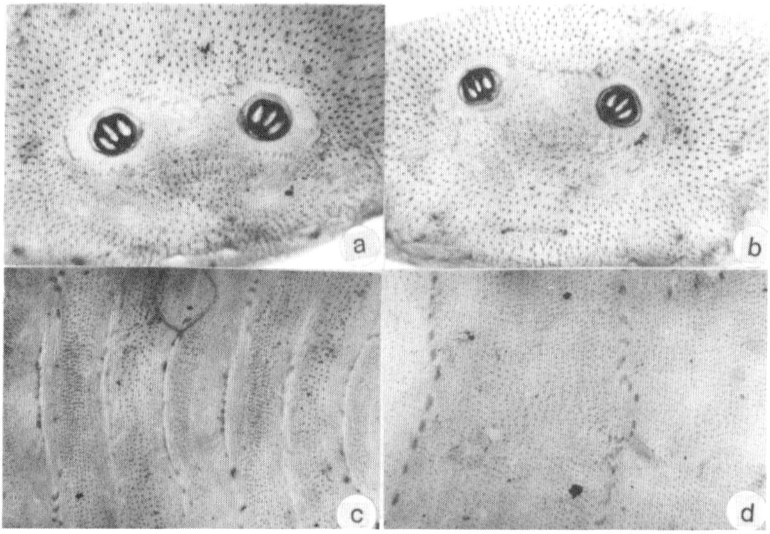

Fig. 42. *Protocalliphora aenea.* a, b, posterior region (note lack of mesostigmatal folds and plaques); c, ventral cuticle; d, dorsal cuticle (note lack of prominent cuticular ridges in c and d).

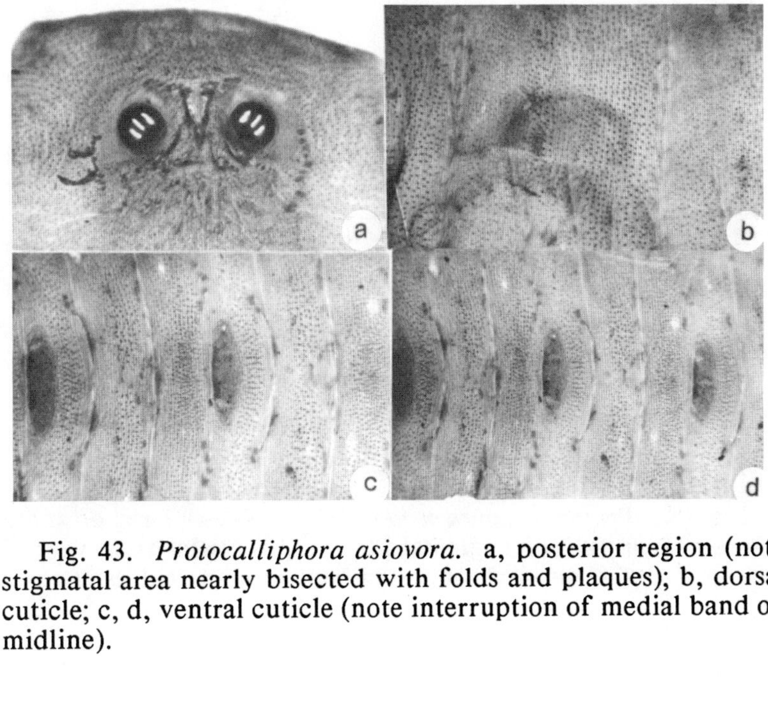

Fig. 43. *Protocalliphora asiovora.* a, posterior region (note stigmatal area nearly bisected with folds and plaques); b, dorsal cuticle; c, d, ventral cuticle (note interruption of medial band on midline).

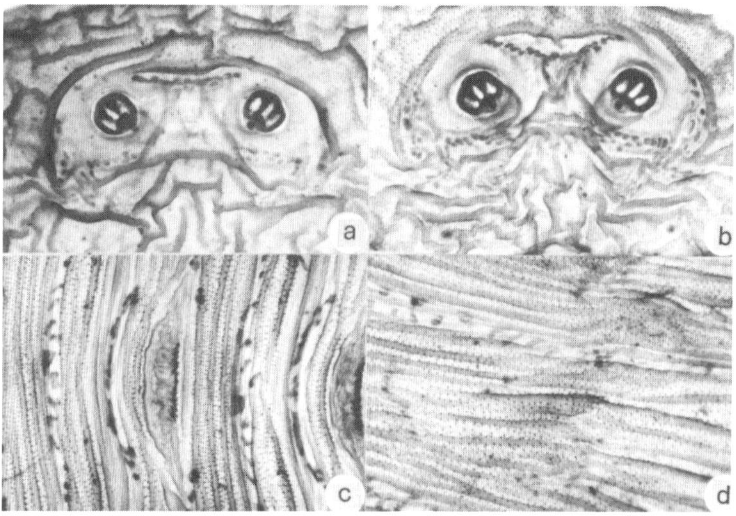

Fig. 44. *Protocalliphora avium*. a, b, posterior region (note marked separation of stigmatal plates by folds in b); c, ventral cuticle; d, dorsal cuticle (note prominent cuticular ridges in c and d).

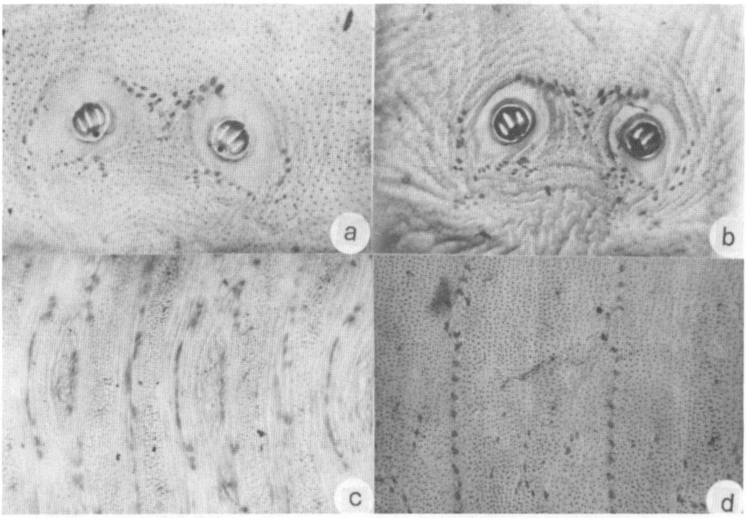

Fig. 45. *Protocalliphora bicolor*. a, b, posterior region (note almost complete bisection of the stigmatal area); c, ventral cuticle; d, dorsal cuticle (note reduction of cuticular ridges in c and d).

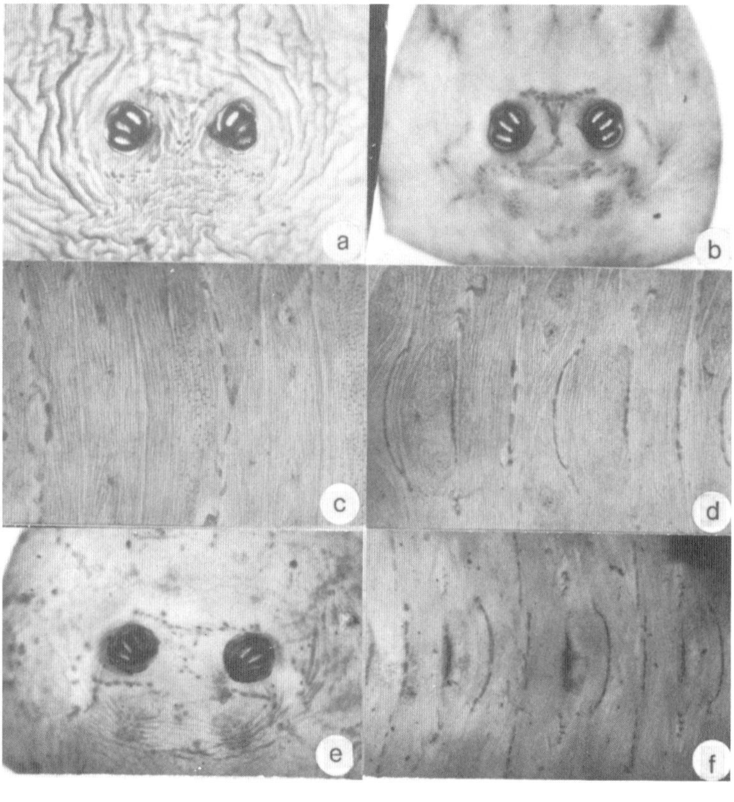

Fig. 46. *Protocalliphora braueri*. Western material: a, b, posterior region; c, dorsal cuticle; d, ventral cuticle. Eastern material: e, posterior region; f, ventral cuticle.

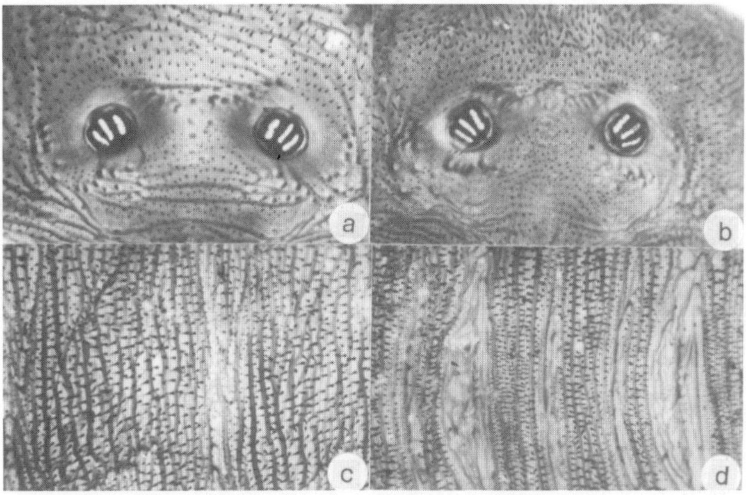

Fig. 47. *Protocalliphora chrysorrhoea.* a, b, posterior region; c, dorsal cuticle; d, ventral cuticle (note prominent cuticular ridges in c and d).

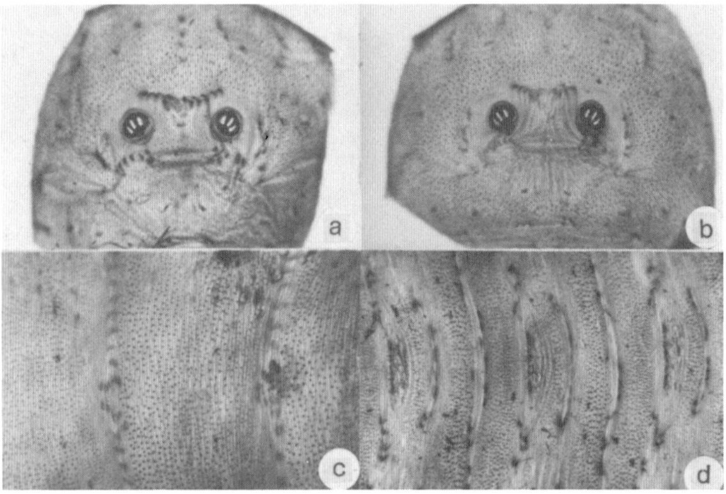

Fig. 48. *Protocalliphora cuprina.* a, b, posterior region; c, dorsal cuticle; d, ventral cuticle.

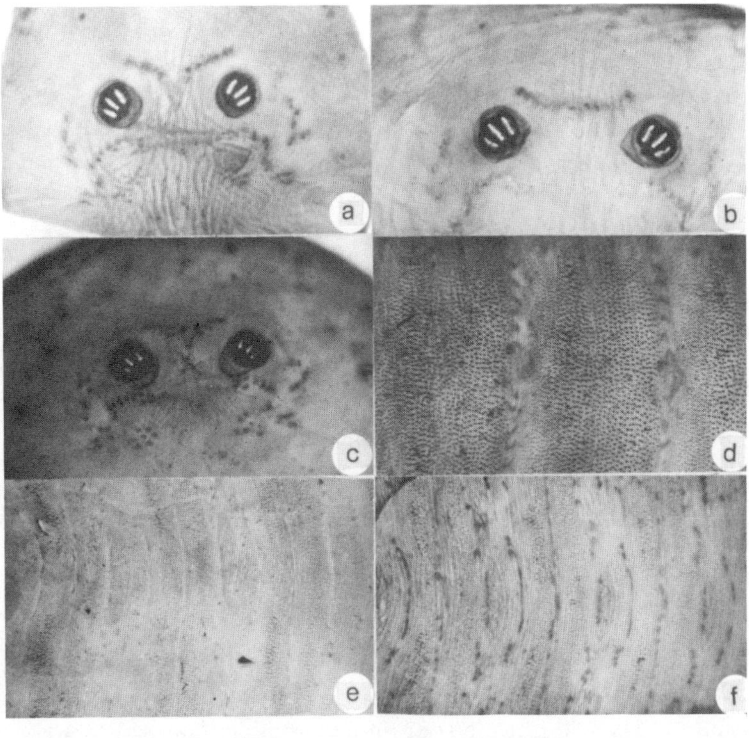

Fig. 49. *Protocalliphora deceptor*. a-c, posterior region (note very small spines and resulting fine-sandpaper appearance; also note lack of circumstigmatal folds; d, dorsal cuticle; e, f, ventral cuticle.

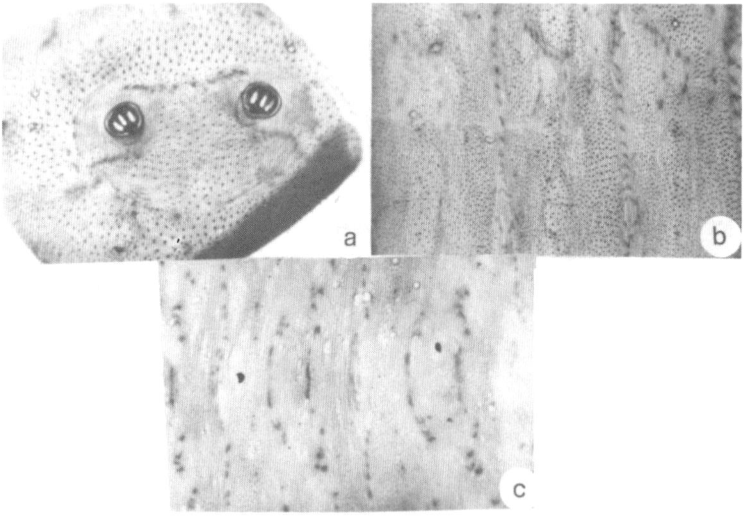

Fig. 50. *Protocalliphora fallisi*. a, posterior region; b, dorsal cuticle; c, ventral cuticle (note much reduced ventral bands).

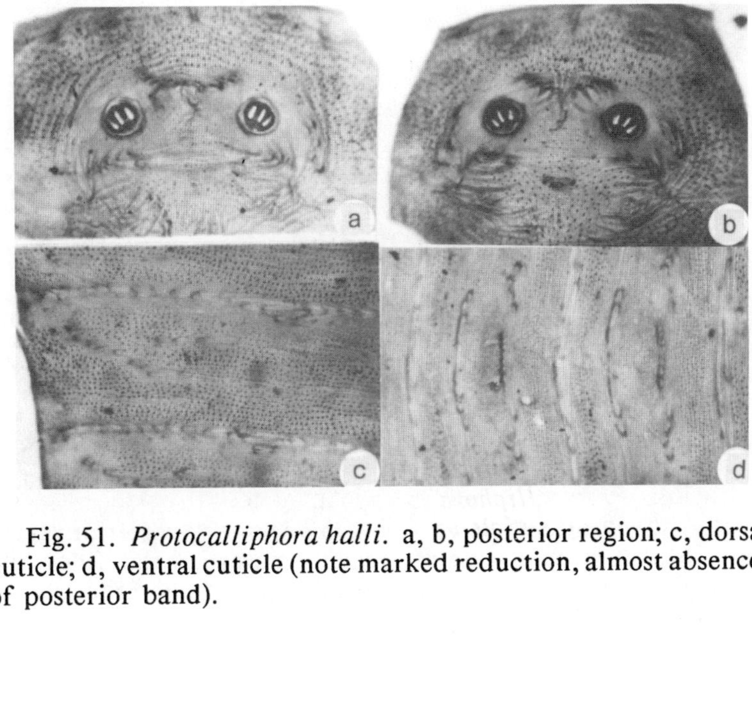

Fig. 51. *Protocalliphora halli*. a, b, posterior region; c, dorsal cuticle; d, ventral cuticle (note marked reduction, almost absence, of posterior band).

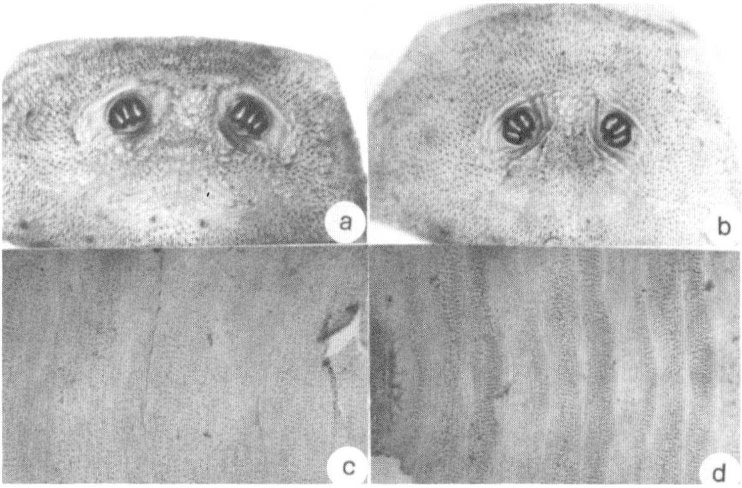

Fig. 52. *Protocalliphora hesperioides*. a, b, posterior region (note bisection of stigmatal area by mesostigmatal folds and plaques); c, dorsal cuticle; d, ventral cuticle (note reduction of cuticular folds in c and d).

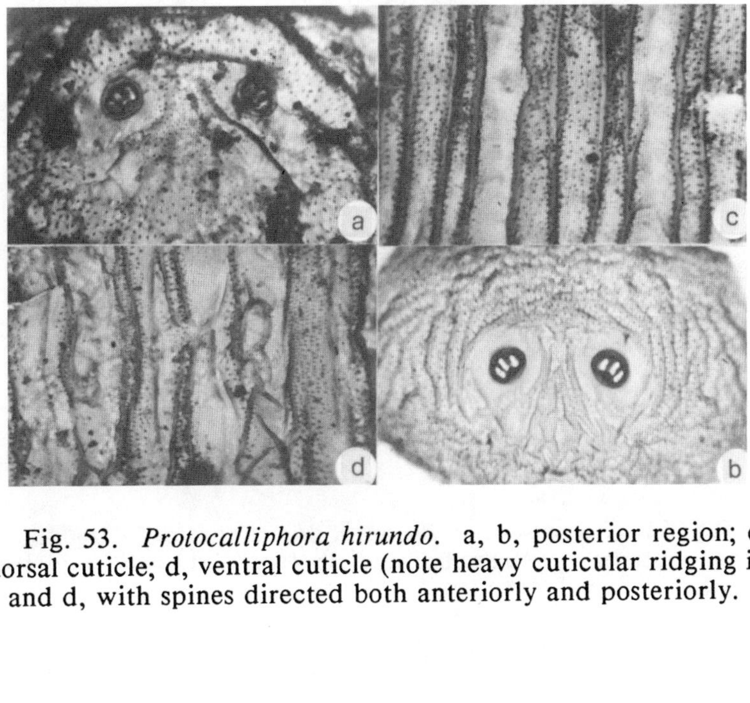

Fig. 53. *Protocalliphora hirundo.* a, b, posterior region; c, dorsal cuticle; d, ventral cuticle (note heavy cuticular ridging in c and d, with spines directed both anteriorly and posteriorly.

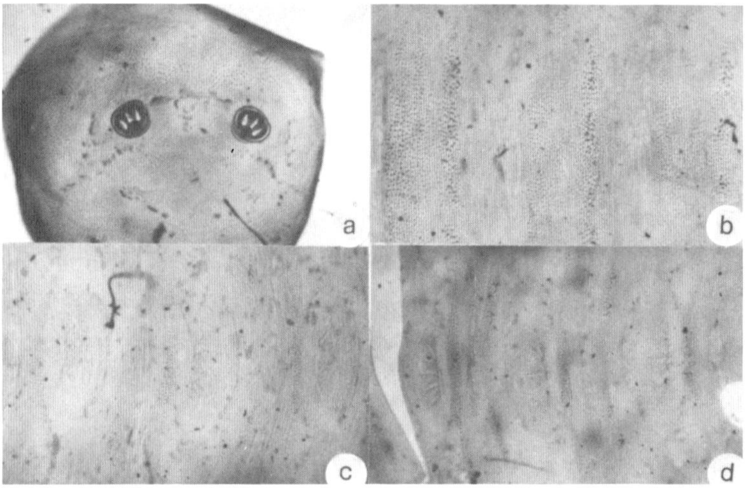

Fig. 54. *Protocalliphora interrupta.* a, posterior region; b, dorsal cuticle; c, d, ventral cuticle (note interrupted medial band).

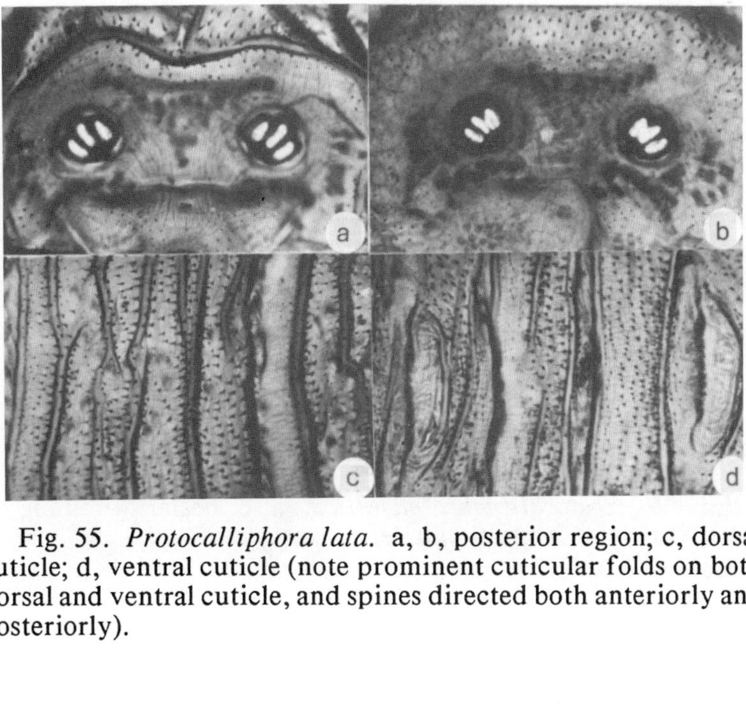

Fig. 55. *Protocalliphora lata.* a, b, posterior region; c, dorsal cuticle; d, ventral cuticle (note prominent cuticular folds on both dorsal and ventral cuticle, and spines directed both anteriorly and posteriorly).

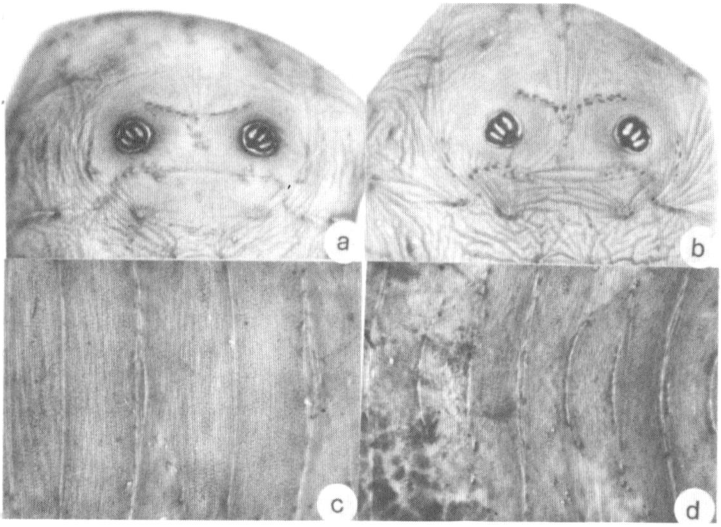

Fig. 56. *Protocalliphora metallica.* a, b, posterior region; c, dorsal cuticle; d, ventral cuticle.

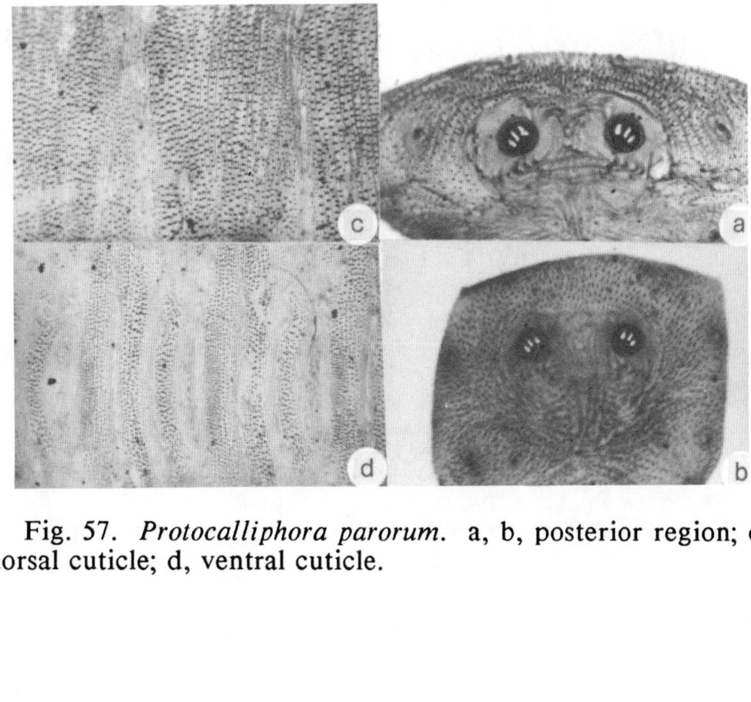

Fig. 57. *Protocalliphora parorum*. a, b, posterior region; c, dorsal cuticle; d, ventral cuticle.

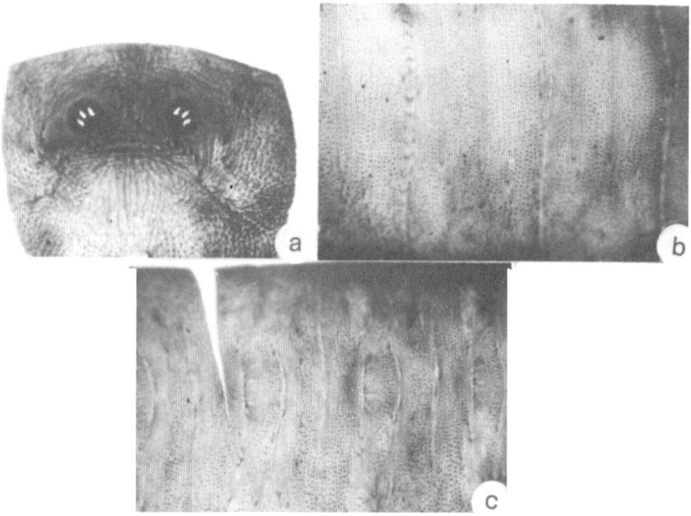

Fig. 58. *Protocalliphora seminuda.* a, posterior region; b, dorsal cuticle; c, ventral cuticle.

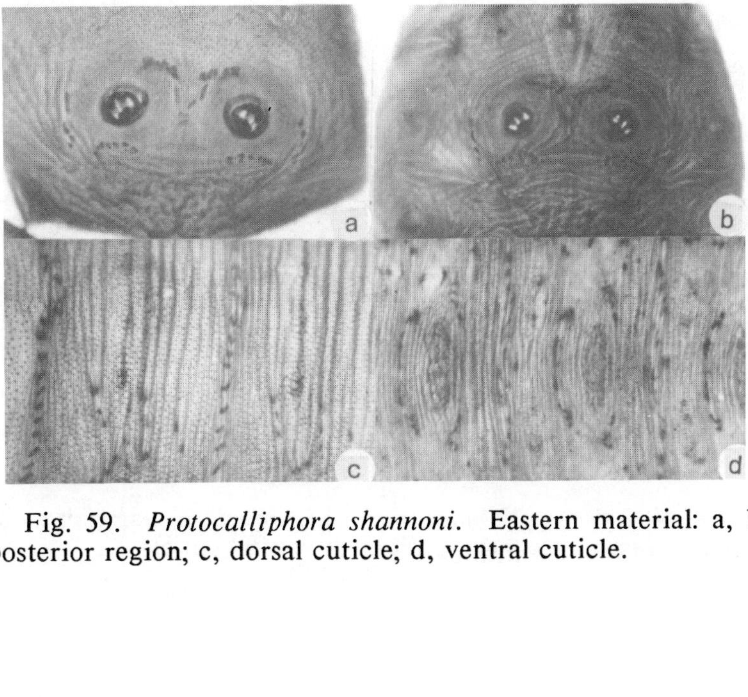

Fig. 59. *Protocalliphora shannoni.* Eastern material: a, b, posterior region; c, dorsal cuticle; d, ventral cuticle.

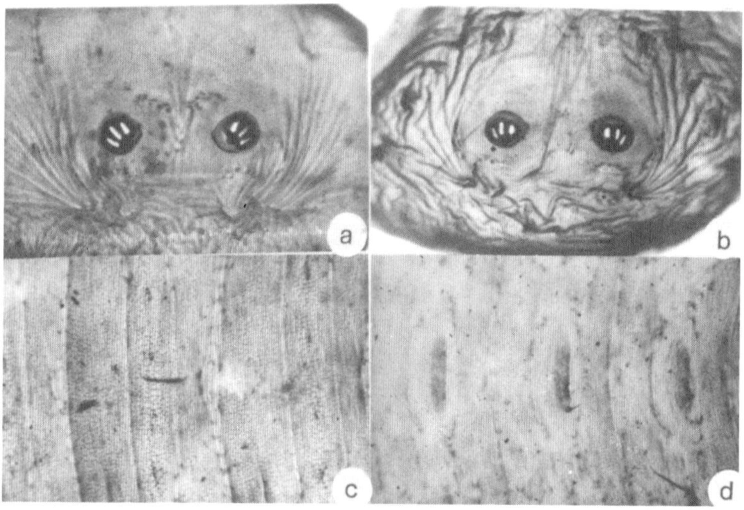

Fig. 60. *Protocalliphora shannoni.* Western material: a, b, posterior region; c, dorsal cuticle; d, ventral cuticle.

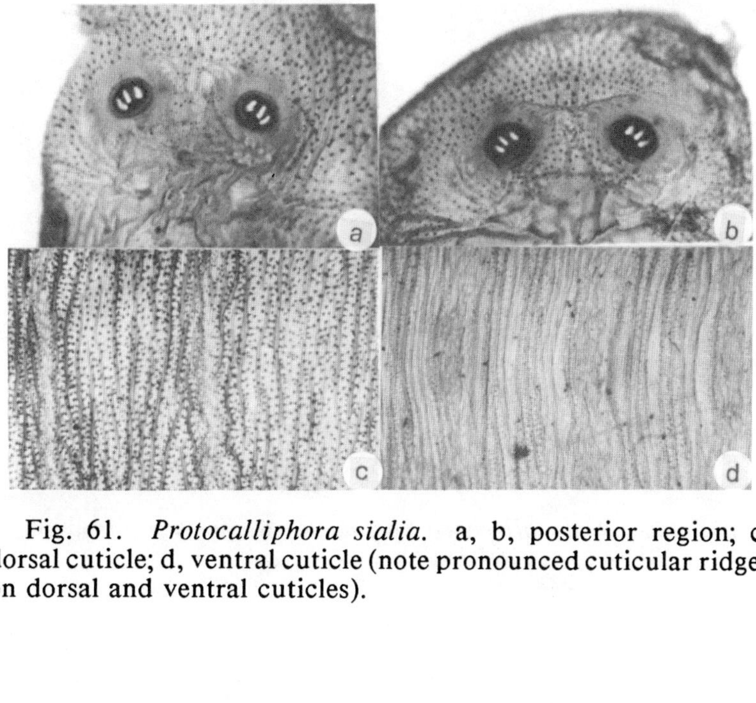

Fig. 61. *Protocalliphora sialia.* a, b, posterior region; c, dorsal cuticle; d, ventral cuticle (note pronounced cuticular ridges on dorsal and ventral cuticles).

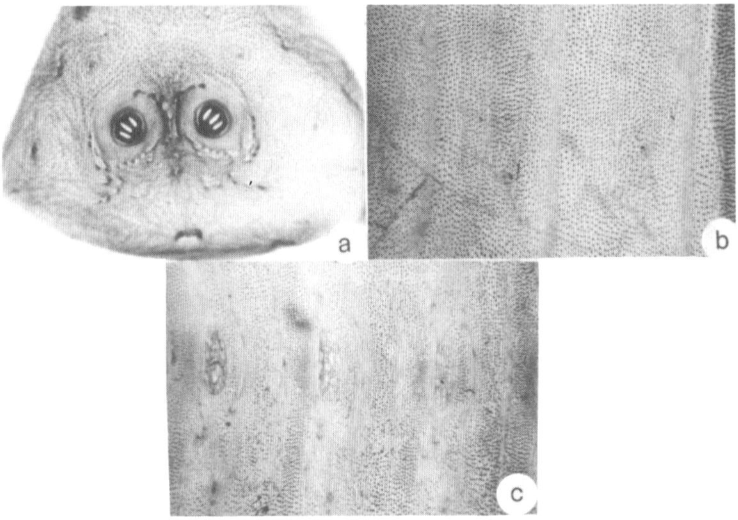

Fig. 62. *Protocalliphora spatulata.* a, posterior region; b, dorsal cuticle; c, ventral cuticle.

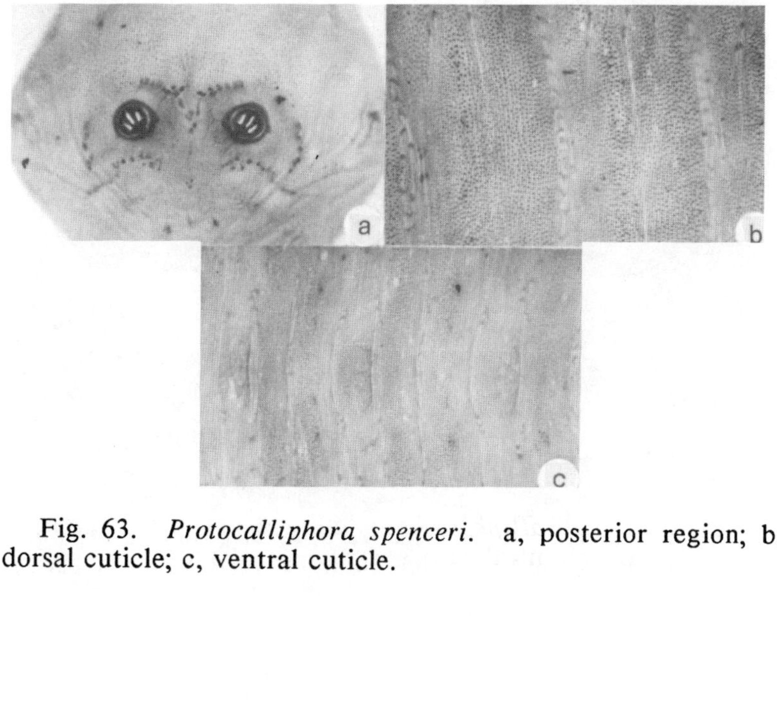

Fig. 63. *Protocalliphora spenceri.* a, posterior region; b, dorsal cuticle; c, ventral cuticle.

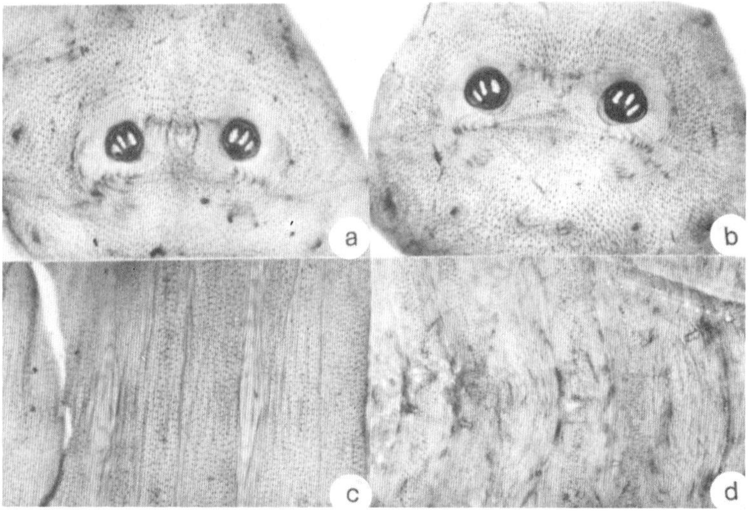

Fig. 64. *Protocalliphora tundrae.* a, b, posterior region; c, dorsal cuticle; d, ventral cuticle (note reduced posterior band).

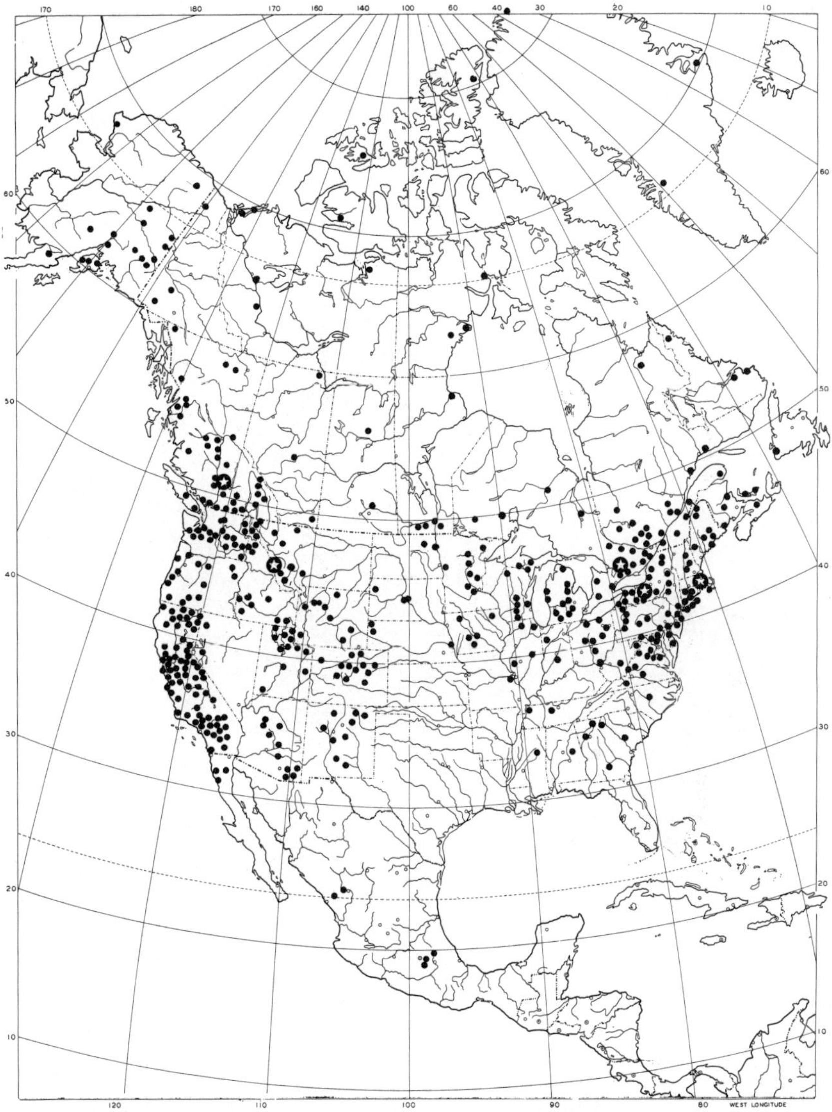

Map 1. Distribution of *Protocalliphora* in North America, as reflected by collecting localities. Major centers of collecting are starred, in addition to concentrations of localities in northern Utah and California.

Map 2. Palearctic distribution of *Protocalliphora*, based on known extreme southern localities and assumed northern limits based on comparable latitudes in North America. Numerous records are known for Europe, but few elsewhere in Eurasia.

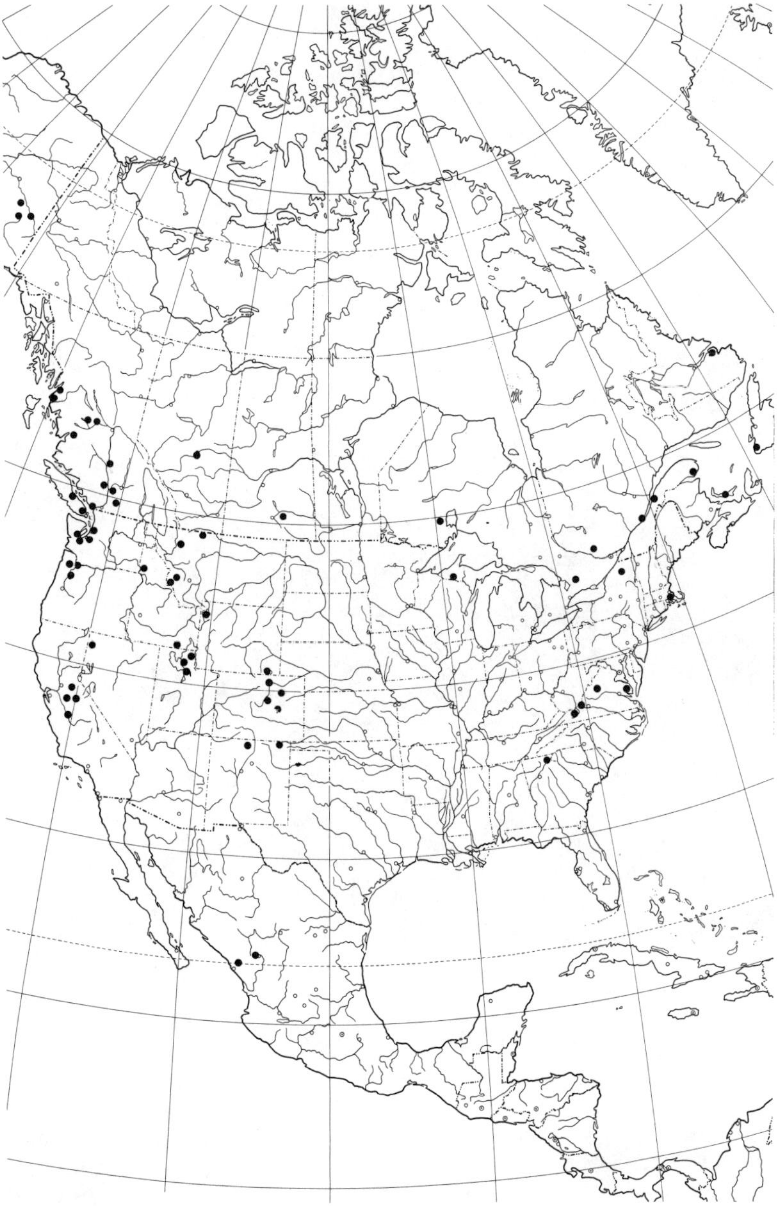

Map 3. Distribution of *Protocalliphora braueri* in North America.

Map 4. Distribution of *Protocalliphora aenea* (stars), *P. beameri* (small circles), and *P. deceptor* (large circles).

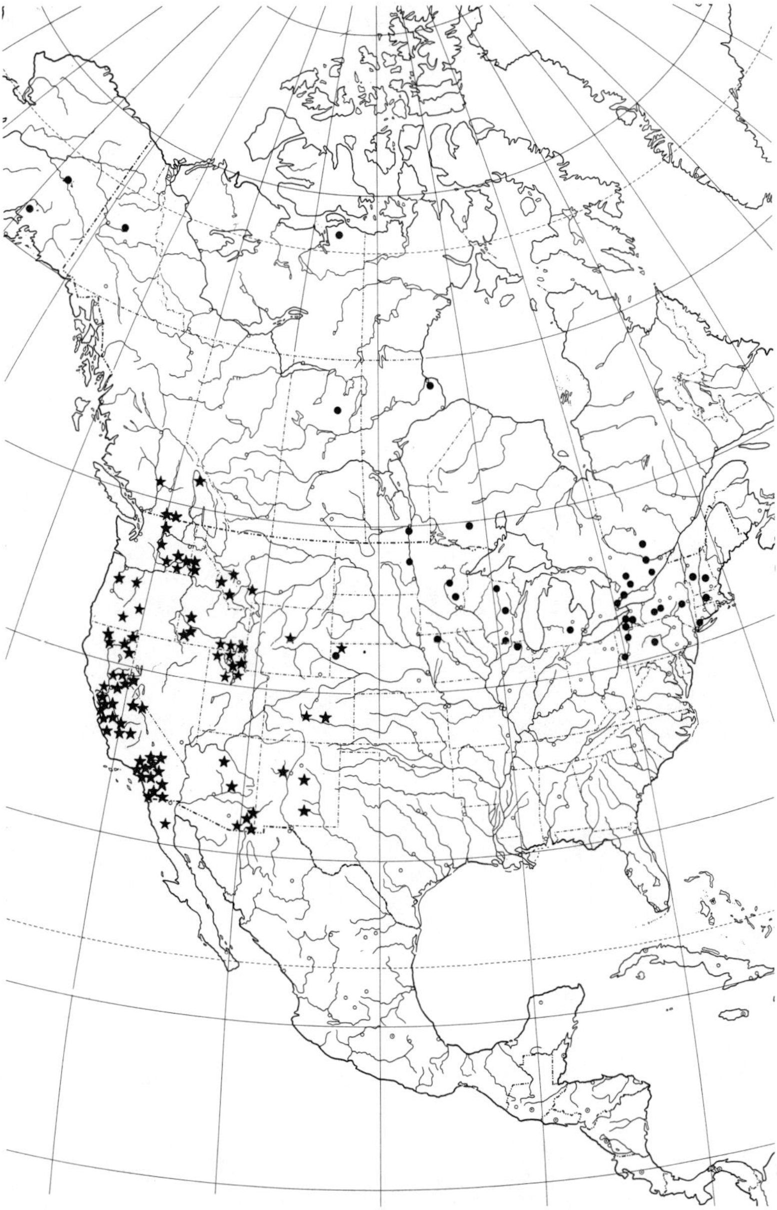

Map 5. Distribution of *Protocalliphora avium* (circles) and *P. asiovora* (stars).

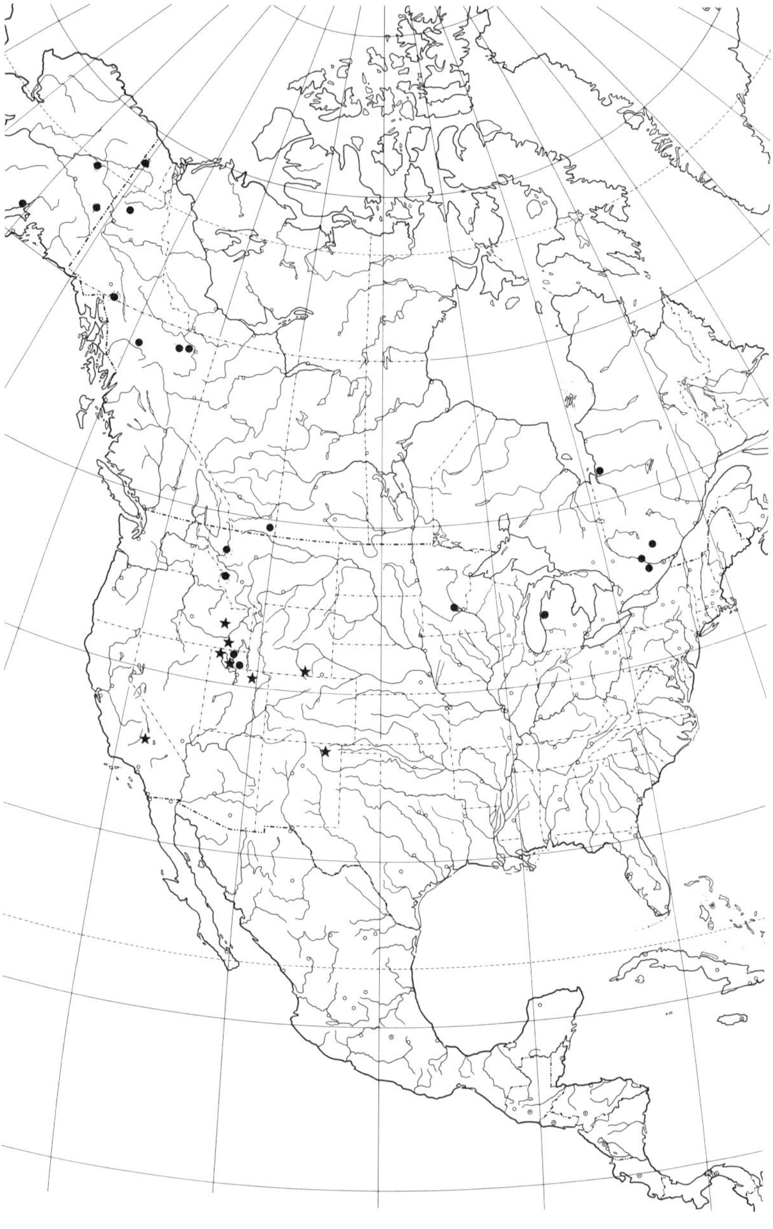

Map 6. Distribution of *Protocalliphora chrysorrhoea* (circles) and *P. seminuda* (stars). The records of *chrysorrhoea* for Michigan and Minnesota are "probables," not verified.

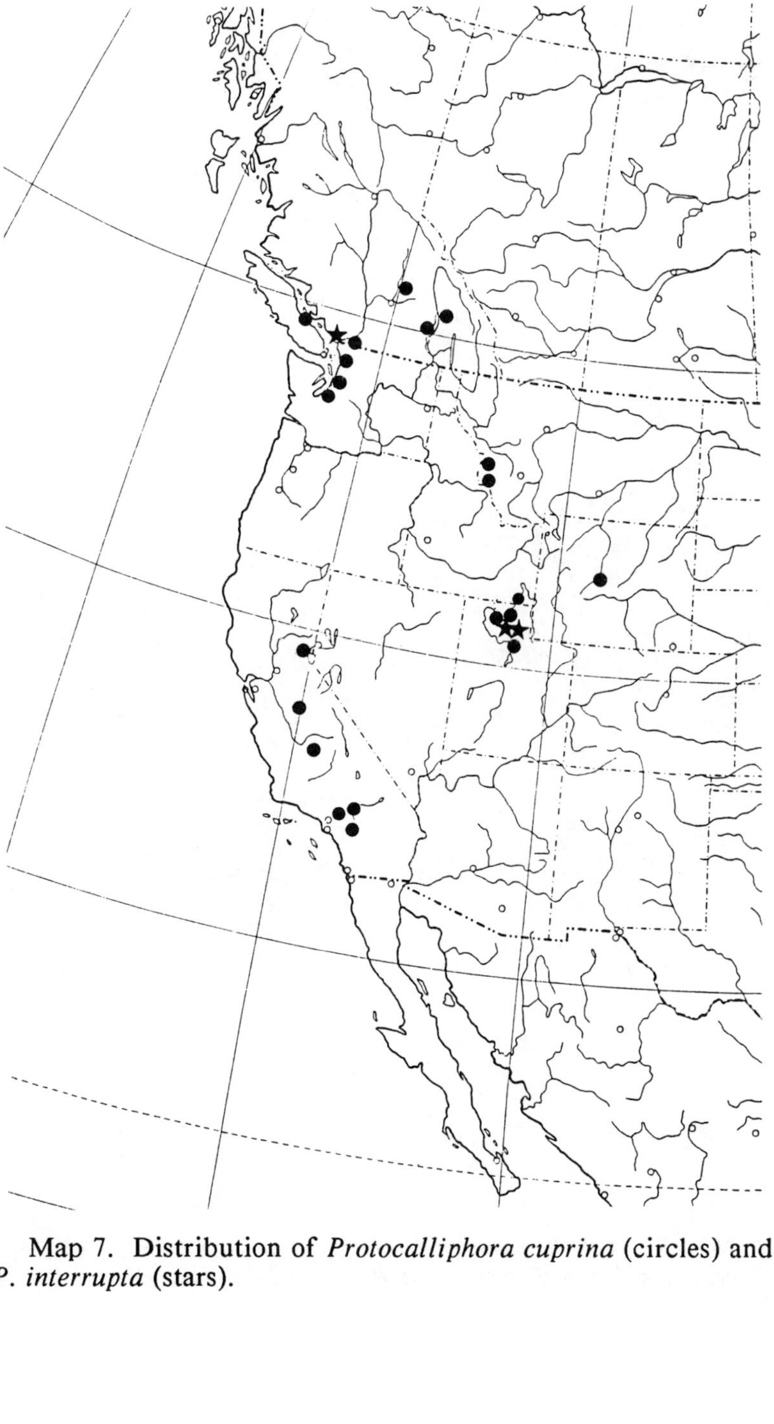

Map 7. Distribution of *Protocalliphora cuprina* (circles) and *P. interrupta* (stars).

Map 8. Distribution of *Protocalliphora halli* (small circles), *P. hesperia* (large circles), and *P. hesperioides* (stars).

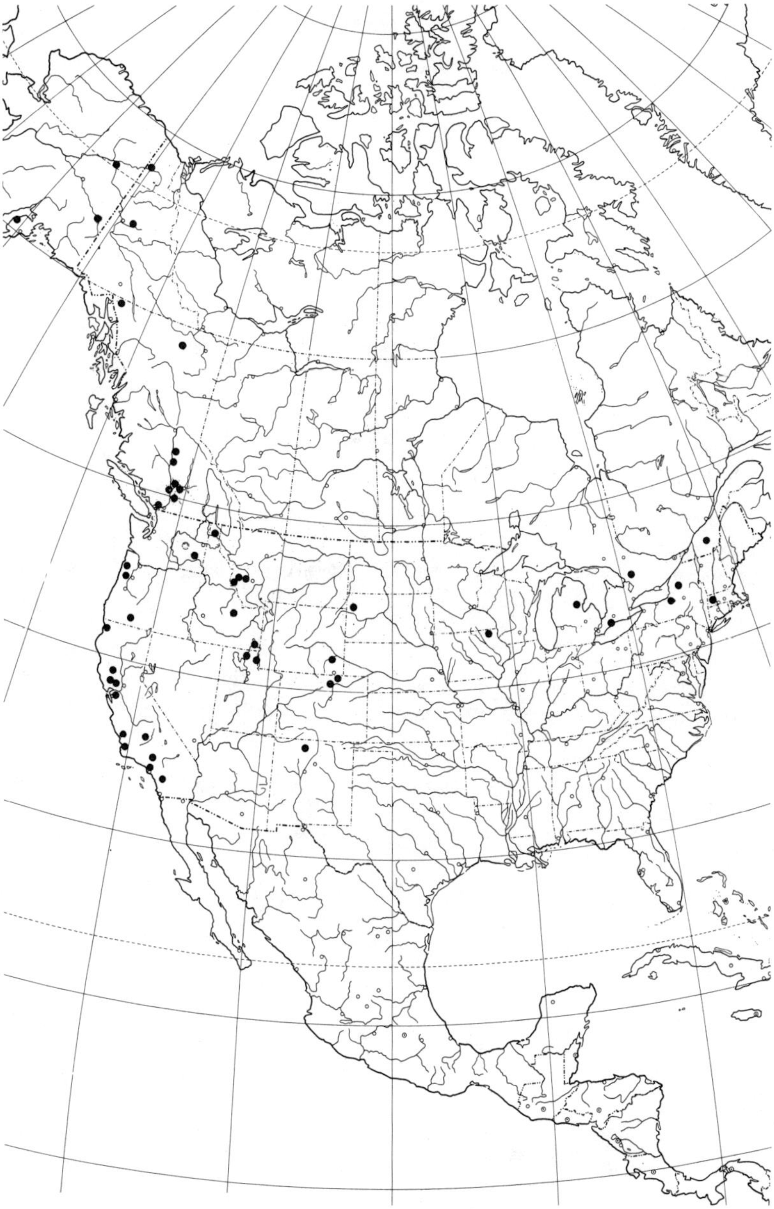

Map 9. Distribution of *Protocalliphora hirundo*.

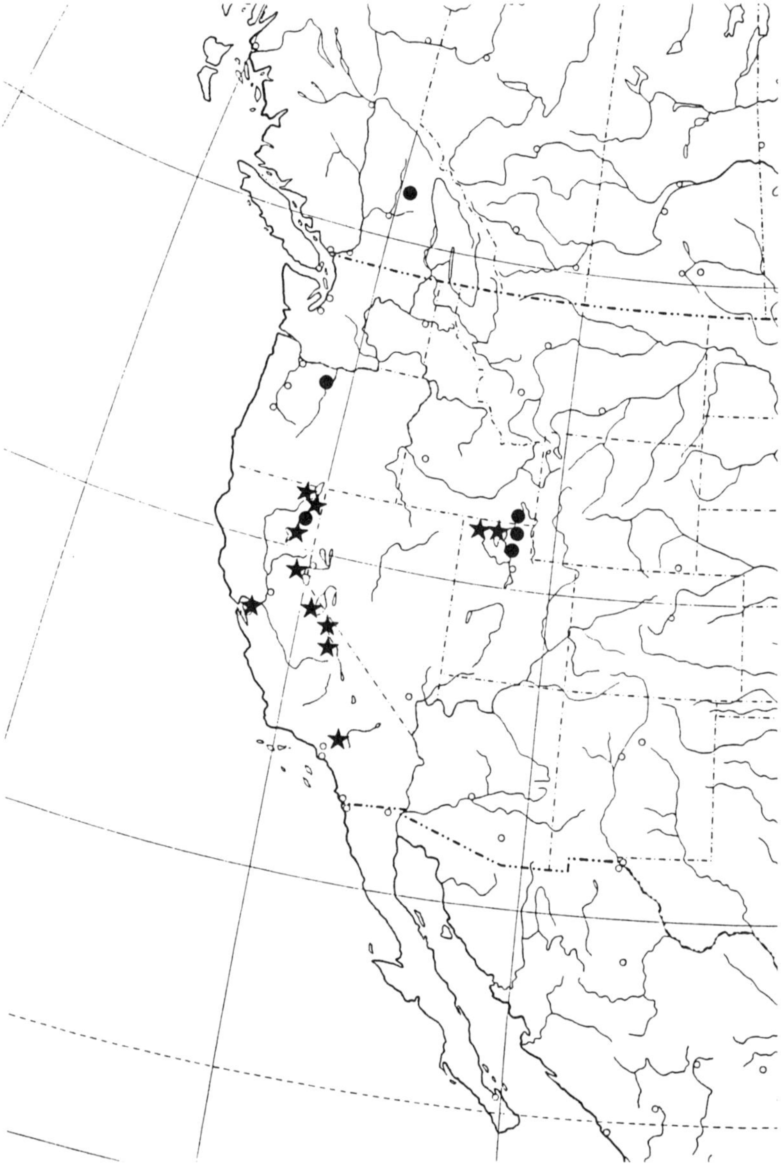

Map 10. Distribution of *Protocalliphora lata* (circles) and *P. parorum* (stars).

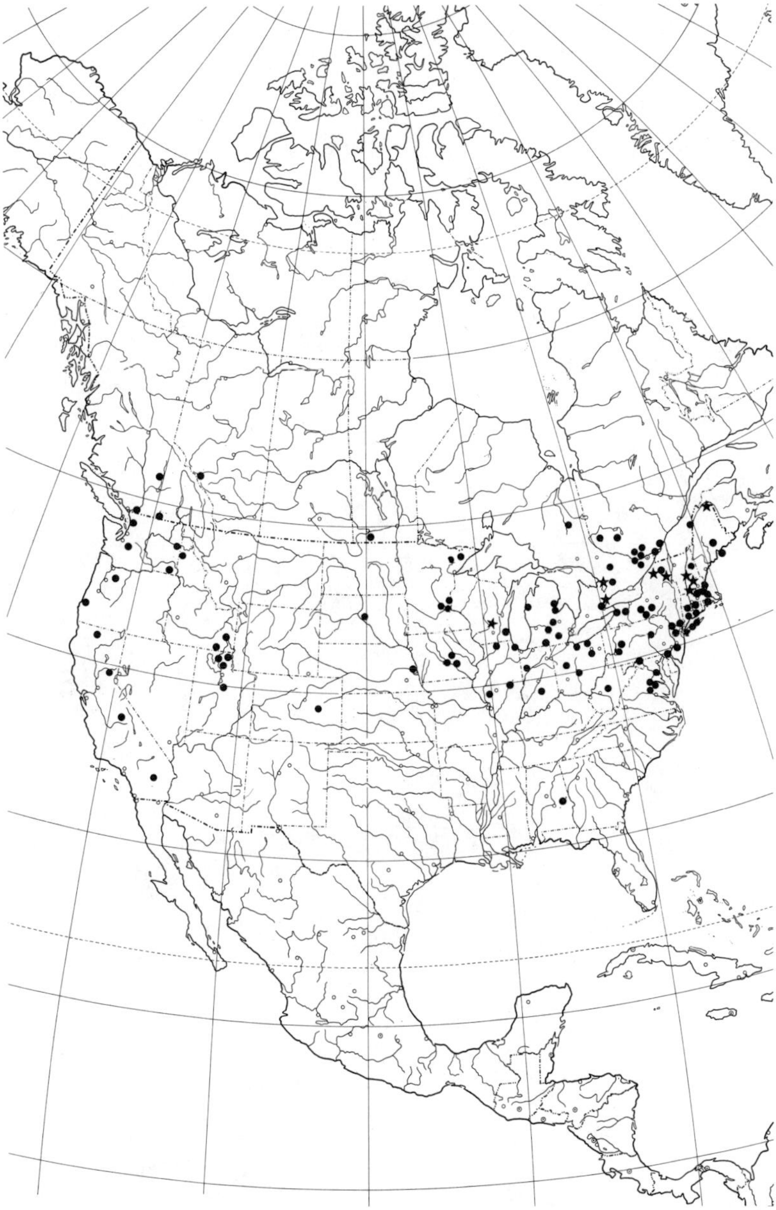

Map 11. Distribution of *Protocalliphora metallica* (circles) and *P. bicolor* (stars). The record of *metallica* for South Dakota is by state only, without definite locality.

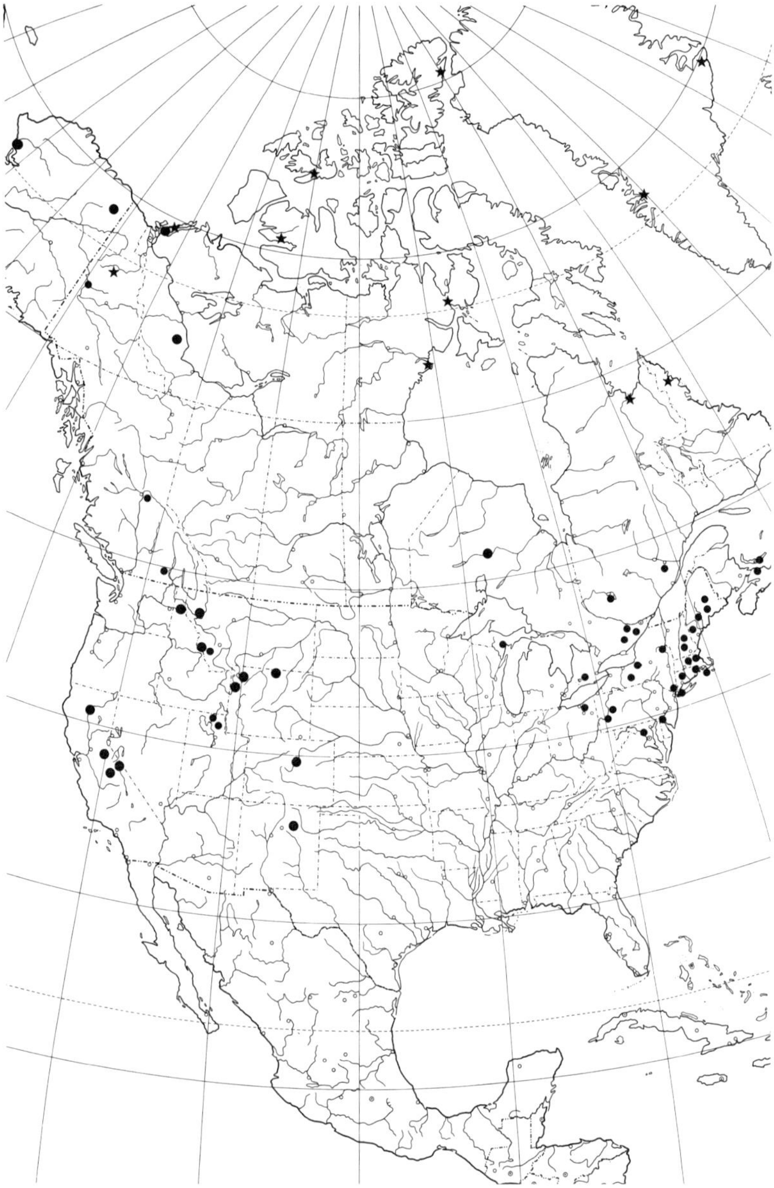

Map 12. Distribution of *Protocalliphora shannoni* (small circles), *P. spatulata* (large circles), and *P. tundrae* (stars).

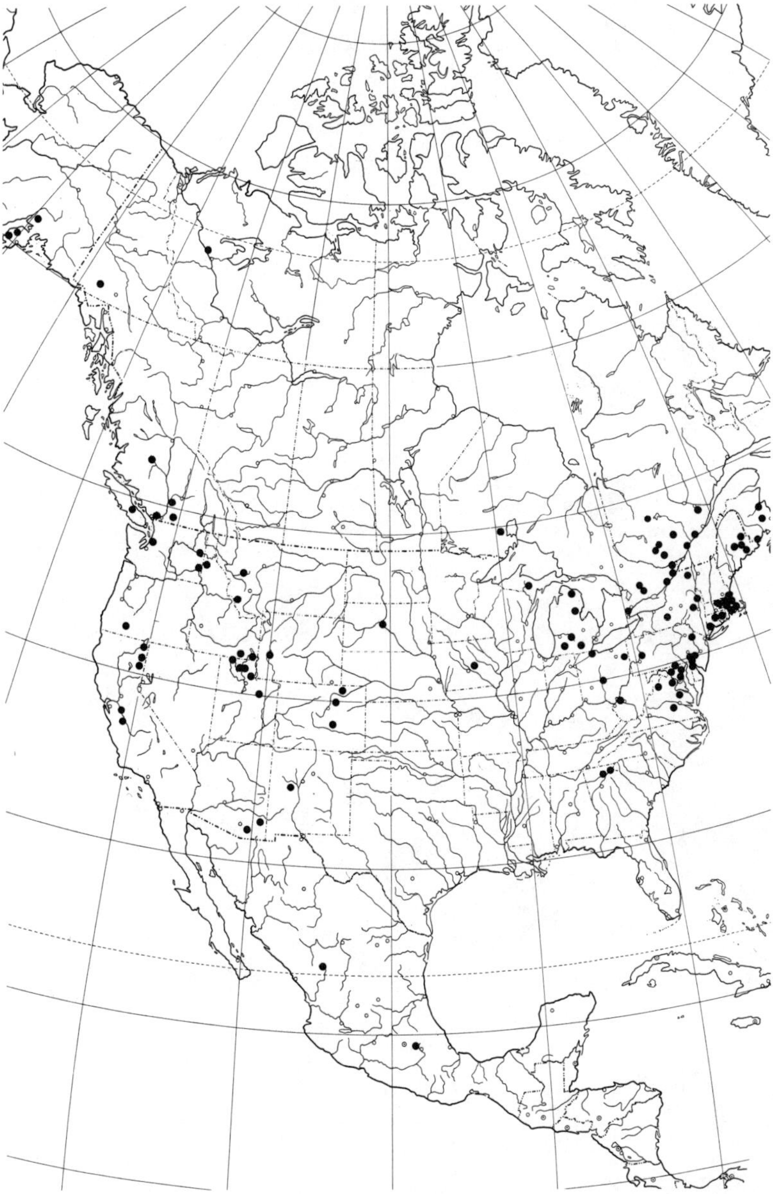

Map 13. Distribution of *Protocalliphora sialia*. The record for South Dakota is by state only, without definite locality.

Map 14. Distribution of *Protocalliphora spenceri* (circles) and *P. brunneisquama* (stars).